MARINE NAVIGATION AND SAFETY OF SEA TRANSPORTATION

Marine Navigation and Safety of Sea Transportation

Maritime Transport & Shipping

Editors

Adam Weintrit & Tomasz Neumann
Gdynia Maritime University, Gdynia, Poland

CRC Press
Taylor & Francis Group
Boca Raton London New York

CRC Press is an imprint of the
Taylor & Francis Group, an **informa** business

A BALKEMA BOOK

Published by:
CRC Press/Balkema
P.O. Box 447, 2300 AK Leiden, The Netherlands
e-mail: Pub.NL@taylorandfrancis.com
www.crcpress.com – www.taylorandfrancis.com

First issued in paperback 2020

ISBN 13: 978-0-367-57641-7 (pbk)
ISBN 13: 978-1-138-00105-3 (hbk)

Visit the Taylor & Francis Web site at
http://www.taylorandfrancis.com

and the CRC Press Web site at
http://www.crcpress.com

Typeset by V Publishing Solutions Pvt Ltd., Chennai, India

List of reviewers

Prof. Roland **Akselsson**, Lund University, Sweden
Prof. Anatoli **Alop**, Estonian Maritime Academy, Tallin, Estonia
Prof. Yasuo **Arai**, Independent Administrative Institution Marine Technical Education Agency,
Prof. Terje **Aven**, University of Stavanger (UiS), Stavanger, Norway
Prof. Michael **Baldauf**, Word Maritime University, Malmö, Sweden
Prof. Michael **Barnett**, Southampton Solent University, United Kingdom
Prof. Eugen **Barsan**, Constanta Maritime University, Romania
Prof. Angelica **Baylon**, Maritime Academy of Asia & the Pacific, Philippines
Prof. Knud **Benedict**, University of Wismar, University of Technology, Business and Design, Germany
Prof. Christophe **Berenguer**, Grenoble Institute of Technology, Saint Martin d'Heres, France
Prof. Tor Einar **Berg**, Norwegian Marine Technology Research Institute, Trondheim, Norway
Prof. Carmine Giuseppe **Biancardi**, The University of Naples „Parthenope", Naples, Italy
Prof. Alfred **Brandowski**, Gdynia Maritime University, Poland
Sr. Jesus **Carbajosa Menendez**, President of Spanish Institute of Navigation, Spain
Prof. Pierre **Cariou**, Word Maritime University, Malmö, Sweden
Prof. A. Güldem **Cerit**, Dokuz Eylül University, Izmir, Turkey
Prof. Adam **Charchalis**, Gdynia Maritime University, Poland
Prof. Andrzej **Chudzikiewicz**, Warsaw University of Technology, Poland
Prof. Kevin **Cullinane**, University of Newcastle upon Tyne, UK
Prof. Krzysztof **Czaplewski**, Polish Naval Academy, Gdynia, Poland
Prof. German **de Melo Rodriguez**, Polytechnical University of Catalonia, Barcelona, Spain
Prof. Decio Crisol **Donha**, Escola Politécnica Universidade de Sao Paulo, Brazil
Prof. Eamonn **Doyle**, National Maritime College of Ireland, Cork Institute of Technology, Cork, Ireland
Prof. Daniel **Duda**, Naval University of Gdynia, Polish Nautological Society, Poland
Prof. Andrzej **Fellner**, Silesian University of Technology, Katowice, Poland
Prof. Börje **Forssell**, Norwegian University of Science and Technology, Trondheim, Norway
Prof. Alberto **Francescutto**, University of Trieste, Trieste, Italy
Prof. Jens **Froese**, Jacobs University Bremen, Germany
Prof. Wiesław **Galor**, Maritime University of Szczecin, Poland
Prof. Avtandil **Gegenava**, Georgian Maritime Transport Agency, Head of Maritime Rescue Coordination Center, Georgia
Prof. Jerzy **Girtler**, Gdańsk University of Technology, Poland
Prof. Stanislaw **Górski**, Gdynia Maritime University, Poland
Prof. Marek **Grzegorzewski**, Polish Air Force Academy, Deblin, Poland
Prof. Andrzej **Grzelakowski**, Gdynia Maritime University, Poland
Prof. Lucjan **Gucma**, Maritime University of Szczecin, Poland
Prof. Stanisław **Gucma**, Maritime University of Szczecin, Poland
Prof. Vladimir **Hahanov**, Kharkov National University of Radio Electronics, Kharkov, Ukraine
Prof. Jerzy **Hajduk**, Maritime University of Szczecin, Poland
Prof. Michał **Holec**, Gdynia Maritime University, Poland
Prof. Qinyou **Hu**, Shanghai Maritime University, China
Prof. Marek **Idzior**, Poznan University of Technology, Poland
Prof. Jung Sik **Jeong**, Mokpo National Maritime University, South Korea
Prof. Mirosław **Jurdziński**, Gdynia Maritime University, Poland
Prof. John **Kemp**, Royal Institute of Navigation, London, UK
Prof. Lech **Kobyliński**, Polish Academy of Sciences, Gdansk University of Technology, Poland
Prof. Serdjo **Kos**, University of Rijeka, Croatia
Prof. Eugeniusz **Kozaczka**, Polish Acoustical Society, Gdansk University of Technology, Poland
Prof. Pentti **Kujala**, Helsinki University of Technology, Helsinki, Finland
Prof. Jan **Kulczyk**, Wroclaw University of Technology, Poland
Prof. Andrzej **Lewiński**, University of Technology and Humanities in Radom, Poland
Prof. Vladimir **Loginovsky**, Admiral Makarov State Maritime Academy, St. Petersburg, Russia
Prof. Mirosław **Luft**, University of Technology and Humanities in Radom, Poland
Prof. Bogumił **Łączyński**, Gdynia Maritime University, Poland

Prof. Zbigniew **Łukasik**, University of Technology and Humanities in Radom, Poland
Prof. Marek **Malarski**, Warsaw University of Technology, Poland
Prof. Francesc Xavier **Martinez de Oses**, Polytechnical University of Catalonia, Barcelona, Spain
Prof. Jerzy **Matusiak**, Helsinki University of Technology, Helsinki, Finland
Prof. Bolesław **Mazurkiewicz**, Maritime University of Szczecin, Poland
Prof. Boyan **Mednikarov**, Nikola Y. Vaptsarov Naval Academy, Varna, Bulgaria
Prof. Jerzy **Merkisz**, Poznań University of Technology, Poznań, Poland
Prof. Daniel Seong-Hyeok **Moon**, World Maritime University, Malmoe, Sweden
Prof. Wacław **Morgaś**, Polish Naval Academy, Gdynia, Poland
Prof. Takeshi **Nakazawa**, World Maritime University, Malmoe, Sweden
Prof. Rudy R. **Negenborn**, Delft University of Technology, Delft, The Netherlands
Prof. Nikitas **Nikitakos**, University of the Aegean, Chios, Greece
Prof. Tomasz **Nowakowski**, Wrocław University of Technology, Wrocław, Poland
Prof. Vytautas **Paulauskas**, Maritime Institute College, Klaipeda University, Lithuania
Prof. Jan **Pawelski**, Gdynia Maritime University, Poland
Prof. Thomas **Pawlik**, Bremen University of Applied Sciences, Germany
Prof. Francisco **Piniella**, University of Cadiz, Spain
Prof. Boris **Pritchard**, University of Rijeka, Croatia
Prof. Jonas **Ringsberg**, Chalmers University of Technology, Gothenburg, Sweden
Prof. Michael **Roe**, University of Plymouth, Plymouth, United Kingdom
Prof. Hermann **Rohling**, Hamburg University of Technology, Hamburg, Germany
Prof. Władysław **Rymarz**, Gdynia Maritime University, Poland
Prof. Aydin **Salci**, Istanbul Technical University, Maritime Faculty, ITUMF, Istanbul, Turkey
Prof. Viktoras **Sencila**, Lithuanian Maritime Academy, Klaipeda, Lithuania
Prof. Shigeaki **Shiotani**, Kobe University, Japan
Prof. Jacek **Skorupski**, Warsaw University of Technology, Poland
Prof. Leszek **Smolarek**, Gdynia Maritime University, Poland
Cmdr. Bengt **Stahl**, Nordic Institute of Navigation, Sweden
Prof. Janusz **Szpytko**, AGH University of Science and Technology, Kraków, Poland
Prof. Leszek **Szychta**, University of Technology and Humanities in Radom, Poland
Prof. Wojciech **Ślączka**, Maritime University of Szczecin, Poland
Prof. Roman **Śmierzchalski**, Gdańsk University of Technology, Poland
Prof. Henryk **Śniegocki**, Gdynia Maritime University, Poland
Prof. Vladimir **Torskiy**, Odessa National Maritime Academy, Ukraine
Prof. Elen **Twrdy**, University of Ljubljana, Slovenia
Capt. Rein **van Gooswilligen**, Netherlands Institute of Navigation
Prof. Nguyen **Van Thu**, Ho Chi Minh City University of Transport, Ho Chi Minh City, Vietnam
Prof. George Yesu Vedha **Victor**, International Seaport Dredging Limited, Chennai, India
Prof. Peter **Voersmann**, Deutsche Gesellschaft für Ortung und Navigation, Germany
Prof. Vladimir A. **Volkogon**, Baltic Fishing Fleet State Academy, Kaliningrad, Russian Federation
Prof. Bernard **Wiśniewski**, Maritime University of Szczecin, Poland
Prof. Krystyna **Wojewódzka-Król**, University of Gdańsk, Poland
Prof. Adam **Wolski**, Maritime University of Szczecin, Poland
Prof. Jia-Jang **Wu**, National Kaohsiung Marine University, Kaohsiung, Taiwan (ROC)
Prof. Hideo **Yabuki**, Tokyo University of Marine Science and Technology, Tokyo, Japan
Prof. Homayoun **Yousefi**, Chabahar Maritime University, Iran

TABLE OF CONTENTS

Maritime Transport & Shipping
Introduction

A. Weintrit & T. Neumann
Gdynia Maritime University, Gdynia, Poland

The monograph is addressed to scientists and professionals in order to share their expert knowledge, experience and research results concerning all aspects of navigation, safety at sea and marine transportation.

The contents of the book are partitioned into eight separate chapters: Pollution at Sea, Cargo Safety, Environment Protection and Ecology (covering the subchapters 1.1 through 1.9), Gas and Oil Transportation (covering the chapters 2.1 through 2.5), Sea Port and Harbours Development (covering the chapters 3.1 through 3.7), Dynamic Positioning and Offshore Technology (covering the chapters 4.1 through 4.5), Container Transport (covering the chapters 5.1 through 5.4), Intermodal Transport (covering the chapters 6.1 through 6.2), Ship's propulsion and Mechanical Engineering (covering the chapters 7.1 through 7.8) and Hydrodynamics and Ship Stability (covering the chapters 8.1 through 8.8).

Each chapter contains interesting information on specific aspects of Maritime Transport & Shipping. The Editors would like to thanks all authors of chapters. It was hard work but worth every minute. This book is the result of years of research, conducted by many people. Each chapter was reviewed at least by three independent reviewers. The Editors would like to express his gratitude to distinguished authors and reviewers of chapters for their great contribution for expected success of the publication. He congratulates the authors for their excellent work.

First chapter is about Pollution at Sea, Cargo Safety, Environment Protection and Ecology. The readers can find some information about overview of the past tanker accidents in the Baltic Sea and chemical related accidents in seas worldwide. The aim of other study is to perform a qualitative research to determine the factors affecting the operational efficiency of ship, berth and warehousing operations in chemical cargo terminals. Chapter also contains information about safe transportation solid bulk cargoes and notice about fire safety assessment concerning nitrates fertilizers in sea transport. The European Union is very active on global market of emission to reduce greenhouse gas emissions from maritime transport. In chapter readers can find information about hovercrafts. There is also notice about disaster preparedness of a maritime university. The new equipment and advantages of the CleanSeaNet System is described and presented as a new method used to protect the marine environment. Authors highlighted problem invasive species travel from one ocean to the other through ballast water from the international shipping industry and survey the changes of diversity and distribution of the gastropods in an important fishing area.

In the second chapter there are described problems related to gas and oil transportation. The readers can find some information about increase in maritime oil transportation in the Gulf of Finland, about possibilities for the use of LNG as a fuel on the Baltic Sea and the general division of ports for the identification of hazards that affect the safety of LNG carrier for port and LNG terminal in Świnoujście located on Pomeranian Bay. In this chapter also presented using natural gas as alternative fuel for vessels sailing in European waters.

The third chapter deals sea port and harbours development. There is a notice about the future of Santos Harbour outer access channel and information about safety management system in sea ports. Presented is method of assessment of insurance expediency of quay structures' damage risks in sea ports. Described are problems in solid waste management, control and compliance measures. In this section also presented are the problems of safety maneuvering of floating unit in yachts ports and application of extruded fenders. Highlighted on the requirements of the application code security and safety of ships and ports and the technical aspects necessary for the application by the Saudi marine Ports.

The fourth chapter is about dynamic positioning and offshore technology. In this chapter readers can found information about a probe of correctness selection of the number and orientation of thrusters in ship's dynamic positioning systems, underwater vehicles' applications in offshore industry, about training for heavy lift and offshore crane loading teams. There is also presented a proposal of international regulations for preventing collision between an offshore platform and a ship, and other than navigation technical uses of the sea space.

The fifth chapter deals container transport. There is described development of container transit from the Iranian south ports and some interesting information about Port Feeder Barge concept. Presented is the concept of modernization works related to the capability of handling E Class container vessels in the Port Gdynia and container transport capacity at the Port of Koper, including a brief description of studies necessary prior to expansion.

In the sixth chapter there are described problems related to intermodal transport. The readers can find some information about intermodal liner passenger connections within Croatian seaports and concept of cargo security assurance in an intermodal transportation.

The seventh chapter deals propullsion and mechanical engineering. There is described diagnostic and measurement system for marine engines', develop a condition based maintenance model for a vessel's main propulsion system. There is also experimental analysis of podded propulsor on naval vessel and presented are the problems of the selection of diesel engines injector nozzles parameters and limitations of the pressure of the fuel injection. There are presented the results of a CFD simulation of marine propeller created with OpenFOAM software. The obtained results were compared with the of the commercial CFD codes simulations and the experimental research. There are described the results of the analysis on the Power Curves and Self Propulsion Factors under various weather and sea conditions. The readers can find some information about engine room simulator training course, information about practicability and essentiality onboard ship.

The eight chapter is about hydrodynamics and ship stability. Presented are information about an approach for preliminary estimating ship's stability when there is a forecast of extreme hydrometeorogical conditions at the area where navigation is supposed. Presented are study about values and locations of the hydrostatic and hydrodynamic forces at hull of the ship in transitional mode and interactions between the model and prototype of boats. The readers can find some information about new methods of measuring the motion and deformation of container vessels in the sea and hybrid Bayesian wave estimation for actual merchant vessels. There is also some information about results of tests of school-ship model's free rolling, the dynamic heeling moment due to liquid sloshing in partly filled wing tanks for varying rolling period of seagoing vessels and about safety for Laker bulker trans-pacific delivery voyage

Each subchapter was reviewed at least by three independent reviewers. The Editors would like to express his gratitude to distinguished authors and reviewers of chapters for their great contribution for expected success of the publication. He congratulates the authors for their excellent work.

Chapter 1

Pollution at Sea, Cargo Safety, Environment Protection and Ecology

Pollution at Sea, Cargo Safety, Environment Protection and Ecology
Maritime Transport & Shipping – Marine Navigation and Safety of Sea Transportation – Weintrit & Neumann (Eds)

Overview of Maritime Accidents Involving Chemicals Worldwide and in the Baltic Sea

J.M. Häkkinen & A.I. Posti
University of Turku Centre for Maritime Studies, Kotka, Finland

ABSTRACT: Transport and handling of hazardous chemicals and chemical products around the world's waters and ports have considerably increased over the last 20 years. Thus, the risk of major pollution accidents has also increased. Past incidents/accidents are, when reported in detail, first hand sources of information on what may happen again. This paper provides an overview of the past tanker accidents in the Baltic Sea and chemical related accidents in seas worldwide. The aim is to find out what can be learned from past accidents, especially from the environmental point of view. The study is carried out as a literature review and as a statistical review.

1 INTRODUCTION

Transport and handling of hazardous chemicals and chemical products has considerably increased over the last 20 years, thus increasing the risk of major pollution accidents. Worldwide, about 2000 chemicals are transported by sea either in bulk or packaged form. Only few hundred chemicals are transported in bulk but these make up most of the volume of the chemical sea-borne trade (Purnell 2009). Chemical releases are thought to be potentially more hazardous than oil. As to marine spills, chemicals may have both acute and long-term environmental effects, and may not be as easily recoverable as oil spills. In addition, public safety risks are more severe in chemical releases (EMSA 2007).

The Baltic Sea is one of the busiest sea routes in the world – 15 % of the world's cargo moves in it. In 2010, the international liquid bulk transports in the Baltic Sea ports contained around 290 million tonnes of oil and oil products, at least 11 million tonnes of liquid chemicals, and 4 million tonnes of other liquid bulk (Holma et al. 2011; Posti & Häkkinen 2012). In addition, chemicals are transported in packaged form, but tonnes are not studied. Navigation in the Baltic Sea is challenging due to the relative shallowness, narrow navigation routes, and ice cover of the sea. Oil and chemicals are a serious threat to the highly sensitive Baltic Sea ecosystems. Recently, both the number and the volume of the transported cargo have increased

significantly in the Baltic Sea (HELCOM 2009), concomitantly raising the spill/ship collision risk in the Baltic Sea areas (Hänninen et al. 2012). The results of previous studies (EMSA 2010, Hänninen & Rytkönen 2006, Bogalecka & Popek 2008, Mullai et al. 2009, Suominen & Suhonen 2007) indicate that both the spill risks and chemical incidents are not as well-defined than those concerning oils. Nevertheless, among the wide range of chemicals transported, the potency to cause environmental damage cannot be overlooked.

The study and analysis of past accidents with consequences to the environment and humans can be a source of valuable information and teach us significant lessons in order for us to prevent future shipping accidents and chemical incidents. The purpose of this study is to provide an overview of the past tanker accidents in the Baltic Sea, and chemical-related accidents in seas worldwide, thus aiming at finding out what can be learned from these past accidents, including e.g. occurrence, causes, general rules and particular patterns for the accidents. The study focuses mainly on chemicals transported in liquefied form, but chemical accidents involving substances in packaged form are also studied. Conventional oil and oil products are observed only on a general level. The special scope in the study is put on environmental impact assessment.

2 MATERIALS AND METHODS

The study was carried out in two stages. First, *a literature review* on maritime accidents involving hazardous substances and especially chemicals was made to find out what kind of studies have previously been conducted on the topic, and what are the main results of these studies. Both scientific articles and research reports were taken into account. The studies were mainly searched by using numerous electronic article databases and a web search engine.

Second, *a statistical review* on maritime tanker-related accidents in the Baltic Sea was carried out to find out the amount and types of tanker accidents that have occurred in the Baltic Sea in recent years, and to examine what kind of pollution these accidents caused and have caused since. All types of tankers (e.g. oil tankers, oil product tankers, chemical tankers, chemical product tankers and gas tankers) were included in the review. An overview of the tanker accidents in the Baltic Sea was made by using maritime accident reports provided by the Helsinki Commission (HELCOM) and by the European Maritime Safety Agency (EMSA). More detailed information about maritime accidents involving a tanker was searched using maritime accident databases and reports provided by the authorities and/or other actors responsible for collecting maritime accident data in each Baltic Sea country. More detailed maritime accident investigation reports on accidents were found from Denmark, Finland, Germany, Latvia and Sweden; basic information about accidents was found from Estonia and Lithuania; and no maritime accident data was found from Poland and Russia.

3 LITERATURE REVIEW ON MARITIME ACCIDENTS INVOLVING CHEMICALS

There are few impact assessment studies for chemical spills in the scientific literature in comparison to those for oil spills. Recently, there have been some good papers and accident analyses concerning chemicals and other hazardous materials (conventional oil omitted), such as Cedre and Transport Canada 2012, EMSA 2007, HASREP 2005, Mamaca et al. 2009, Marchand 2002 and Wern 2002. In addition, the Centre of Documentation, Research and Experimentation on Accidental Water Pollution (Cedre) collect information about shipping accidents involving HNS for an electric database by using various data sources (Cedre 2012). None of those aforementioned sources are, or even try to be, exhaustive listings of all accidents involving chemicals and other hazardous materials, but they have gathered examples of well-known accidents with some

quality information. By compiling accident data from aforementioned sources, 67 famous tanker/bulk carrier accidents involving chemicals and/or other hazardous materials were detected. These accidents frequently involved chemicals or chemical groups like acids, gases, vegetable oils, phenol, ammonia, caustic soda and acrylonitrile. Using the same information sources, 46 accidents involving packaged chemicals or other hazardous materials were listed. In comparison to bulk chemicals, it can be seen that the variety of chemicals involved in accidents is much higher in the case of packaged chemicals. In this section, key findings and lessons to be learned from in relation to vessel chemical accidents are discussed in more detail, the analysis being based on original key studies.

3.1 *Overview of maritime chemical accidents worldwide*

Marchand (2002) presented an analysis of chemical incidents and accidents in the EU waters and elsewhere, and stated that 23 incidents had information written down on related facts, such as accident places and causes, chemical products involved, response actions and environmental impacts. The study categorized the accidents into five groups according to how the substance involved behaved after being spilled at sea: products as packaged form; dissolvers in bulk; floaters in bulk; sinkers in bulk; and gases and evaporators in bulk. Based on Marchand's (2002) analysis, most of the accidents happened in the transit phase at sea, that is, while the vessel was moving. Only four accidents happened in ports or in nearby zones. Most of the accidents happened with bulk carriers (62 per cent of all the incidents), and less often with vessels transporting chemicals in packaged form (38 %). Bad weather conditions and the resulting consequences were the main cause of the accidents (in 62 per cent of all the cases). Marchand (2002) highlighted several issues concerning human health risks in the case of maritime chemical accidents. He also pointed out that in most accident cases the risks affecting human health come usually from reactive substances (reactivity with air, water or other products) and toxic substances. The evaluation of the chemical risks can be very difficult if a ship is carrying diverse chemicals and some of those are unknown during the first hours after the accident. A more recent study, Manaca et al. (2009) weighted the same chemical risks as Marchand (2002). Certain substances such as chlorine, epichlorohydrine, acrylonitrile, styrene, acids and vinyl acetate are transported in large quantities and may pose a very serious threat to human health being highly reactive, flammable and toxic. Both Marchand (2002) and Mamaca et al. (2009) pointed out that consequences and hazards to the environment have varied a lot, considering chemical tanker accidents. Both studies stated that, in light of accidents, pesticide products are one of the biggest threats for the marine environment. If pesticides enter the marine environment, consequences for the near-shore biota, and simultaneously for the people dependent on these resources could be severe. On the other hand, even substances considered as non-pollutants, such as vegetable oils (in accidents like Lindenbank, Hawaii 1975; Kimya, UK 1991; Allegra, France 1997), can also have serious effects for marine species like birds, mussels and mammals (Cedre 2012, Marchand 2002).

By surveying 47 of the best-documented maritime transport accidents involving chemicals in the world from as early as 1947 to 2008, Mamaca et al. (2009) gathered a clear overview of lessons to be learned. Even though the data was too narrow for it to be used in making any statistical findings, the study presented some good examples of maritime chemical accidents. 32 of those accidents occurred in Europe. The list of chemicals that were involved in the accidents more than one time included sulphuric acid (3), acrylonitrile (3), ammonium nitrate (2), and styrene (2). Only 10 of the 47 accidents occurred in ports or in nearby zones. Moreover, 66 per cent of the accidents involved chemicals transported in bulk, whereas 34 per cent involved hazardous materials in packaged form. Primary causes for the reviewed accidents were also studied. Improper maneuver was most frequently the reason for the accident (in 22 per cent of all the cases), shipwreck came second (20 %), and collision was third (13 %), closely followed by grounding and fire (11 % each).

Based on past accident analysis considering packaged chemicals, Mamaca et al. (2009) pointed out that, in light of packaged goods, as a consequence of high chemical diversity present on the vessel, responders must know environmental fates for different chemicals individually as well as the possible synergistic reactions between them. Even though smaller volumes are transported, packaged chemicals can also be extremely dangerous to humans. This could be seen when fumes of epichlorohydrine leaking from the damaged drums on the Oostzee (Germany 1989) seriously affected the ship's crew and caused several cancer cases that were diagnosed years after (Mamaca et al. 2009). However, these types of accidents involving packaged chemicals have only a localized short-term impact on marine life. As to accidents caused by fire, there are difficulties in responding to the situation if the vessel is transporting a wide variety of toxic products. It is important yet difficult to have a fully detailed list of the transported products for the use of assessing possible dangers for rescue personnel and public. Based on the analyses of the reviewed accidents,

Mamaca ct al. (2009) showed that the highest risk for human health comes mainly from reactive substances (reactivity with air, water or other products). They also noted that many chemicals are not only carcinogenic and marine pollutants, but can form a moderately toxic gas cloud which is often capable of producing a flammable and/or explosive mix in the air. Acrylonitrile is a toxic, flammable and explosive chemical, and if it is exposed to heat, a highly toxic gas for humans (phosgene) is formed. Vinyl acetate, in turn, is a flammable and polymerizable product that in the case of Multi Tank Ascania incident (in United Kingdom, in 1999) caused a huge explosion. Little is known about the actual marine pollution effects of most of these substances. If hazardous chemicals and oil are compared, it can be said that the danger of coastline pollution is a far greater concern for oil spills than it is for chemical spills. On the other hand, the toxic clouds are a much bigger concern in the case of chemical accidents (Mamaca et al. 2009).

In their HNS Action Plan, EMSA (2007) reviewed past incidents involving a HNS or a chemical. About 100 HNS incidents were identified from 1986 to 2006. These incidents included both those that resulted in spill and those that did not. EMSA (2007) stated that caution should be applied to the data concerning the total sum of the incidents as well as the amount of spills, because there is variability in the reports from different countries. Statistics showed that the principle cause for both release and non-release incidents were foundering and weather (in 22 per cent of all the incidents), followed by fire and explosion in cargo areas (20 %), collision (16 %) and grounding (15%). Majority of the accidents involved single cargoes (73 %), in which most of the material was carried in bulk form (63 %). Moreover, 50 % of all studied incidents resulted in an HSN release. As to these release accidents/incidents, most of them happened in the Mediterranean Sea (40 %); some in the North Sea (22 %) and Channel Areas (20 %), whereas only 8 per cent occurred in the Baltic Sea. The foundering and weather was again the principle cause of these release incidents in 34 per cent of the cases, followed by fire and explosion in cargo areas (18 %), collision (14 %), and grounding (10 %). The majority of the incidents resulting in HNS release involved single cargoes (78 %) of which 61 per cent was in bulk form (EMSA 2007).

HASREP project listed major maritime chemical spills (above 70 tonnes) in the EU waters from 1994-2004 (HASREP 2005). The project found 18 major accidents altogether, and most of them happened in France or Netherlands. Interestingly, 8 accidents listed in HASREP (2005) were not mentioned in the study of Mamaca et al. (2009). The average occurrence of a major maritime chemical accident in the European Union was nearly 2 incidents per year

(HASREP 2005). By comparison, the statistical study made by the U.S. Coast Guard (USCG) in the United States over 5 year-span (1992–1995) listed 423 spills of hazardous substances from ships or port installations, giving an average of 85 spills each year. The 9 most frequently spilled products were sulfuric acid (86 spill cases), toluene (42), caustic soda (35), benzene (23), styrene (20), acrylonitrile (18), xylenes (18), vinyl acetate (17) and phosphoric acid (12). Over half of the spills were from ships (mainly carrier barges), and the rest from facilities (where the spill comes from the facility itself or from a ship in dock). A complementary study made over a period of 13 years (1981–1994) on the 10 most important port zones reported 288 spills of hazardous substances, representing on average, 22 incidents each year (US Coast Guard 1999). Small spillages in Europe were not recorded with a similar care because they were not detected and/or there was a lack of communication between environmental organizations and competent authorities (HASREP 2005).

Cedre and Transport Canada (2012) analyzed a total of 196 accidents that occurred across the world's seas between 1917 and 2010. The substances that were most frequently spilled and that had the greatest quantities were sulphuric acid, vegetable oils, sodium hydroxide solutions and naphtha. Quite surprisingly, the study showed that structural damage (18 %) was the main cause of accidents involving hazardous materials, followed by severe weather conditions (16 %), collision (13 %), and grounding (11 %). Loading/unloading was the cause for only 7 per cent of the accidents (Cedre and Transport Canada 2012).

3.2 *Animal and vegetable oils*

Even though vegetable oil transport volume remains 200 times smaller than the volume of mineral oil transport, it has increased dramatically (Bucas & Saliot 2002). Thus, the threat of a vegetable oil spill due to a ship accident or accidental spill is presently increasing. Even though vegetable oils are regarded as non-toxic consumable products, they may be hazardous to marine life when spilled in large quantities into the marine environment. Bucas & Saliot (2002) observed that there are 15 significant cases of pollution by vegetable or animal oils that have been reported during the past 40 years worldwide. Rapeseed oil was involved in five cases, soybean oil and palm oil in three cases each, coconut oil, fish oil and anchovy oil in one case each, and in two cases the product was unknown. The largest amount of vegetable oil was spilled in Hawaii in 1975 when M.V. Lindenbank released 9500 tonnes of vegetable oils to coral reef killing crustaceans, mollusks and fishes. It also impacted green algae to grow excessively as well as caused tens of birds to die. Similarly, the fish oil accident had also a serious effect on marine environment, killing lobsters, sea urchins, fishes and birds (Bucas & Saliot 2002).

Based on past cases, Bucas & Saliot (2002) described the environmental fate of vegetable oil spills. The specific gravity of vegetable oils is comprised between 0.9 and 0.97 at 20° Celsius. After spilled into the sea, these oils remain at the surface of the sea and spread forming slicks. The further fate of these oils depends on the nature of the oil, the amount spilled, the air and sea temperatures etc. In open seas or in ports, the consequences are often severe because of local and tidal current movements. The slick can easily spread over several square kilometers. Few hours or days after a spill, the slick is usually no longer regular. A part of the oil may be mingled with sand, some of it may have polymerized and sunk, and in the open sea, mechanical dispersion of the oil slick makes it more available to bacterial degradation. Overall biological degradation can be achieved within 14 days, whereas it takes 25 days for a petroleum product to degrade. If the accident happens in a shallow bay, this bacterial degradation may result in lack of oxygen in the water column (Bucas & Saliot 2002).

Bird loss is usually a major consequence of vegetable oil spills. Slicks are often colorless with a slight odor, and thus they are not easily detected by birds. Several mechanisms lead birds to death after oiling: For example, the loss of insulating capacity of wetted feathers makes birds die from cold; the loss of mobility makes them as easy catch; the loss of buoyancy due to coated feathers results in drowning; the laxative properties of the oil ingested during self-cleaning cause lesions; and the clog of nostrils and throat can result to suffocation. As to crustaceans, the invertebrates have died, for instance, from asphyxiation of clogging of the digestive track. Anoxia of the whole water column may also be the cause of these deaths, and there is also evidence that e.g. sunflower oil can be assimilated on tissues of mussels, as it has happened in the case of the Kimya accident (Bucas & Saliot 2002, Cedre 2012). Bucas & Saliot (2002) stated that it is necessary to quickly collect the oil after spillage by using usual methods like booms and pumps.

3.3 *Risk assessment of different chemicals*

Risk posed by maritime chemical spill depends also on accident scenario and environmental conditions besides inner properties of the spilled chemical. Basically, accidents involving chemical tankers can be classified into four groups. Offshore, in the open sea area, chemical spill has space to have a larger effect or to dissolve and be vaporized. This mitigates the negative effects of the spill. On the other hand, response actions can take a longer time and environmental conditions can be challenging, as well. The incident occurring closer to shoreline can be easier or faster to reach, even if the impact to the environment can potentially be more disastrous. The third scenario portrays a casualty that happens in a closed sea area, like in a port or in a terminal area. In these cases, the spill is usually localized and effectively restricted. However, even smaller spill may elevate toxicity levels in a restricted area. Ports are also situated near city centers, and there is an elevated risk for the health of the public and workers in the area. The fourth possibility is an accident during winter in the presence of ice and snow (Hänninen & Rytkönen 2006). The properties of the chemicals may change in cold water. Some chemicals may be more viscous or even become solids, and thus, easier to recover. On the other hand, hazardous impacts of some chemicals may multiply in the cold environment because the decomposition of the chemicals becomes slower. Thus, chemicals may drift to larger areas. They may also accumulate to the adipose tissues in animals which decreases the probability of an animal to survive beyond winter (Riihimäki et al. 2005).

The marine pollution hazards caused by thousands of chemicals have been evaluated by, for example, the Evaluation of Hazardous Substances Working Group which has given GESAMP Hazard Profile as a result. It indexes the substances according to their bio-accumulation; bio-degradation; acute toxicity; chronic toxicity; long-term health effects; and effects on marine wildlife and on benthic habitats. Based on the GESAMP evaluation, the IMO has formed 4 different hazard categories: X (major hazard), Y (hazard) and Z (minor hazard) and OS i.e. other substances (no hazard) (IMO 2007). Over 80 per cent of all chemicals transported in maritime are classified as belonging to the Y category (GESAMP 2002; IMO 2007). This GESAMP categorization is very comprehensive, but different chemicals having very different toxicity mechanisms, environmental fate and other physico-chemical properties may end up to same MARPOL category. The GESAMP hazard profile, although being an excellent first-hand guide in a case of a marine accident, will not answer the question of which chemicals belonging to the same Y category are the most dangerous ones from an environmental perspective.

Many risk assessment and potential worst case studies exist to help find out what impacts different chemicals might have if instantaneous spill were to happen (Kirby & Law 2010). For example, Law & Campell (1998) made a worst case scenario of circa 10 tonnes insecticide spill (pirimiphos-ethyl), and concluded that it might seriously damage crustacean fisheries in an area of 10,000 km^2 with a recovery time of 5 years. In the case of marine accidents, the greatest risk to the environment is posed by chemicals which have high solubility, stay in the water column, and are bioavailable, persistent and toxic to organisms. Based on the analysis of chemicals transported in the Baltic Sea, Häkkinen et al. (2012) stated that nonylphenol is the most toxic of the studied chemicals and it is also the most hazardous in light of maritime spills. The chemical is persistent, accumulative and has a relatively high solubility to water. Nonylphenol is actually transported in the form of nonylphenol ethoxylates but it is present as nonylphenol when spilled to the environment, and in the aforementioned study the worst case scenario was evaluated. Other very hazardous substances were sulphuric acid and ammonia (Häkkinen et al. 2012). Similarly, the HASREP (2005) project identified top 100 chemicals which are transported between major European ports and involved in trade through the English Channel to the rest of the World. The assessment was based both on transport volumes and the GESAMP hazard profile. The project highlighted chemicals such as benzene, styrene, vegetable oil, xylene, methanol, sulphuric acid, phenol, vinyl acetate, and acrylonitrile. It was concluded that these chemicals were the ones that have high spillage probability but may not result in significant environmental impact. Similarly, French McKay et al. (2006) applied a predictive modeling approach for a selected range of chemicals that are transported by sea in bulk and concluded that phenol and formaldehyde present the greatest risks to aquatic biota. Harold et al. (2011) evaluated human health risks of transported chemicals, based on the GESAMP ratings for toxicity and irritancy. This gives more weight to chemicals that are floaters; form gas clouds; or are irritable and toxic like chlorine (Harold et al. 2011). It is clear that different weightings have a certain impact on the difference in results in these studies. However, the chemicals of real concern vary depending on the sea area for which the risk assessment is conducted since the amounts and types of chemicals differ in different sea areas as do marine environment and biota (Kirby & Law 2010).

The impacts of a release or a spill depend on the behavior of the chemical or chemicals in question. It can be concluded that the most harmful chemicals for human health have quite opposite properties to those that are most hazardous for water biota. For human health, the most hazardous chemicals are those that are very reactive, form either very toxic or irritating (or explosive) gas clouds, and also have possible long-term effects, such as carcinogenic effects. From the environmental point of view, the most hazardous chemicals are those that sink, have a high solubility, possibly stay at the water column, are persistent, bioavailable and very toxic and can have possible long-term effects (French McKay et al. 2006, Häkkinen et al. 2012, Harold et al. 2011).

3.4 Response actions in case of maritime chemical spills

There are many excellent reviews (e.g. Marchand 2002, EMSA 2007, Purnell 2009), based on lessons learned from past accidents, which also contain data about response actions in case of chemical spills. Even if response actions taken differ in every accident case according to special conditions and chemicals involved, it is nevertheless possible to demonstrate certain significant or specific elements valid in all chemical incidents at sea (Marchand 2002).

Firstly, like the information concerning the ship cargo, an evaluation of chemical risks is of primary importance before any operational decisions are to be made, especially if the ship is carrying a wide variety of chemicals (Marchand 2002). Following the chemical spill at sea, the response authorities must immediately take measures in order to minimize the chemical exposure to the public as well as contamination of the marine environment. The primary factors which determine the severity and extent of the impact of the accident are related to the chemical and physical properties of the chemicals in question. It should be noted that in the case of oil spills, the hazard to human health is generally considered to be low, and the more toxic and lighter fractions often evaporate before response actions are able to be started. However, in case of chemical accidents, an initial assessment and monitoring of potential hazards should be undertaken first in order to ensure a safe working environment. In that stage, the primary hazards and fate of the chemical in that marine environment are evaluated. The monitoring techniques need to be designed to measure the key parameters that could give rise to a hazard. It should also be noted that in some cases doing nothing might be the best option, as long it happens under observation (Marchand 2002, Purnell 2009). Le Floch et al. (2010) stated that in case of an instantaneous chemical spill, response usually follows three accepted scenarios: 1) response is not possible, because the spill occurred in a geographical environment that is incompatible with reasonable response times, 2) response is not possible due to reactivity of the substances (major, imminent danger), and 3) response is possible. Gases and evaporators, very reactive substances, and explosives are the biggest concern for human health and safety. Several monitoring devices and dispersion models exist which may aid decision making and help protect responders and the public. The floaters can be monitored by using the same techniques that are used for oil spills. Chemicals that prove to be the most difficult to be monitored are sinkers and dissolvers (such as acrylonitrile in the case of Alessandro Primo in Italy in 1991), even if some techniques e.g. electrochemical methods and acoustic techniques exist (EMSA 2007, Purnell 2009).

Several international, regional and national authorities have published operational guides to describe the possible response options in case of a chemical spill. For example Cedre and IMO have made manuals providing information about different response techniques that can be used in case of chemical spills (Cedre 2012, HELCOM 2002, IMO 2007). Usually response techniques depend on the behavior of a chemical in the environment, and on whether it is released or still contained in packaged form. In practice, the response action varies substantially. Techniques that are applicable in case of oil accidents may be suitable for only some floating chemicals. However, it should not be forgotten that some floating chemicals can also potentially create toxic and maybe explosive vapor clouds (e.g. diesel, xylene and styrene). If this happens, the spark/static-free equipment should be used. Moreover, foams or sorbent materials can also be used near the spill source. Risks associated with evaporators or gases, such as ammonia and vinyl chloride, could be diminished by diluting or using release methods (Purnell 2009). In shallow water areas, neutralizers, activated carbon, oxidizing or reducing agents, complexing agents, and ion-exchangers can be used. Chemicals that are heavier than seawater, in turn, may contaminate large areas of the seabed. Recovery methods that are used include mechanical, hydraulic or pneumatic dredges, but the recovery work is time-consuming and expensive and results in large quantities of contaminated material. Other option is capping the contaminated sediment in-situ (Purnell 2009).

As Marchand (2002) listed, the time involved in response operations can vary from 2–3 months (Anna Broere, Holland; Cason, Spain; Alessandro Primo, Italy); to 8 months (Fenes, France); to 10 months (Bahamas, Brazil); or to even several years as in the case of the research carried out on a sunken cargo (Sinbad, Holland). Cold weather and ice cover may create further problems to response actions in the Baltic Sea in the winter. The viscosity of chemicals may change in cold, and they can be more persistent. Collecting techniques based on fluid-like masses are no longer effective, if fluids change and act more like solid masses. Moreover, it is difficult for a recovery fleet to operate, if it is surrounded by ice and snow. If chemicals have spread under the ice cover, detecting the spill is more difficult, and the use of dispersing agents is ineffective. However, ice breakers may be used to break the ice cover and to improve mixing chemicals with larger water masses (Hänninen & Rytkönen 2006).

4 STATISTICAL REVIEW ON TANKER ACCIDENTS IN THE BALTIC SEA

4.1 *Accident statistics by HELCOM and EMSA*

The Helsinki Commission (HELCOM) has reported that during the years 1989–2010 approximately 1400 ship accidents happened in the Baltic Sea. Most of the accidents were groundings and collisions, followed by pollutions, fires, machinery damages and technical failures (Fig. 1). One in ten of the accidents are defined as other types of accidents (HELCOM 2012).

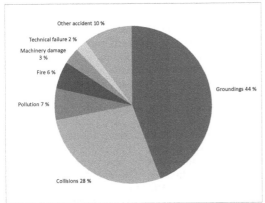

Figure 1. Vessel accidents in the Baltic Sea in 1989–2010 by accident types. (HELCOM 2012)

According to HELCOM (2012), 1520 vessels in total have been involved in the accidents occurred in the Baltic Sea during the years 1989–2010. Almost half of the vessels were different types of cargo vessels excluding tankers (Fig. 2). Large number of other vessel types (e.g. pilot vessels, tugs, dredgers) was also involved in the accidents. One in seven of the accidents involved a tanker and a passenger vessel.

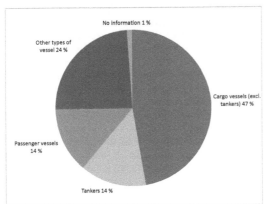

Figure 2. Vessel accidents in the Baltic Sea in 1989–2010 by vessel types. (HELCOM 2012)

Based on the HELCOM's accident statistics, 210 tankers (including crude oil tankers, chemical tankers, oil/chemical product tankers, gas carriers and other types of vessels carrying liquid bulk cargoes) were involved in the accidents that occurred in the Baltic Sea during the years 1989–2010. During this period, 28 of all tanker accidents in the Baltic Sea led to some sort of pollution. Due to these 28 pollution cases, approximately 3100 m³ of harmful substances in total spilled in the sea. In almost all of the pollution cases, spilled substance was conventional oil or an oil product (e.g. crude oil, gasoline oil, fuel oil, diesel oil) (Fig. 3). In one pollution case only, the spilled substance was a chemical (a leakage of 0.5 m³ of orthoxylene in Gothenburg on 13 February 1996). 13 out of the 28 tanker pollution cases in the Baltic Sea that were reported by HELCOM have been classified as spills/pollutions; 5 were classified as collisions; 3 as groundings; 2 as technical failures; 1 as machinery damage; 1 as contact with bollard; 1 as hull damage; 1 as loading accident; and 1 as an accident caused by broken hose. Over one-third (11) of all these tanker pollution accidents happened on the Swedish coast; 4 accidents happened in Lithuania; 3 accidents in Latvia; 2 accidents in Estonia; 2 accidents in Russia; 1 accident in Finland; 1 accident in Poland; 0 accidents in Germany; and 4 accidents in other areas of the Baltic Sea. The largest pollution case involving a tanker in the Baltic Sea during the period of 1989–2010 happened in the Danish waters on 29 March 2001 when approximately 2500 m³ of oil spilled into the sea as a result of a collision between a tanker and a bulk carrier (HELCOM 2012).

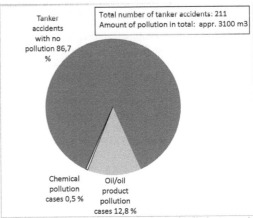

Figure 3. Tanker accidents and the share of pollution cases in the Baltic Sea in 1989–2010. (HELCOM 2012)

Based on the EMSA's Maritime Accident Reviews (EMSA 2007, 2008, 2009, 2010), the annual number of accidents in the Baltic Sea has varied between 75 and 120 accidents over the period of 2007–2010. In each of these years approximately 15 per cent of all maritime accidents in the EU happened in the Baltic Sea. During the reviewed period, the main causes of the accidents have been groundings (32–52 per cent of all accidents), followed by collisions/contacts (23–35 %), fires and explosions (10–17 %) and sinkings (2–5 %). In every year, the largest proportion of accidents happened in the south-western approaches off the Danish and Swedish coasts, with these accounting for around 70–77 per cent of the regional total. Groundings off the Danish and Swedish coasts accounted for around 80–88 per cent of the total Baltic Sea region groundings in the years 2007–2010. Most of the accidents in the region happened in the heavily trafficked approaches around eastern Denmark, which can be more difficult to navigate than many other areas. The recorded figures show that the Finnish and Estonian coastlines accounted for around 15–17 per cent of the total number of accidents happened in the Baltic Sea in this 4 year period. Accidents recorded by EMSA in the years 2007–2010 include 4 significant pollution events in total. As a consequence of these pollution events, at least 695 tonnes of oil/oil products spilled into the Baltic Sea (the size of pollution in one accident was not available). No significant chemical accidents happened in the Baltic Sea during the reviewed period. In addition to these significant pollution events, some smaller accidental spills were recorded by EMSA in the years 2007–2010. For example, in 2007 EMSA's daily research recorded about 30 accidental oil spills of different sizes in and around EU waters (EMSA 2007).

HELCOM and EMSA mainly provide coarse-level information about each maritime accident. Therefore, more detailed information on maritime accidents involving a tanker was searched using maritime accident databases and reports provided by the authorities and/or other actors who are responsible for collecting maritime accident data in each Baltic Sea country. More detailed maritime accident investigation reports were found about Denmark, Finland, Germany, Latvia and Sweden, and basic information about accidents was found about Estonia and Lithuania. There was no maritime accident data found about Poland or Russia.

4.2 National accident statistics

According to the Danish Maritime Authority's (DMA) annual marine accident publications (Danish Maritime Authority 2009), the total of 42 accidents involving a tanker registered under the Danish or Greenlandic flag happened during the period of 1999–2008. When examining foreign vessels, it can be seen that 63 foreign tankers in total were involved in the accidents that happened in Denmark's territorial waters in the reviewed period.

51 of these foreign tankers are classified as oil tankers, 9 as chemical tankers, and 3 as gas tankers. In addition to the DMA's annual marine accident publications, Danish Maritime Authority and the Danish Maritime Accident Investigation Board (DMAIB) have published, on their Internet sites, 142 maritime accident investigation reports or investigation summary reports on merchant ships during the years 1999–2011 (Danish Maritime Authority 2012, Danish Maritime Accident Investigation Board 2012). Study of these investigation reports revealed that 21 accidents involving a tanker in total were investigated by the DMA and the DMAIB. 9 of these accidents can be classified as personal accidents, 6 as collisions, 4 as groundings, 1 as an explosion, and 1 as an oil spill. Over half (11) of the accidents occurred in the Baltic Sea, 1 accident in the North Sea, and the rest of the accidents in other sea areas around the world. Only 2 of the investigated accidents led to pollution: 1) 2700 tonnes of fuel oil spilled in the sea as a consequence of a collision between two vessels in Flensburg Fjord in 2001 and 2) 400–500 litres of heavy fuel oil spilled into the sea during bunkering near Skagen in 2008.

Accident investigation reports provided by the Finnish Safety Investigation Authority shows that 10 tanker-related accidents in total happened to vessels in Finland's waters and to those that were sailing under Finnish flag during the period of 1997–2011. 4 of these accidents were groundings, 3 collisions, 2 spills and 1 personal injury. Two of the accidents led to spill: 1) on 20th July 2000 in the Port of Hamina, about 2 tonnes of nonyl phenol ethoxylate leaked on the quay area and into sea during loading, and 2) on 27th February 2002 in the port of Sjöldvik, about 2 m^3 of flammable petrol leaked into sea during unloading (Finnish Safety Investigation Authority 2012).

The study of the marine casualty statistics (BSU 2012a) and maritime casualty investigation reports (BSU 2012b) provided by the Federal Bureau of Maritime Casualty Investigation (BSU) revealed that during 2002–2011 the BSU recorded 27 marine casualties involving a tanker that happened in Germany's territorial waters or to vessels sailing under the German flag. 16 of these casualties were collisions, 7 personal accidents, 2 groundings, 1 water contamination, and 1 carbon monoxide exposure. 17 chemical tankers, 10 tankers, 1 river tanker and 1 motor tanker in total were involved in the accidents. Most of the accidents occurred in the Kiel Canal, in the Elbe River, in the Port of Hamburg, or outside Germany's waters. Only one of the accidents happened in the Baltic Sea, north of Fünen. Information about possible pollution as a consequence of an accident was not available in all cases. However, at least 18 of 27 accidents involving a tanker did not cause pollution and only 1 of the accidents was reported to have led to pollution (appr. 960 tonnes of sulphuric acid in the Port of Hamburg on 6 June 2004).

According to the maritime accident statistics of the Latvian Maritime Administration, the total of 30 accidents involving a liquid bulk vessel happened in Latvia's territorial waters or to vessels sailing under the Latvian flag during the period of 1993–2010. 17 of these accidents were classified as collisions, 3 as groundings, 3 as personal injuries, 2 as fires/explosions, 2 as pollutions, and 3 as other types of accidents. Unfortunately, the Latvian Maritime Administration's accident statistics do not provide information on whether the accidents caused pollution or not (Latvian Maritime Administration 2012).

The Swedish Transport Agency's annual maritime accident/incident reports (Swedish Transport Agency 2012a) revealed that the total of 90 accidents and 14 incidents involving a tanker occurred in the Swedish territorial waters during the period of 2002–2010. Machine damages (24 per cent of all the tanker accidents), groundings (22 %), collisions with other object than a vessel (19 %), and collisions between vessels (17 %) have been the most common reasons for tanker accidents. Approximately 51 per cent of the tankers involved in the accidents were vessels sailing under the Swedish flag and 49 per cent were foreign vessels. There was some lack of information, but it could be determined that at least 4 of these accidents led to pollution (Swedish Transport Agency 2012a, 2012b): 1) 500 litres of fuel oil spilled from a fuel tank during bunkering in Gothenburg in 2005; 2) 100 litres of gas oil spilled into the sea as a consequence of a collision between two vessels in Gothenburg in 1998; 3) approximately 45 m^3 of gas oil spilled from a fuel tank due to vessel grounding in Brofjorden in 1999; and 4) approximately 600 tonnes of hydrochloric acid were released into the sea under the control of the Swedish Maritime Administration near Öresund in 2000 as a consequence of a collision between two vessels.

According to the Estonian Maritime Administration, the total of 16 accidents involving a tanker happened to vessels in Estonia's territorial waters, or to vessels which have been sailing under Estonia's flag during the period of 2002–2011. 7 of these accidents were groundings, 3 fires, 4 contacts with a quay, and 2 collisions. None of the accidents have caused pollution (Estonian Maritime Administration 2012).

According to the maritime accident statistics of the Lithuanian Maritime Safety Administration, 12 accidents involving a liquid bulk vessel happened in Lithuania's territorial waters or to vessels sailing under the Lithuanian flag during the period of 2001–2010. 4 of these accidents can be classified as spills, 3 as collisions, 2 as contacts with a quay/other

vessel, 1 as fire, and 2 as other types of accidents. As a consequence of the 4 spill types in the accidents, at least 3.5 tonnes of oil and 0.06 tonnes of diesel fuel leaked into the sea in the Lithuanian waters. The amount of oil spilled in the water is probably higher since regarding the 2 oil spill cases, there was no information available about the level of pollution (Lithuanian Maritime Safety Administration 2012).

5 SUMMARY AND CONCLUSIONS

This paper provided an overview of the past tanker accidents in the Baltic Sea and HNS accidents in seas worldwide. It also aimed at finding out what can be learned from past accidents, especially from the environmental point of view.

The results of this study showed that chemical tanker accidents are very rare, even though there is always the possibility that such incident may happen. Many other studies have shown that the most commonly transported chemicals are the ones most likely to be involved in an accident. Moreover, the risks are different and vary in different sea areas. The risk of an accident is the highest in water areas where the largest amounts of chemicals are transported, the density of the maritime traffic is at its highest point, where bad weather conditions exists, as well as the ship-shore interface in ports where unloading/loading take place. Incidents involving chemical spills are statistically much less likely to occur than oil spills.

Actually, very little is known about the actual marine pollution effect of most of highly transported substances. From the environmental point of view, the previous studies have highlighted accidents in which pesticides were released to water, but also substances considered as non-pollutants (vegetable oils) seem to have a negative effect on biota in the water environment. When comparing hazardous chemicals with oil, it can be said that the danger of coastline pollution is a far greater concern in oil spills than in chemical spills. It is very difficult to evaluate chemical risks if a ship is carrying diverse chemicals and some of those substances are unknown during the first hours after the accident. This aforementioned situation is often faced when a vessel is carrying packaged dangerous goods. The most important difference between chemical and oil spill may be related to response actions. The air quality or the risk of explosion does not usually cause concern for response personnel in case of oil spills, but for chemical spills, it should be carefully evaluated if some response actions are made. In case of chemical spills, the response may be limited, in most cases, to initial evaluation, establishing exclusions zones, modeling and monitoring, followed by planning of a controlled release, recovery or leaving in-situ. This process will take many weeks or even months.

Both literary and data mining showed that neither major chemical spills nor oil spills, such as Erika or Prestige, have happened in the Baltic Sea. However, every year over 100 shipping accidents (all cargoes included) take place in the Baltic Sea. Collisions and groundings are the main types of accident/incidents in the Baltic Sea. Human factor is the main cause for the accidents, followed by technical reasons. The largest proportion of accidents happens in the south-western approaches off the Danish and Swedish coasts. Annually, on average, 15 per cent of all shipping accidents in the Baltic Sea have involved a tanker. Less than 5 per cent of the tanker accidents have led to spill/pollution. The spilled substance has in most cases been oil or an oil product – only very few chemical spill cases have been reported in the Baltic Sea. Considering both chemical and oil tankers, only very small spills have happened and their environmental impact has been neglected. Since there have been no major accidents in the Baltic Sea, it is not possible to learn about accident cases. However, there are some excellently described international tanker accidents which give valuable lessons to be learned from by different stakeholders and rescue services.

There are many parties in the Baltic Sea Region, including e.g. HELCOM, EMSA and the national authorities, which are collecting/producing data on the maritime accidents that have occurred in the Baltic Sea. In addition, some European or worldwide databases (e.g. Cedre) contain data of accidents that have occurred in the Baltic Sea. However, in the future, the maritime accident databases on the Baltic Sea Region should be improved and harmonised. Regarding accident investigation reports, each Baltic Sea country should publish these reports publicly in electronic format. It would be worth to contemplate whether all accident investigation reports concerning accidents that have occurred in the Baltic Sea waters or to vessels sailing under a Baltic Sea country's flag could be gathered under one public information service.

ACKNOWLEDGEMENTS

This study is made as a part of the Chembaltic (Risks of Maritime Transportation of Chemicals in Baltic Sea) project. Special thanks to the European Regional Development Fund (ERDF), the Finnish Funding Agency for Technology and Innovation (Tekes), companies supporting the research project, and all the research partners being involved in the project.

REFERENCES

Bogalecka, M & Popek 2008. Analysis of Sea Accidents in 2006. TransNav, International Journal on Marine Navigation and Safety of Sea Transportation 2(2):179–182.
BSU 2012a. The Federal Bureau of Maritime Casualty Investigation's (BSU) statistics about marine casualties and serious marine incidents on sea and an account of its activities in the last accounting year. Available at: http://www.bsu-bund.de/cln_030/nn_101790/EN/publications/Annual__Statistics/annual__statistics__node.html?__nnn=true (accessed 10 August 2012).
BSU 2012b. Investigation Reports 2003–2012. Available at: http://www.bsu-bund.de/cln_030/nn_101790/EN/publications/Investigation_20Reports/investigation__report__node.html?__nnn=true (accessed 13 August 2012).
Bucas, G. & Saliot, A. 2002. Sea transport of animal and vegetable oils and its environmental consequences. Marine Pollution Bulletin 44: 1388–1396.
Cedre 2012. The Internet site of Centre of Documentation, Research and Experimentation on Accidental Water Pollution. Available at: www.cedre.fr (accessed 10.08.2012).
Cedre and Transport Canada 2012. Understanding Chemical Pollution at Sea. Learning Guide. Brest: Cedre, 2012. 93 pp.
Danish Maritime Accident Investigation Board 2012. Casualty reports from the years 1999–2011. Available at: http://www.dmaib.com/Sider/CasualtyReports.aspx (accessed 21 August 2012).
Danish Maritime Authority 2009. Marine Accidents 2009. Available at: http://www.dma.dk/SiteCollectionDocuments/Publikationer/Maritime-accidents/Accidents%20at%20Sea%202009.pdf (accessed 16 August 2012).
Danish Maritime Authority 2012. About the Division for Investigation of Maritime Accidents. Available at: http://www.dma.dk/Investigation/Sider/Aboutus.aspx (accessed 14 June 2012).
EMSA 2007. Maritime Accident Review 2007. Available at: http://emsa.europa.eu/emsa-documents/download/374/216/23.html (accessed 8 August 2012).
EMSA 2008. Maritime Accident Review 2008. Available at: http://emsa.europa.eu/emsa-documents/latest/download/373/216/23.html (accessed 8 August 2012).
EMSA 2009. Maritime Accident Review 2009. Available at: http://emsa.europa.eu/emsa-documents/latest/download/308/216/23.html (accessed 8 August 2012).
EMSA 2010. Maritime Accident Review 2010. Available at: http://emsa.europa.eu/implementation-tasks/accident-investigation/download/1388/1219/23.html (accessed 8 August 2012).
Estonian Maritime Administration 2012. Laevaõnnetuste juurdluskokkuvõtted [Marine casualty reports]. In Estonian. Available at: http://www.vta.ee/atp/index.php?id=720 (accessed 17 July 2012).
Finnish Safety Investigation Authority 2012. Vesionnettomuuksien tutkinta [Investigation of water accident]. In Finnish. Available at: http://www.turvallisuustutkinta.fi/Etusivu/Tutkintaselostukset/Vesiliikenne (accessed 17 July 2012).
French McKay, D.P., Whittier, N. Ward, M. & Santos, C. 2006. Spill hazard evaluation for chemicals shipped in bulk

using modeling. Environmental Modelling and Software, vol 21, pp. 156–159.

GESAMP (2002). The revised GESAMP hazard evaluation procedure for chemical substances carried by ships, GESAMP reports and studies No 64, No 463/03, 137 pp.

HASREP 2005. Response to harmful substances spilled at sea. Task 2 Risk assessment methodology for the transport of hazardous and harmful substances in the European Union maritime waters. Cedre. 32 pp.

Harold, P., Russell, D. & Louchart 2011. Risk prioritization methodology for hazardous & noxious substances for public health, ACROPOL, The Atlantic Regions'Coastal Pollution Response, Pembrokeshire County Council and Health Protection Agency.

HELCOM 2009. Overview of the shipping traffic in the Baltic Sea. Available at: http://www.helcom.fi/stc/files/shipping/Overview%20of%2 0ships%20traffic_updateApril2009.pdf (accessed 21 March 2011).

HELCOM 2002. Response to accidents at sea involving spills of hazardous substances and loss of packaged dangerous goods. HELCOM Manual on Co-operation in Response to Marine Pollution within the framework of the Convention on the Protection of the Marine Environment of the Baltic Sea Area (Helsinki Convention), Volume 2, 1 December 2002.

HELCOM 2012. Accidents and response – Compilations on Ship Accidents in the Baltic Sea Area. Available at: http://www.helcom.fi/shipping/accidents/en_GB/accidents/ (accessed 5 October 2012).

Holma, E., Heikkilä, A., Helminen, R. & Kajander, S. 2011. Baltic Port List 2011 – Annual cargo statistics of ports in the Baltic Sea Region. A publication from the Centre for Maritime Studies, University of Turku. 180 p.

Häkkinen, J., Malk, V., Penttinen, O.-P., Mäkelä, R. & Posti, A. 2012. Environmental risk assessment of most transported chemicals in sea and on land. An analysis of southern Finland and the Baltic Sea. In: Töyli, J., Johansson, L., Lorentz, H., Ojala, L. and Laari, S. (Ed.), NOFOMA 2012 – Proceedings of the 24th annual Nordic logistics research network conference, 7–8 June 2012, Naantali, Finland.

Hänninen, S. & J. Rytkönen 2006. Transportation of liquid bulk chemicals by tankers in the Baltic Sea. Technical Research Centre of Finland. VTT publications 595. 121 p. Espoo, Finland. Available at: http://www.vtt.fi/inf/pdf/publications/2006/P595.pdf (accessed 14 February 2012).

Hänninen, M., Kujala, P., Ylitalo, J., & Kuronen, J. 2012. Estimating the Number of Tanker Collisions in the Gulf of Finland in 2015. TransNav, International Journal on Marine Navigation and Safety of Sea Transportation 6(3): 367–373.

IMO 2007. Manual on Chemical Pollution.2007 edition.

Kirby Mark F. & Law R. J. 2010. Accidental spills at sea - risk, impact, mitigation and the need for co-ordinated post-incident monitoring. Marine Pollution Bulletin 60: 797–803.

Latvian Maritime Administration 2012. The Internet site of Latvian Maritime Administration. In Latvian. Available at: http://www.jurasadministracija.lv/index.php?action=145 (accessed 4 June 2012).

Law, R.J. & Cambell, J.A. 1998. The effects of oil and chemical spillages at sea. The Journal of the Chartered Institutions of Water and Environmental Management 12, 245–249.

Le Floch, S., Fuhrer, M., Slangen, P. & Aprin, L. 2012. Environmental Parameter Effects on the Fate of a Chemical Slick. Chapter: 02/2012; ISBN: 978-953-51-0161-11In book: Air Quality - Monitoring and Modeling.

Lithuanian Maritime Safety Administration 2012. Laivų avarijų ir avarinių atvejų 2001–2010, išnagrinėtų Lietuvos saugios laivybos administracijoje, statistika [Ship accidents and emergency situations in 2001–2010, investigated by the Lithuanian Maritime Safety Administration, the statistics]. In Lithuanian. Available at: http://msa.lt/download/1406/avariju_statistika.pdf (accessed 9 August 2012).

Mamaca, E., Girin, M. le Floch, S. & le Zir R. 2009. Review of chemical spills at sea and lessons learnt. A technical append. to the Interspill 2009 conference white paper. 39 pp.

Marchand, M. 2002. Chemical spills at sea. In M. Fingas (ed.), The handbook of hazardous materials spills technology. McGraw-Hill, New York, 2002.

Mullai, A., Larsson, E. & Norrman, A. 2009. A study of marine incident databases in the Baltic Sea region. TransNav, International Journal on Marine Navigation and Safety of Sea Transportation 3(3): 321–326.

Posti, A. & Häkkinen, J. 2012. Survey of transportation of liquid bulk chemicals in the Baltic Sea. Publications from the Centre for Maritime Studies University of Turku, A 60.

Purnell, K. 2009. Are HSN spills more dangerous than oil spills? A white paper for the Interspill Conference & the 4th IMO R&D Forum, Marseille, May 2009.

Riihimäki, V., L. Isotalo, M. Jauhiainen, B. Kemiläinen, I. Laamanen, M. Luotamo, R. Riala & A. Zitting 2005. Kemikaaliturvallisuuden tiedonlähteet. (Sources of information about the chemical safety). In Finnish. 2. ed. 151 p. Finnish Institute of Occupational Health. Helsinki, Finland.

Suominen, M. & Suhonen, M. 2007, Dangerous goods related incidents and accidents in the Baltic Sea region, DaGoB publication series, vol. 7:2007.

Swedish Transport Agency 2012a. Sjöolyckor i svenska farvatten – Sammanställning av rapporterade sjöolyckor i svenska farvatten med svenska och utländska handels- och fiskefartyg, årliga redovisningar 2002–2010 [Maritime accidents in Swedish waters – Summary of reported marine casualties in Swedish waters with Swedish and foreign merchant and fishing vessels, annual reports 2002–2010]. In Swedish. Available at: http://www.transportstyrelsen.se/sv/Sjofart/Olyckor--tillbud/Statistiksammanstallning (accessed 23 August 2012).

Swedish Transport Agency 2012b. Publicerade haverirapporter i 1997–2011 [Published accident investigation reports in 1997–2011]. In Swedish. Available at: http://www.transportstyrelsen.se/sv/Sjofart/Olyckor--tillbud/Haverirapporter/Publicerade-haverirapporter/ (accessed 27 August 2012).

US Coast Guard 1999. Hazardous Substances Spill Report, Vol. II no 8.

Wern, J. 2002. Report on incidents involving the carriage of hazardous and noxious substances (HNS) by sea. Department for Transport. London.

Factors Affecting Operational Efficiency of Chemical Cargo Terminals: A Qualitative Approach

T.A. Gülcan, S. Esmer, Y. Zorba & G. Şengönül
Dokuz Eylul University, Maritime Faculty, Izmir, Turkey

ABSTRACT: Chemical cargo terminals constitute are a special terminal form where high and international levels of safety and quality elements applied. Unlike conventional bulk cargo and container cargo operations, chemical cargo operations include own priorities, applications, and the evaluation criteria. The aim of this study is to perform a qualitative research to determine the factors affecting the operational efficiency of ship, berth and warehousing operations in chemical cargo terminals.

1 INTRODUCTION

The influence of the chemicals, mineral oils and petrochemicals industry in daily life and in industry is well known – chemical and petrochemical products go into the manufacture of soaps, pharmaceuticals, plastics, tires and other objects vital to the onward march of civilization as well as mineral oils are both used by public and industry. However, before consumers can reap the benefits of these products, a great deal of logistical planning goes into the manufacture, transport and processing (Gaurav Nath & Brian Ramos, 2011, Marine Dock Optimization for a Bulk Chemicals Manufacturing Facility). Today there are three kinds of terminals; the ones having their own refineries, terminals that only rent storage tanks for their customers only and the ones which include the both. The logistics part of these terminals deal with loading, unloading and also transporting these products via truck, train, pipeline and ships in which operation activities play the most important role. To become a global and regional terminal, today's ports should always be in improvement process about operational efficiency of their terminals in accordance with the regional and international rules and manuals.

2 IMPORTANCE OF SEA TERMINALS

In today's global economic conditions, there is worldwide storage need for chemical mineral oil and petrochemical industry producers and customers. Port of Rotterdam offers more than 30 million cubic meter of tank storage capacity for all types of liquid bulk. Products handled include crude oil, mineral oil products such as petrol, diesel, kerosene and naphtha, all kinds of bulk chemicals and edible oils and fats. In Port of Rotterdam region there are now five oil refineries, which process the imported oil, and over 45 chemical companies which have large-scale facilities. There is also 1500 km of pipelines interconnecting oil and chemical companies (http://www.portofrotterdam.com).

These liquid raw materials and products are commonly transported by maritime transportation mode because of its lowest cost per ton mile and amount efficiency. Also pipelines as mentioned above play another important role for the transfer of the raw materials and products between refineries and terminals, especially located in the same geographical area or where maritime transportation is not cost/effective like Baku-Tiflis-Ceyhan pipeline.

Truck and railway transportation modes are mostly used domestically for shipping the products from the terminals to the manufacturers.

All of these facilities require a terminal with its berth or jetties for the ships and also for the barges, railway for the trains, locomotives for the wagons, roads and stations for the trucks, pipelines between the terminals and/or refineries, tank farms for the storage of the raw materials and products, hoses or pipelines between the berth/jetty, wagon and truck loading/unloading stations.

During loading and unloading of the liquid chemicals, operational safety is another important factor. Spills and accidents can be seen in many ways e.g. (Duffey and Saull,185:2009); while filling, in storage, during transport, at process and transfer facilities; plus failures of vessels and pipeline. Safe and efficient operational procedures should include design, control and management with together considering all relevant factors in chemical terminals. Therefore "The Operational Efficiency of the Terminals" is a very important component on top of the facilities mentioned above.

3 METHODOLOGY

In this work "In-depth Interview" method was used face to face with the authorized Operational Manager/Staff of the companies as listed below. Because of all manager and staff do not want to disclose their names, the table do not include name of the participants.

Terminals	Staff Positions	Date
OIL Tanking / Hamburg / Germany	Terminal Manager	Nov. 2012
VOPAK / Hamburg / Germany	Operations Manager	Nov. 2012
DOW International / Hamburg / Germany	Dock Operations Leader	Nov. 2012
SOLVENTAŞ / İzmit / Turkey	General Manager	Dec. 2012
LİMAŞ / İzmit / Turkey	Tank Terminal Manager	Dec.2012

The research questions were about the following topics:
– Jetty capabilities of the companies,
– The intermodal logistics capabilities of the companies,
– Loading and unloading automatic system/tools they use,
– The software systems they benefit during the operations and their tools.
– The watch systems for the operational staff the companies apply (number of personnel at operation stations, working hours, watch system etc.),
– The training systems,
– The inspections of the terminals,
– The Risk analyses procedures.

4 RESEARCH FINDINGS

4.1 *Jetty capabilities of the companies;*

Numbers of Jetties of the terminals are as listed below.

	Vopak	Oil Tanking	Dow	Solventaş	Limaş
# of Jetties	9	5	3	2	2
Drafts (m)	3.5-12	3.6-12.8	7-14	10-25	11-22

The products handled in the jetties of VOPAK and OIL Tanking are mostly mineral oils and this is the reason why these jetties are convenient for ships between 2.000 and 200.000 dwt. VOPAK is also handling sulfuric acid as chemicals. In the inside parts of the jetties of these two terminals, handling operations are usually realized with the barges and only hoses are used in handling operations. The mineral oils can be handled up to 2000 cbm/hour in OIL Tanking and also 1000 cbm/hour in VOPAK with loading arms according to the receiving capacity of the ships and to the property of the products. Although, pipelines used in mineral oil handlings are generally produced for a maximum pressure of 12-13 bars, they're usually used under pressures of between 6-7 bars due to safety and material lifetime.

DOW is handling only chemical products in its terminal with its jetties between 155 meter and 270 meter long. The loading arms on the jetties can be remote controlled which prevents the possible delays caused by the ship maneuvers.

SOLVENTAŞ uses one of its jetties for chemical liquids and the other one for fuel and gas oil handlings which are 250 and 275 meter long. There is real-time fuel oil and gas oil blending capability on the jetty as loaded to the barges for bunkering. On chemical jetty, 42 separate products can be handled at the same time with 4 or 8 ships according to their tonnages. LİMAŞ can handle 10 separate chemicals simultaneously on its 165 meter long jetty with two ships.

As described "The Physical Oceanographic" effect, tidal level in the Elbe River reaches up to 5 meter which causes delays in ship operations in connection with the drafts of the ships sometimes.

4.2 *Intermodal logistics capabilities of the companies;*

The European railway network is directly connected to the terminals in Hamburg and therefore is a very flexible instrument for transports leaving Hamburg and arriving at the terminals from the hinterland. All three companies in Hamburg have their own locomotives and railway inside their terminals. The yearly average number of wagons handled in OIL Tanking is 20.000. Also this number in VOPAK is daily between 100-200 wagons. As a result, the amount of handled liquid by railway is more than seaborne transports in these two companies 26% of the products leave DOW / Hamburg terminal by railway.

VOPAK and OIL Tanking has pipeline connection between their terminals and also with other refineries in their region. DOW international has a 380 km. long Ethylene pipeline inside Germany to its other refineries.

Tanker loading capabilities allow these three companies serious amounts of product handling and transporting them via trucks inside Germany and Europe. OIL Tanking handles average 65.000 tankers yearly and DOW / Hamburg forwards its 21% of chemical products by road transport by tankers.

The firms located in İzmit/TURKEY use seaborne and tanker transportation modes in common.

LİMAŞ has pipeline connections with two companies producing chemical products in its region. The average Tanker loading number in SOLVENTAŞ is daily 250 and has 43 loading stations which allows a yearly handling amount 1.400.000 tons in average. The loading stations number in LİMAŞ is 16 with a daily average 100 tankers loading capacity.

4.3 *The Automatic Loading and Unloading system/tools the companies use;*

All the terminals use automatic handling systems in accordance with their capacities. In this case, VOPAK and OIL Tanking can control all the handling cycle with the help of the software by which they realize the planning and handling that includes from which station and line number the product loading is going to be realized or which tank is going to be unloaded/loaded, in the "Control Rooms" they use. The staff working in these control rooms can control the level of the products in the Tanks and also the physical conditions of the products real time as well. Handling operations with ships and wagons are completed under the auspices of terminal staff.

The three Hamburg located terminals use full automatic loading systems for the tankers. This loading process is realised under the terminal's safety and security rules only by the tanker drivers who pass the tests made at the entrance of the terminal and who are experienced in automatic loading at least for a specific time that the company defines.

If the driver makes some mistakes during the loading process, then the system doesn't let him to go on with loading and warns the staff in the control room for helping the driver with the communication system or personally.

SOLVENTAŞ is realizing all the handling operations, including the ones that are completed under nitrogen cover automatically with help of the software the company created. The handling planning should be done by using this program and

it doesn't let the planner to do this over the lines or valves that malfunction or under construction which inhibits the accident possibilities by the material. In loading process of tankers, it starts automatically by entering the number of "Loading Conformity Paper" by the staff to the system at the loading station which is brought by the tanker driver and ends automatically when the volume of the product reaches the required amount as it should be.

4.4 *The Software systems the terminals benefit during the operations;*

The examined terminals are all using various software according to their capacities during their operational facilities, connected within the framework of delegated limitations to the other departments such as technical and commercial.

After the clients order, handle planning is realized via these Decision Support System software including the variables like ETA of the vehicles or ships, the line numbers going to be used during handling, the necessary tank levels at the beginning and at the end etc. Additionally by the Local Area Network, operators can achieve ship's information, essential manuals, and procedures and check lists for the operations which they're assigned for with these software's. During the operations if operator does something wrong than the program automatically stops the handling process and informs the control room or quality management departments of the terminals.

Further the stated tools, some terminals like SOLVENTAŞ enable tank leaseholders, owners of the products and freight forwarders to achieve with in competence of they are allowed to its software database to check out the real time information about their products, the bureaucratic works status etc. This software tool capability enables the freight forwarders make their loading and shipping plans by entering all the information about the tanker and also the drivers to the system.

After the freight forwarders' handling planning are loaded in the system, if traffic or other issues don't let the plan get realized at the terminal then the related staff inform the forwarders about the situation and guide them.

4.5 *The operational staff working systems;*

In the Hamburg terminals, the handling process continues 24 hours for ships, barges and wagons. Tanker operations are 24 hours only in OIL Tanking terminal. In SOLVENTAŞ and LİMAŞ terminals ship handling processes are also 24 hours. Tanker operations in this two terminals are only daytime available.

Although, all terminals have various watch systems according to their personnel numbers, they apply daily 8 hour working with 3 watches (LİMAŞ has 2 watches). Some of them support the day time watches with staff who works only at day times on working days. Every watch except DOW has Watch Leaders. The watch leaders at SOLVENTAS should be ship engineers in principle.

The watch leaders assign their watch staff to the stations according to their skills and experience after they analyze the Planning Department's daily operational plans. Except operational problems, OIL Tanking doesn't assign any staff to the tanker loading area.

According to the GERMAN rules, during the handling operations at jetties, one staff should always be on duty on jetty. Additionally on jetties, in all terminals in HAMBURG there are always enough numbers of staff at train loading stations and in tank farm area. The terminals in Hamburg and also IZMIT principle about their staff are their having the skills to work on every station inside the terminal. In DOW and OIL Tanking terminals in every watch there are a few locomotive drivers who are trained and licensed by Deutsche Bahn.

4.6 *The training systems for the Operational staff;*

All operational staff both in Germany and Turkey are well trained by internal and also external trainers as well. According to the international and national rules, all of the staff should be trained in specific issues like IMDG Code, ISPS Code, Fire Fighting and First Aid. These trainings are generally given by licensed internal trainer in the terminal. SOLVENTAS and LİMAŞ are also trains it's staff about "Emergency Response Against Marine Pollution".

Additionally these trainings, simulators are used in some terminals for training the operators especially to build up their visual memories. OIL Tanking is using a wagon simulator from an external training company to train its staff and is planning to do this with a ship simulator next year.

4.7 *The inspections of the terminals;*

Today's global economic circumstances, safety and security rules forces the terminals to have certificates which are valid worldwide to subsist in the market. All terminals in this work have the technical and quality (ISO) certificates according to their capabilities and are inspected frequently to keep these standards.

Today, intuitions like CDI or SGS imposed themselves worldwide and the terminals which work with their standards and have their certificates are always one step forward to the others in the competition.

Some companies like OIL Tanking creates an inspection team with its employees who work at the other terminals worldwide an inspects it's terminals with this teams.

4.8 *The risk analyze procedures to minimize accidents during the operations activities;*

Analyzing all risks, accidents and taking precautions principle is implemented by all the terminals in this work. Although the analyzing methods are various, the managers and watch leaders determines the possible risks during the operations and after analyzing them with coefficients, bring out measures to minimize them.

5 CONCLUSION

Almost all terminals included in this work primary subject is to convert the manual handling systems to full automatic systems by the time to prevent the accident possibilities caused by human mistakes and to save up from labor force and leeway.

Especially railway intermodal mode affects the operational efficiency positively in terminals and doesn't require labor force like road mode. Investments on upper structure in this case by Eastern European countries and Turkey and integration with Western European countries would increase the capacity seriously.

Determining the specific criteria for the tanker drivers to enable them to do loading operations in automatic stations without terminal staff and applying them widely would affect the operational efficiency positively.

Making use of simulators by training the operational staff would give the personnel a visual memory which would be helpful them during the operational activities.

Allowing the customers to enter the terminals software within the framework and to make their own handle plan with the terminals planning department can help the planning department in making operational plans.

REFERENCES

Duffey, R.B., Saull, J.W., 2009, Managing and Predicting Maritime and Off-shore Risk, International Journal on Marine Navigation and Safety of Sea Transportation, Vol.3, Number 2, pp.181-188

Gaurav Nath & Brian Ramos, 2011, Marine Dock Optimization for a Bulk Chemicals Manufacturing Facility

http://www.portofrotterdam.com, 08.10.2012.

http://www.dowstade.de, 08.10.2012.

Interviews:

DOW International, Hamburg, Germany. Dock Operations Leader, November 2012.

LİMAŞ, İzmit/Turkey, Tank Terminal Manager, December 2012.

SOLVENTAŞ, İzmit/Turkey, General Manager, December 2012.

OIL Tanking, Hamburg/Germany, Terminal Manager, November 2012.

VOPAK, Hamburg/Germany, Operations Manager, November 2012.

Pollution at Sea, Cargo Safety, Environment Protection and Ecology
Maritime Transport & Shipping – Marine Navigation and Safety of Sea Transportation – Weintrit & Neumann (Eds)

The Parameters Determining the Safety of Sea Transport of Mineral Concentrates

M. Popek
Gdynia Maritime University, Gdynia, Poland

ABSTRACT: Solid bulk cargoes belong to two major groups of goods classified in sea transportation. The safe transportation of these cargoes is a responsible task. When the wet granular materials, such as mineral concentrates and coals lose their shear strength resulted from increased pore pressure, they flow like fluids. Too high humidity of cargo leading to its liquefaction may cause the shift of the cargo . In consequence, it may cause ship's heel and even its capsizing and sinking. The oxidation of mineral concentrates, under certain circumstances, leads to spontaneous combustion which can cause many serious problems during storage and transportation.
The results of the investigation on possibility of using starch as absorber (hydrophilic) material are presented. Biodegradable materials, composed of starch are added to the ore to prevent sliding and shifting of ore concentrates in storage. The role of starch materials in properties of mineral concentrates from the point of view of safe shipment was investigated.

1 SEA TRANSPORT AS A SECTOR OF MARITIME ECONOMY

1.1 *Sea transport*

Carriage of goods by sea is the most important form of transport in the world. In 2011, it was accounted for more than 80% of global freight. The volume of transportation of goods by sea is an indicator of the global economy. Global reduction in the gross world product and reduction of the trade in goods drastically affected the value of maritime transport. Positive macroeconomic phenomena have a direct impact on the state of maritime transport and structure. An important factor determining the state of the maritime industry is the structure of maritime transport [Grzelakowski 2009].

Te United Nations Conference on Trade and Development, making an annual analysis of the maritime economy, divided all the cargoes according to the following groups:
– oil and oil products,
– dry cargo - divided into two groups:
1 five basic solid bulk cargoes: concentrates and iron ore, coal, bauxite, phosphates, and grain;
2 other bulk cargoes: metal ore concentrates. agricultural goods and construction materials.

The volume of the cargo transportation of these groups is an indicator of the global economy [UN 2010].

1.2 *Transport of bulk cargoes*

Ore concentrates and other similar fine – grained materials shipped by sea are mostly loaded in bulk without packing and considered as bulk cargoes. Over the 1990 - 2011 period major dry bulk volumes moved, growing at an average rate over 5%. They represent over 25% of the volume of cargo transported by sea. Two loads – coal and iron ore concentrate predominate in this group. The demand for iron ore remains in the global economy at a high level, thus a steady increase in volumes is observed, for example: in the 2010 approximately 8,6%. The demand for coal remained at the same level, as in 2010 only a slight increase in volume of approximately 2,1% was observed.

The year 2010 was positive for major and minor dry bulks and other dry cargo as total volumes bounced back and grew by 8,4%. In the group of the remaining goods solid, especially mineral resources, in the year 2010 there was an increase in carriage by about 10 %, and the increase of demand for oil, which revived the freight market in this area [UN 2011].

2 SEA TRANSPORTOTION OF MINERAL CONCENTRATES

2.1 *Hazards*

The transportation safety of mineral concentrates is dependent upon the measurement and control of its properties as well as the behavior of bulk on a macro-scale. Because there is a great variety in the solids properties and characteristics of such assemblies, there is a need to understand the mechanico-physical properties of solid particles as well as the physico-chemical interaction of adsorbed solid-water boundaries on particles.

The comparison of qualitative behavior of solid materials gives a very general characterization of the cargoes. These terms may be sufficient to ensure the ability for safe shipment by sea. Solid bulk cargoes as being either three–phase (solids–water–air) or two (solids–water) structure can be treated as a single continuum.

These cargo may liquefy when contains at least some fine particles and some moisture, although it need not be visible wet in appearance. They may liquefy if shipped with a moisture content in the excess of its Flow Moisture Point [IMO 2011]. Too high humidity of cargo leading to its liquefaction may cause the shift of the cargo and in consequence , ship's heel and even its capsizing and sinking.

Self-heating is a catalytic process resulting in the accumulation of heat in the load, which is released during the oxidation of sulphide minerals with oxygen from the air. Consequently, the local temperature rise in the load causes a further acceleration of the process of oxidation, generating a faster increase in temperature [Zarrouk & O'Sullivan 2006].

The oxidation of mineral concentrates, under certain circumstances, leads to self-heating and, under exceptional conditions, to combustion. Associated phenomena, such as atmospheric oxygen depletion, emission of toxic fumes, corrosion, agglomeration and sintering also arise during transport and the storage of concentrates. In conditions of maritime transport, this process poses a threat to human life and health and the environment.It lowers the quality of the cargo as well as consequently substantial economic losses [Bouffard & Senior 2011].

Many factors would determine the liquefaction and oxidation such as the cargo environment and the physical variables.

2.2 *Liquefaction*

When the wet granular materials lose their shear strength resulted from increased pore water pressure, they flow like fluids. The flow state is a state that occurs when a mass of granular material is saturated with liquid to an extent that, under the influence of prevailing external force such as vibration, impaction or ship's motion [Zhang 2005].

To minimize the risk of liquefaction, the IMSBC Code introduces the upper bound of moisture content of cargo called the Transportable Moisture Limit (TML). The TML is defined as 90% of the Flow Moisture Content (FMP), which depends on the characteristics of cargo and should be measured experimentally.

2.2.1 *The method of estimation TML*

Many methods dealing with determination of the moisture content which simulates transition of fine-grained bulk cargo from solid into liquid state in sea trade conditions can be found in the scientific literature. IMO approved the following assessment methods of safe moisture content:
- Flow Table Test,
- Japanese Penetration Method,
- Proctor C/Fagerberg Method.

The results of estimationn TML obtained using Proctor C/Fagerberg Method are higher in all cases than those given by remaining methods. The results of FMP determination obtained using Penetration Method are consistent with those got from the Flow Table Test [Popek & Rutkowska 2001]. Statistical parameters calculated for the measured values confirmed the conclusion.

2.3 *Self–heating of concentrates*

Typical moisture contents of concentrates under shipment vary from 3 to 10% by weight. Thus, mineral concentrates are moist and, in fine particle size–conditions, conducive to oxidation. Mineral concentrates may exhibit self-heating to various degrees during storage or shipping. The worst cases present a serious fire to SO_2 emission hazard. It has been recognized that oxygen and moisture was of great importance in the triggering of self–heating. Bone dry material will neither heat nor moist the material in the absence of a supply of oxygen. The onset of self–heating leading to combustion in metal sulphide concentrates has been taken to consist of an initiating event, a low temperature, aqueous oxidation ($<100^0C$) and high–temperature oxidation. At low temperatures, aqueous exothermic sulphation reaction, similar to the process of weathering, is exhibited by many sulphids in the initial stages of the oxidation. The accumulation of the heat can lead to autothermic reaction associated with active combustion and evaluation of sulphur dioxide, which leads, in turn, to oxide products.

2.3.1 *Methods of estimation ability to self-heating*

Using different methods research have been made focusing on the propensity of sulphide concentrates to spontaneous combustion. To guard against the risk of self-heating during overseas transport, there is usually testing used and approved by protocol such as the United Nations Transportation of Dangerous Goods Protocol. [UNECE 2008]. According to the Protocol, concentrate is subjected to the "*cage*" test which classifies it as self-heating or not. The most likely exothermic reactions are the oxidation of elemental sulphur to sulphur dioxide in the 140-300^{0}C and the oxidation of sulphide minerals above 400^{0}C. If the concentrate temperature does not increase to over 200^{0}C over the 24 h period, then the concentrate is classified as a non-self–heating material. [Rosenblum et al. 2001].

This test has some inconveniences: requires a large mass of concentrate, a minimum of 24 h, the cage and furnace are not the equipment available in most laboratories.

That is the reason, why a fast, inexpensive and quantitative method to measure the self –heating character of sulphide concentrates has been developed. The new method, intended for quick diagnosis, consists of measuring the amount of sulphur evolved at 300^{0}C. Elemental sulphur is a product of oxidation of sulphide minerals. The content of elemental sulphur in the concentrates has proven to be a good indicator of self –heating potential of a sulphide concentrate. The sulphur content is correlated to the UN self –heating protocol that measures the rise of temperature that occurs when a concentrate is heated. The sample that contains less than 2% elemental sulphur is not to experience a temperature rise of more than 60^{0}C, considered to be not prone to self–heating [Bouford & Senior 2011]

The used "*hot - storage*" test allows to determine the activation energy, but it is a time-consuming method and therefore research are made on the development of alternative methods [Malow & Krause 2004].

This development resulted in the introduction of the heat realase (HR) rate method [Jones et al. 1998], and the crossing – point temperature (CPT) method [Nugroho et al. 2000]. The advantage of these method is that only one sample size needs to be investigated providing a faster and less expensive method.

3 EXPERIMENTAL PROCEDURES

The behavior of a mineral concentrate is liable to liquefy and its threat to the ship's stability is closely related to the effect of a liquid free surface. The liquefaction is created by moisture migration – the water content of a cargo to rise to the bottom of a hold. The wetter bottom layer may therefore be prone to liquefaction and provoke instability of the entire cargo, even though the average moisture content of the whole cargo is less than the TML [Eckersley 1997].

The specific behavior of mineral concentrates, when being transported by sea, makes it necessary to find a new solution to prevent movement of these cargoes by liquefaction.

The purpose of this work was investigation on possibility of using starch materials as moisture absorbers.

Spontaneous combustion of sulphide concentrates can cause many serious problems during storage and transportation. The next aim of the work was estimation of the possibility of the self-heating of selected mixtures mineral concentrates with starch materials by using crossing – point method and determination corresponding to the apparent activation energies.

3.1 *Materials*

Two types of concentrates: zinc (zinc blende), and iron sulfide concentrates were sampled and used for the investigation. These materials are a typical bulk cargoes "*which may liquefy*". For the liquefaction to occur, the mineral concentrates need to have a permeability low enough that excess pore pressures cannot dissipate before sliding occurs. It is controlled by grain size distribution, expressed by requirement that 95% or more, the cargo should be coarser than 1 mm to prevent liquefaction. The results of grain size analysis are presented in Fig. 1.

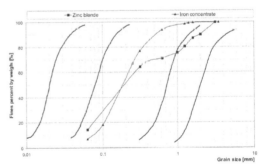

Figure 1. The grain size distribution in zinc blende and iron concentrate

In zinc blende, the content of particle with a diameter greater than 1 mm is about 22%. In iron concentrate, the content of particle with a diameter greater than 1 mm is about 2%. These cargoes are typical materials which are able to liquefy. The concentrates should be tested for liquefaction and shall only be accepted for loading when the actual moisture content is less than its TML.

It may be concluded from the results given in Figure 1 that zinc blende is composed of finer grains than iron concentrates. The values of TML depend on grain size, so if content of smaller grains in mineral concentrates increases, the value of TML increases too. The higher TML values of zinc blende (9,2%) than iron TML values (7,4%) is connected with the degree of concentrates grinding.

The results of the grain size analysis of mixtures concentrates with selected starch material presented in Figure 2.

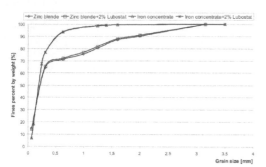

Figure 2. The grain size distribution in mixture of zinc blende and iron concentrate with Lubostat

The starch material does not significantly change grain size distribution. Based on the effective size D_{10} it can be said that the tested mixtures are the materials, which may liquefy during shipment.

The chemical composition of concentrates influences the transport – technological properties, particularly the ability to oxidize. The study found also that among many parameters that determine susceptibility to self-heating, the most important role is of the chemical composition [Iliyas et al. 2010]. In addition, differences in elemental composition, especially in the content of useful part which is metal as well as the presence of sulfur, can cause variations in chemical activity. Table 1 presents the percentages of major elements – sulphur and metals in tested minerals.

Table 1. Chemical compositions of the samples of concentrates (wt.%)

| Sample types | Elements | |
	Metal	S
Zinc blende	Zn = 61,57	31,49
Iron concentrate	Fe = 61,20	3,20

Source: based on data obtained from manufacturers.

Following starch materials were tested (potato starch obtained from Potato Industry Company at Luboń):
– distarch phosphate -"Lubostat"
– acetylated distarch adipate -"AD"
– granulated product - granulated starch.
Selected starch materials have different ability to absorb moisture. The greatest capacity to collect moisture from the atmosphere is distinguished by an acetylated distarch adipate. Hygroscopicity distarch

phosphate is comparable with the properties of granulated starch.

High ability to absorb moisture is a very desirable phenomenon in the conduct of research. AD and Lubostat are characterized by a similar water absorption and water absorption of granulated starch is lower.

3.2 Methods

The influence of adding starch materials to the ores on its parameters determining ability for safe shipment by sea was assessed on the basis of determination of the following parameters: Flow Moisture Point and reaction activity. The evaluation of FMP was performed with the use of the Flow Table Test according to the recommendations given in IMSBC Code. The samples with Lubostat, AD and granulated starch were tested for estimation TML at several time intervals.

The self-heating property of sulfide concentrates was estimated according to the procedure presented by Yang [Yang et al. 2011]. The constant temperature in the chest was maintained stable at 140^0C, 150^0C, 160^0C and 180^0C, respectively over a long period of the experiments to determine the crossing – point temperature at four different ambient temperature.

4 RESULTS AND DISUSSION

The results of estimation TML for zinc blende and iron concentrate and for their mixtures with starch materials are presented in Figures 3 and 4.

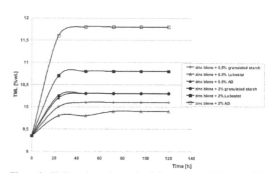

Figure 3. TML values determined by means of Flow Table Test –zinc blende +starch materials

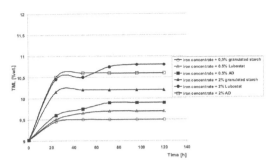

Figure 4. TML values determined by means of Flow Table Test –iron concentrate +starch materials

Starch material absorbed water from the mixtures at the amount approximately proportional to the starch material content in the mineral concentrates. It can be noticed that modified starch presents higher solubility than granulated starch. In general, the higher values of TML were observed in the case of testing concentrates + 2 % of starch material. For the mixtures containing 2 % Lubostat and AD, greater increasing of TML was observed than for concentrate containing 2 % granulated starch.

In the case of zinc blende, the highest expected changes of TML were observed for a mixture with 2% share of AD. The time, in which half of the maximum absorption was obtained, shows a high absorption rate that is a highly desirable in the conditions of transport.

In the case of iron concentrate, determined values of the parameters indicates that 2% share in mixture of the Lubostat with the concentrate results in the largest increase of the TML. The time, in which half of the maximum value of TML was obtained , indicates that the rate of moisture absorption by this starch materials is rapid, the waiting time to improve parameters relevant to the transport is short. The equilibrium absorption of water by starch materials is reached in four days.

The results of estimation activation energy for concentrates and mixtures with starch materials are presented in Table 2.

Table 2. Activation energy for concentrates and mixtures with starch materials

Type of mixture	Activation energy [kJ/mol] Type of concentrate	
	Zinc blende	Iron concetrate
Concentrate	84,90	86,75
Concentrate + 2% Lubostatu	84,95	86,57
Concentrate + 2% AD	84,60	86,53
Concentrate + 2% granulated starch	84,36	86,90

The ability to oxidize and self-heating depends primarily on the chemical composition of concentrates. Sulfide minerals contain non – stoichiometric sulfur that would readily be decomposed into stoichiometric sulfide with sulfur liberation to precede the exothermic oxidation. Sulfide–reach mineral samples exhibit accelerated self-heating behavior compared to lower sulfide samples, because of spontaneous reaction with oxygen following desulfuration [Ilias et al. 2010].

As presented in Table 1, tested concentrates differ in sulfur content which may significantly influence the susceptibility to oxidation.

The greatest ability to self-heating showed a zinc concentrate that probably contains the largest non-stoichiometric amount of sulfur. The research results indicate that iron concentrate exhibits the higher activation energy. The presence of starch materials in mixtures does not increase the ability of concentrates to self-heating.

5 CONCLUSIONS

It may be concluded that there are several criterions which influence the transportation safety of mineral concentrates.

The ability to absorb water is related primarily to the composition of starch material and the percentage of starch in mixture. The higher content starch material in mixtures contributes to the increase value of FMP estimated by Flow Table Test. It may be concluded that applied starch materials can be used for decreasing moisture content of mineral concentrate before shipment. Due to presence starch materials, the risk of passing mineral concentrate into liquid state is lower.

Physical and chemical characteristics of sulphide concentrates and their behavior during oxidation are very important for sea transportation. The activation energy is the critical input – parameter and its accurate experimental determination is of primary importance for estimation of self-heating. The experimental data confirm the relationship between sulfur content and the ability to self-heating of the concentrate. The information gained by using the crossing – point temperature (CPT) method in the present investigation will prove to be useful when simulating the self–heating behavior of a large pile of sulfide concentrates during shipping and storage.

REFERENCES

Bouffard, S.& Senior, G. D. 2011. A new method for testing the self –heating character of sulphide concentrates. Mineral Engineering 24: 1517 -1519.
Eckerley, J.D. 1997. Coal cargo stability, The AusIMM Proceedings. No 1: 33 – 41.
Grzelakowski, A.S. 2009. Maritime transport development in the Global Scale-the Main Chances, Threats and Challenges. International Journal on Maritime Navigation and safety of sea Transportation 3(2): 197-205.

Iliyas, A. Hawboldt., Khan, F. 2010. Thermal stability investigation of sulfide minerals in DSC. *Journal of Hazardous Materials* 178: 814-822.

Jones, J. Henderson, K. Littlefair, J. Rennie, S. *1998*. Kinetic parameters of oxidation of coals from heat release measurement and their relevance to self-heating tests. *Fuel* 77 (1-2): 19-22.

IMO. 2011. *International Maritime Solid Bulk Cargoes Code*, London.

Malow, M. Krause, U. 2004. The overall activation energy of the exothermic reactions of thermally unstable materials. *Journal of Loss Prevention in the Process Industries* 17: 51-58.

Nugroho, Y. McIntosh, A. Gibbs, B. 2000. Low-temperature oxidation of single and blended coals. *Fuels* 79 (15): 1951-1961.

Popek, M. Rutkowska, M. 2002. The Methods for Determination of flow Moisture Point in Bulk Cargoes. *Commodity Science in Global Quality Perspective*. Proc. intern. Symp., Maribor, 2-8 September 2001.

Popek, M. 2010. The Influence of Organic Polymer on Parameters determining ability to Liquefaction of Mineral concentrates. *International Journal on Maritime Navigation and safety of sea Transportation*. 4(4):435-440.

Shitaram, T. 2003. Descrete element modeling of cyclic behavior of granular materials. *Geotechnical and Geological Engineering* 21: 297 -329.

Rosemblum, F. Nesset, J. Spira, P. 2001. Evaluation and control of self -heating in sulphide concentrates. CIM Bulletin 94(1056): 92-99.

UN. 2010. Review of Maritime Transport. United Nations Conference on Trade and Development.

UN. 2011. Review of Maritime Transport. United Nations Conference on Trade and Development.

UNECE. 2008. Manual of Tests and Criteria for Self-Heating Substances (Part III, Classification, Procedures, Test Method and Criteria Relating to Class 3, Class 4, Division 5.1 and Class 9). Fifth Edition: 357-359.

Yang, F. Wu, Ch. Li, Z. 2011. Investigation of the propensity of sulfide concentrates to spontaneous combustion in storage. *Journal of Loss Prevention in the Process Industries* 24: 131-137.

Zarrouk, S. J. O'Sullivan, M.J. 2006. Self-heating of coal the diminishing reaction rate. *Chemical Engineering Journal* 119: 83-92.

Zhang, M. 2005. Modeling liquefaction of water saturated granular materials under undreined cyclic shearing. *Act. Mech. Sinica* 21: 169-175.

Determination of the Fire Safety of Some Mineral Fertilizers (3)

K. Kwiatkowska-Sienkiewicz, P. Kutta & E. Kotulska
Gdynia Maritime University, Department of Chemistry and Industrial Commodity Science, Poland

ABSTRACT: This paper provides an outlook on fire safety assessment concerning nitrates fertilizers in sea transport. The investigation was aimed at comparison of two methods of classification and assignment to a packing group of solid fertilizers of class 5.1 of International Maritime Dangerous Goods Code. First research was conducted in accordance with the Manual of Test and Criteria, "Test for oxidizing solids" described in the United Nations Recommendations on the Transport of Dangerous Goods. The second method was the differential thermal analysis (DTA), where the basis was the determination of the temperature change rate during thermal reaction. According to two used tests, the investigated three fertilizers belong to 5.1 Class and to packaging group III of the International Maritime Dangerous Goods Code. Two fertilizers do not belong to dangerous goods. The DTA method gives more quantitative information about fire risk on the ship than the method recommended in the International Maritime Dangerous Goods Code.

1 INTRODUCTION

The Chemical Abstract Service (CAS) lists over 63 000 chemicals outside the laboratories environments and the number increases each year. The United States Department of Transportation and International Maritime Organization (IMO) regulate over 3800 hazardous materials in transportation. The European Union and IMO regulations of dangerous goods are similar. Sea transport of hazardous substances is regulated in International Maritime Dangerous Goods Code – IMDG Code (for packaging goods) and International Maritime Solid Bulk Cargoes Code - IMSBC Code.

According to UN Directive dangerous chemicals are classified as: T+ – very toxic, T – toxic, Xn – harmful. Xi – irritant, C – corrosive, N - dangerous to the environment, E – explosive, O – oxidizing, F+ – extremely flammable, F – highly flammable [Emergency Guide Book 2002, Bruke 2003].

Dangerous goods in the sea transport classification is regulated by IMO Codes [IMDG-Code 2011, IMSBC Code 2012]. IMDG Code classification is based on physical and chemical dangerous properties. Dangerous goods are divided into nine classes: 1 – explosives, 2 - gases (where 2.1 - flammable gases, 2.2 – non-flammable gases, 2.3 – toxic gases), 3 – flammable liquid, 4 – flammable solids (where: 4.1 – flammable solids, self- reactive substances and solid desensitizes explosive, 4.2 - substances liable to spontaneous combustion, 4.3 – substances which, in contact with water , emit flammable gases), 5 – oxidizing substances and organic peroxides (5.1 – oxidizing substances, 5.2 – organic peroxides), 6 – toxic and infectious substances (where: 6.1 – toxic substances, 6.2 – infectious substances). 7 – radioactive materials, 8 – corrosive substances, 9 – miscellaneous dangerous substances and articles.

A major fire aboard a ship carrying these materials may involve a risk of explosion in the event of contamination by combustible materials. An adjacent detonation may also involve a risk of explosion.

During thermal decomposition nitrate fertilizers give toxic gases and gases which support combustion. Dust of fertilizers might be irritating to skin and mucous membranes.

There are at present no established good criteria for determining packaging groups of dangerous goods Class 5.1. Substances of great danger belong to packaging group I, of medium danger to packaging group II - medium danger, or minor danger to packaging group III. In this paper are presented two methods of classification oxidizers as the dangerous goods.

2 MINERAL FERTILIZERS

According to International Fertilizer Industry Association the global consumption of nitrogen, potassium and phosphate mineral fertilizers (also called NPK) in 2009 amounted to 165 millions ton of pure NPK component.

With commercial types of fertilizers the following are identified as necessary for plant growth: Primary nutrients - nitrogen (N), phosphorus (P), and potassium (K). Secondary nutrients: calcium (Ca), magnesium (Mg), sulfur (S). Trace minerals – (boron (B), chlorine (Cl), copper (Cu), iron (Fe), manganese (Mn), molybdenum (Mo), zinc (Zn), and selenium (Se).

Poland ranks second among manufactures of phosphoric and third as to the size of nitrogen fertilizer production. On a world scale the Polish production of nitrogen and phosphoric minerals constitutes per about 1,5% of world production. Modern, ecologic and friendly for the work environment base to storage and handling of fertilizers there is at Gdynia in Baltic Bulk Base.

Transports of mineral fertilizers resist changes affected by conditions during successive stages of transport processes, i.e. carriage, loading and storage. Mineral fertilizers are generally mixtures of salts and physically appear in granulated form. There are hygroscopic, liquefy, and have dangerous properties like: oxidation, corrosion, susceptibility to self-heating, self-ignition and explosion. They have tendency to shifting on board and can lost stability of boat, particularly if the angle of repose is greater than 30° and after exceeding relative humidity [IMSBC Code 2012]. Mixtures of ammonium nitrate, potassium and sodium nitrates also potassium sulphate are prone to shifting

Mineral fertilizers generally concern nitrates. Nitrates belong to oxidizers.

Oxidizers are dangerous goods in accordance with International Maritime Dangerous Goods Code, belong to 5.1 Class, they are not necessarily flammable, but able to intensify the fire by emission of oxygen. Some oxidizing substances have toxic or corrosive properties, or have been identified as harmful to the marine environment. They will react in contact with reducing reagents. Hence oxidizing agent will invariably accelerate the rate of burning of combustible material. The National Fire Protection Association in United Stated classified oxidizing substances according the stability in four classes [Burke, 2003].

Chile, nire, lime saltpeter are nitrates. Pure ammonium nitrate, the base of fertilizers, belongs to compounds transported in limited quantities (UN 0222 –ammonium nitrate, with more than 0,2% combustible substance). Ammonium nitrate UN 1942 with not more than 0,2% total combustible substances, calculated as carbon, during transport

the temperature of material should not be above 40°C. Do not ventilate this cargo.

Ammonium nitrate based fertilizers UN 2067, UN 2071, UN 2067 may be transported in bulk. Fertilizers: potassium nitrate – UN 1486, sodium nitrate UN 1498 and sodium and potassium nitrate mixtures UN 1499, calcium nitrate UN 1454 may be also transported in bulk [IMSBC Code 2012]. Fertilizers are highly hygroscopic and will cake if wet. They belong to cargo group A and B. Group A consists of cargoes which may liquefy if shipped at a moisture content in excess of their transportable limit. Group B consists of cargoes which possess chemical hazard which could give rise to dangerous situation on the ship [Appendix A and B IMSBC Code, 2012].

3 CLASSIFICATION OF OXIDIZERS

Classification of oxidizing substances to class 5.1 and the assignation criteria to the packaging groups are based on physical or chemical properties of goods and are based on the test described in the IMDG Code [Manual Tests and Criteria -UN Recommendations Part III]. In this test, the investigated substances were mixed with combustible material - cellulose in ratios of 1:1 and 4:1, by mass, of substance to cellulose. The mixtures were ignited and the burning time was noted and compared to a reference mixture, in ratio 3:7, by mass, of potassium bromate(V) to cellulose.

If a mixture of the tested substance and cellulose burns equal to or less than the reference mixture, this indicates that the substance is classified in class 5.1. This also means that oxidizing substance is assigned to a packing group III (if the criteria of packing group I and II are not met). Next the burning time is compared with those from the packing group I or II reference standards, 3:2 and 2:3 ratios, by mass, of potassium bromate(V) and cellulose.

Any substance which, in both the 4:1 and 1:1 sample-to-cellulose ratio (by mass) does not ignite and burn, or exhibits mean burning times greater than that of a 3:7 mixture (by mass) of potassium bromate(V) and cellulose, is not classified as class 5.1.

Using these criteria we tested a big mass sample of a component which involves larger volumes of toxic gases, as opposed to a differential thermal analysis, where the basis is the determination of the temperature change rate during thermal decomposition.

Using differential thermal analysis (DTA) we can register quality and quantity changes during dynamic heating of investigated materials in time.

The self-heating or thermally explosive behavior of individual chemicals is closely related to the appearance of thermogravimetry-differential thermal analysis (TG-DTA) curve with its course.

In previous examinations of mixtures of oxidizers with flour wood [Michałowski, Barcewicz 1997] the temperature change rates [^{0}C/s] were calculated into 1 milimole of an oxidizer and tested oxidizing substances were blended with combustible substance in mass ratio 5:1. Then some fertilizers are investigated using recommended test in IMDG Code and differential thermal analysis [Kwiatkowska-Sienkiewicz 2008, Kwiatkowska-Sienkiewicz 2011, 2011(a)].

According to later DTA experiments any substances which temperature change rate are lower than 0,96 [^{0}C/s] are not classified to class 5.1.

Substances, which during thermal analysis of mixtures of oxidants with cellulose, the temperature change rate values are between 0,96 to 1,82 [^{0}C/s] should be assigned to packaging group III.

Substances blended with cellulose of with the temperature change rate values between 1,82 to 5,0 [^{0}C/s] should be assigned to packaging group II.

Mixtures of oxidizers with cellulose which temperature change rate values exceed 5,0 [^{0}C/s] belong to packaging group I

Cellulose as combustible material is used in experiments. Cellulose belongs to polysaccharides which develop free radicals on heating [Ciesielski, Tomasik 1998, Ciesielski, Tomasik, Baczkowicz 1998]. Free radicals exposed during thermal reaction of polysaccharides mixed with nitrates(V) give possibility of self-heating and self-ignition.

In practice, during long transport combustible materials and commodities containing polysaccharides we can observe self-heating effect, specially, if polysaccharides are blended with oxidizers.

In this paper fertilizers, containing nitrates, blended with cellulose were investigated.

That examinations were based on potassium bromate(V) blends with cellulose in mass ratio 2:3 and 3:7 as a standard. Class 5.1 includes substances which temperature change rate was greater than temperature change rate of mixture of potassium bromate(V) with cellulose, in mass ratio 3:7 – 0,96 [°C/s].

Substances, which during thermal analysis of mixtures of oxidants with cellulose, the temperature change rate values between 0,96– 1,82 [°C/s] should be assigned to the III packaging groupII packaging group involves the crossing value of the temperature change rate under 1,82 [°C/s].

In this paper we concentrated on comparison of two methods of assignation to class 5.1 and classification to packing groups. The first one is recommended by United Nations [UN Recommendations] and second one, differential thermal analysis used in chemistry.

4 EXPERIMENTAL

Determination of oxidizing substances according to IMDG Code and UN Recommendations test and DTA method was carried out using the same blends of fertilizer and cellulose.

The following substances were blended with cellulose in mass ratio 1:1 or 1:4:
– ammonium nitrate (60%) + calcium carbonate (40%),
– ammonium nitrate (5%) + ammonium sulphate (95%)
– ammonium nitrate (55%) + ammonium sulphate (45%),
– Canwile - nitrogen is as a sum of ammonium and nitrate nitrogen (27%),
– WAP-MAG CaO (28%), MgO(16%), SO$_3$ (10%).

As reference material - potassium bromate(V) pure was blended with cellulose in mass ratio 3:7 and 2:3 for analysis

In experiments microcrystalline cellulose, grade Vivapur type 101, particle size >250 μm (60 mesh), bulk density 0.26 – 0.31 g/ml was used

Chemical reaction course during the heating can be investigated by means of differential thermal analysis (DTA) method. Using thermal analysis, the changes of mass, temperature and heating effects curves are recorded.

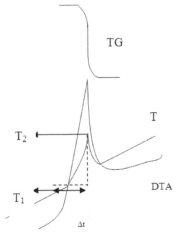

Figure 1. Self –ignition effect monitored by DTA method
Source: [Kwiatkowska-Sienkiewicz, Barcewicz 2001]

The following outputs were recorded during measurements using DTA method.
– T the temperature change curve which is a straight line till the mixture flash point is reached, with a district?/distinct peak in the self-ignition region, especially during reaction of very active oxidizers,
– DTA curve – gives information about heat effects,

– TG curve is of mass change during the reaction.

The temperature increase value was determined from the temperature change curve T on the basis of its deflection out of the straight line, in the peak region. On the ground of the above mentioned data the temperature change rates [°C/s] were calculated by dividing the temperature increase by the time of self-ignition effect, counted into 1g of a fertilizer. Thermal treatments of the blends were heated from room temperature to 500°C. The procedure was run in the air, under dynamic condition. The rate temperature increase was 10°C/min. Ceramic crucibles were taken. Paulik-Paulik-Erdley 1500 Q Derivatograph (Hungary) was used. The measurements were carried out three times. Decomposition initiating temperatures of the blends fertilizers with cellulose were read from the recorded curves.

5 RESULTS

The results of performed thermal reactions of fertilizers and cellulose are presented in the Tables 1 and 2.

On the basis of results of the test described in Manual of Test and Criteria three of examined fertilizers belong to class 5.1 of dangerous goods and require packaging group III. Two fertilizers do not belong to dangerous goods because mixtures of those fertilizers with cellulose had shorter burning time (0,81cm/min.) than standard. Only Canvile was burning in mass ratio 1:1 with cellulose. Four fertilizers were burning with cellulose only in mass ratio 4:1. According to this test WAP-MAG and ammonium nitrate (5%) with ammonium sulfide (95%) do not belong to dangerous goods of 5.1 Class. Those fertilizers belong to Class 9 of IMDG Code. A mineral fertilizer containing about 28% of ammonium nitrate belongs to 5.1 Class dangerous goods and has III packaging group.

The results of the second method of performed thermal reactions between cellulose and selected fertilizers are presented in Tables 2 and 3. Blends of potassium bromate(V) and cellulose in mass ratio 3:7 and 2:3 are the standards in classification using differential thermal analyses tests (we used the same standards as in Manual Test recommended by IMDG Code).

Table 1. Determination risk of fire fertilizers according to Manual Test and Criteria IMDG Code

Fertilizer	Sample to cellulose.	Burn rate [cm/min.] Sample	Standard	Proposed Class IMDG	Packagingg roup
Ammonium nitrate (60%) + calcium carbonate (40%)	1:1	-	0,83	5.1	III
	4:1	3,58			
Ammonium nitrate (55%) + ammonium sulphate (45%)	1:1	-	0,83	5.1	III
	4:1	3,69			
Canwile - nitrogen as a sum of ammonium and nitrate nitrogen (27%)	1:1	0,70	0,83	5.1	III
	4:1	4,75			
WAP-MAG CaO (28%), MgO(16%) SO₃ (10%)	1:1	-	0,83	Not belong to 5.1 Class	
	4:1	0,81			
NH_4NO_3 (5%) $(NH_4)_2$ SO_4 (95%)	1:1		0,83	Not belong to 5.1 Class	
	4:1				
Potassium bromate (V) (p.a.) (standard)	2:3	10	10	5.1	II
	3:7	0,83	0,83	5.1	III

The ignition temperature and temperature change rate make it possible to assess packaging group of investigated fertilizers, belongs to class 5.1 of dangerous goods.

The blends of fertilizers and cellulose had lower ignition temperature then pure oxidizers. Hence those fertilizers will invariably accelerate the rate of burning with combustible materials.

The mixtures of mineral fertilizers with combustible material – cellulose were decomposed in temperature about 105 - 242 °C.

During reactions exothermic processes and weight losses (Canwile -57%) were observed

Loss of mass is very dangerous in shipping, especially of bulk cargo. Fertilizers containing nitrates and ammonium salts, during fire on board, lose stowage mass about 1/3 to ½ .

The results of differential thermal analysis suggest similar effects like in the tests recommended by IMDG Code.

Table 2. Thermal decomposition oxidizers/fertilizers and his blends with cellulose using DTA method

Fertilizer	Sample to cellulose	Temperature [°C] of Self-heating	Self-ignition	Temperature change rate [°C/s]	Loss of mass [%]
Ammonium nitrate (60%) + calcium carbonate (40%)	1:1	-	242	1,01	29,5
	4:1	-	228	1,11	13
Ammonium nitrate (55%) +	1:1	-	199	0,92	15,5

ammonium sulphate (45%)	4:1	127	197	0,96	6,5
Canwile - nitrogen as a sum of ammonium and nitrate nitrogen (27%)	1:1	153	207	1,75	57
	4:1	117	182	1,12	27
WAP-MAG CaO (28%), MgO(16%) SO$_3$(10%)	1:1	276	-	-	6
	4:1	105	-	-	8
NH$_4$NO$_3$ (5%) (NH$_4$)$_2$ SO$_4$ (95%)	1:1	128	-	-	36
	4:1	122	-	-	21,5
Potassium bromate (V) (p.a.)	2:3		329,3	0,96	
(standard)	3:7		190	1.82	

Table 3. Assignment of the fertilizers to the packaging group based on temperature change rate

Name of the fertilizer	Temperature change rate [°C/s]		Assigned packaging group	Proposed class of IMDG Code
	1:1	4:1		
Ammonium nitrate (60%) + calcium carbonate (40%)	1,01	1,11	III	5.1
Ammonium nitrate (55%) + ammonium sulphate (45%)	0,92	0,96	III	5.1
Canwile - nitrogen as a sum of ammonium and nitrate nitrogen(27%)	1,75	1,12	III	5.1
WAP-MAG CaO (28%), MgO(16%) SO$_3$(10%)	-	-		Not belong to 5.1 Class
NH$_4$NO$_3$ (5%) (NH$_4$)$_2$ SO$_4$ (95%)	-	-		Not belong to 5.1 Class
Potassium bromate (V) (Standard)	0,96÷1,81		III	5.1

During Manual Test (according to IMDG Code) we have only qualitative data of burning time of fertilizers blends with cellulose.

The ignition temperature and temperature change rate make it possible to assess packaging group of investigated fertilizers, belongs to class 5.1 of dangerous goods.

After comparison of these two methods of assignation to class 5.1 and packaging group's data, thermal analysis gives quantitative information about thermal effects (melting, self-heating, self-ignition) and loss mass during heating. During Manual Test (according to IMDG Code) we have only qualitative data burning time of blends of oxidizers with cellulose. In Manual Test big probe - 30g blends of fertilizer and cellulose was used, in DTA method only - 300 ÷500 mg.

Differential thermal analysis is an objective chemical method which could make it possible to determine the criteria of assignment of oxidizers to packaging groups, required for sea transport. Data DTA method gives more information about fire risk assessment than Manual Test recommended by IMDG Code.

6 CONCLUSION

Manual Test recommended in IMDG Code provides qualitative data about burning time.

Using DTA method, during heating we can register changes of temperature, loss mass, melting point temperatures mixtures, before self-heating, self-ignition and explosive effects.

During experiments in DTA method were use 300 – 500 mg of materials, in Manual Test – 30 g.

Data based on the differential thermal analysis gives more information about fire risk assessment than Manual Test recommended by IMDG Code.

The comparison of two methods of classification and assignment to a packing group of solid substances of class 5.1 of IMDG Code indicates, that differential thermal analysis (DTA method) gives objective, quantitative information about fire risk on board a ship.

REFERENCES

Appendix A in the IMSBC Code 2012: London, Printed by International Maritime Organization.

Appendix B in the IMBC Code 2012: London, Printed by International Maritime Organization.

Brudulk H. Rynek usług transportowo-spedycyjno-logistycznych (TSL) w Polsce, Rzeczpospolita, dodatek LTS Logistyka Transport Spedycja 2009 nr 2, 19 czerwca 2009.

Bruke R.: Hazardous Materials Chemistry for Emergency Responders. Lewis Publishers, Boca Raton, Florida 2000, 255 -257.

Ciesielski W., Tomasik P. Starch radicals. Part III: Semiartifical complexes. 1998. Z. Lebensm. Untes. Forsch. A. 207:pp. 292-298.

Ciesielski W., Tomasik P. Baczkowicz M. Starch radicals. Part IV: Thermoanalitical studies. 1998. Z. Lebensm. Untes. Forsch. A. 207:pp. 299-303.

Emergency Response Guide Book, 2002, New York, Printed by The United States Department of Transportation.

International Maritime Dangerous Goods Code (IMDG Code) 2012: London, Printed by International Maritime Organization.

International Maritime Solid Bulk Cargoes Code (IMSBC Code) 2012: London, Printed by International Maritime Organization.

Kwiatkowska-Sienkiewicz K., Barcewicz K.: 2001;The new criteria of separation of oxiedizers from ammonium salts. Proceedings: European Safety & Reliability Conference, ESREL, Turyn, Italy, Ed. Technical University in Turyn, pp .959-963.

Kwiatkowska-Sienkiewicz K., 2008; Application of thermal analyze in studies on levels of separation of oxidizers from ammonium chlorate(VII). Proc.The 16 th Symposium of IGWT, Suwon, Korea, pp. 737-739.

Kwiatkowska-Sienkiewicz K., 2011 Fire Safety Assessment of Some Oxidizers in Sea Transport,Transport Systems and Processes – Marine Navigation and safety Transportation, Taylor & Francis Group, Boca Radon, London, New York, Leiden, pp. 33-37.

Kwiatkowska-Sienkiewicz K., (a) 2011, Fire safety assessment of some nitrate fertilizers in sea transport, Zeszyty Naukowe Akademii Morskiej w Szczecinie, 27(99) z.2 pp. 24-27,

Manual of Tests and Criteria Part III, 34.4.1.: 2012: London, United Nations Recommendations on the Transport of Dangerous Goods.

Michałowski Z., Barcewicz K., 1997. A new proposal for assignment criteria fo the oxidizers into 5.1 class and packaging categories of IMDG Code. Polish Maritime Research, , 3 (13) 4-25.

The Ecological Hovercraft – Dream or Reality!

Z.T. Pagowski & K. Szafran
Institute of Aviation , Warsaw, Poland

ABSTRACT: The European Union is very active on global market of emission to reduce greenhouse gas emissions from maritime transport. There is currently no international regulation of emissions from this transport, but the new generation biofuel technology is critical for reducing emission of wide range of transport resources including a small water transport. The paper presents the results of preliminary biofuel tests of ecological fuel BioE85 on the hovercraft PRP 600 M/10, which was manufactured in Poland by Institute of Aviation. A hovercraft transport is all known a hybrid transport vehicle and the tests confirmed significant changes in emission using biofuel with 85% of ethanol and will help to support the idea of new generation of biofuels that reduce the environmental pollution on land, sea and air.

1 INTRODUCTION

40 mm in sea level and mean global temperature about 1 °C in the last 100 years, disappearing arctic ice and other local vibration with climate indicates problem of reducing emissions from all mode of sea transport including in future also hovercrafts. Hovecrafts are operating on sea and water, all ground condition like wet sand, swamps, soft mud, tundra, permafrost, sandbars and ice. Depending from military, civil, commercial or sport applying and size hovercrafts are using all types of engines from aviation turbine engines to car engines including hybrid version or electric engines powered by Lithium battery. Breakdown of transport emissions indicates that sea and water transport is responsible for about 2-5% of emission and of course depends from the country and still increases (see below). The European Union is very active on global market of emission to reduce greenhouse gas emissions from maritime transport. There is currently no international regulation of emissions from this transport, but the new generation biofuel technology is critical for reducing emission of sea and water transport including hovercrafts. For this type of maritime vehicle are not known different statistical data, but generally published information for ships indicates on growing ecological problem to reduce main emission factors like CO, HC, NOx and PM.

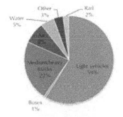

Source : http://www.c2es.org/technology/overview/transportation
Figure 1. US transport greenhouse gas emissions by mode

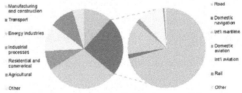

Source: http://ec.europa.eu/clima/policies/transport/index_en.htm
Figure 2. EU27 greenhouse gas emissions by sector and mode of transport, 2007

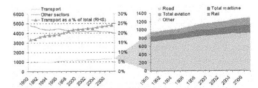

Source: http://ec.europa.eu/clima/policies/transport/index_en.htm
Figure 3. EU greenhouse gas emissions from transport by sectors and mode

Emissions (tonnes)	NO$_x$		SO$_2$		CO$_2$		PM	
	Absolute	%	Absolute	%	Absolute	%	Absolute	%
At Sea	156 521	89%	96 288	99%	6 091 920	85%	15 006	87%
In Ports	20 296	11%	677	1%	1 076 411	15%	2 277	13%
Total	176 817	100%	96 965	100%	7 168 331	100%	17 283	100%

Source: Policy Research Corporation

Source: Policy Research Corporaton
Figure 4. Emissions from cruise tourism at sea and in EU ports in 2009

Beginning from the sixties specialists from Institute of Aviation have designed some experimental hovercrafts (Jerzy Bień, Andrzej Moldenhawer). Next beginning from 1997 Institute of Aviation elaborated and produced new type of construction, versatile tested hovercraft PRP 560, known under name "Ranger". Certified under no 2/War/12 by the Polish Register of Shipings - Polski Rejestr Statków hovercraft IL - PRC – 600 M/10 (Fig.6), redesigned lately version of " Ranger" and used in tests with biofuel Bio E85 is equipped with a propulsion system consisting of 2l motor engine GM l DOHC 130 hp, belt transmission and fan Multi Wing.

Figure 5. Hovercraft IL- PRC - 600M/10 on Zegrze lake in tests

2 COMPARISON TESTS WITH BIOFUEL E85

Actual and future on board diagnostic systems and procedures used in car engines on hovercraft IL-PRC-600/M10 were no used according different mode of working cycle between the same engine in car or hovercraft, which is charged by fan and ground level conditions. Hovercraft was tested from 1000 to 5000 rev/min of engine charged on fan full power characteristic using automatic tester in the same temperature and pressure of ambient air and water. Like the tester was used last generation emission multigas analyzer M-488 Plus for measurement of the concentration of automobile emission gas CO, CO2, HC, O2 and NOx based on infrared (NDIR) method. Comparison tests, without any change in regulation of engine system, were

performed using standard unleaded petrol Eurosuper 95 and ethanol biofuel BIO E 85 from "Orlen" with about 85% ethanol acc. Dz.U. 2009 nr 18 poz. 98 (Summer 75%, Winter 70%) with RON/MON Octane Nr min. 85/95. Comparison preliminary results of the tests are shown below on fig. 7-9.

Figure 6. Certificate of hovercraft IL-PRC-600/10

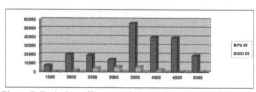

Figure 7. Emission of hovercraft - CO v/v [ppm]

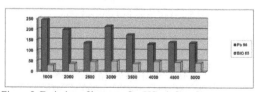

Figure 8. Emission of hovercraft – HC v/v [ppm]

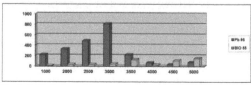

Figure 9. Emission of hovercraft – NOx v/v [ppm]

The tests confirmed significant changes in emission using biofuel E85 and will help to support the idea of biofuels that reduce the environmental pollution. Next hovercraft was tested dynamically on Zegrze lake and confirm advantages of new biofuel in dynamic flights. Additional advantage of this fuel will be quick biodegradation on water environment. Noise of the hovercraft made mechanical moving parts such as motors, fans, propellers and vibration of curtain and airbags. It have an impact on comfort inside the cabin and outside of the hovercraft. In the Institute of Aviation designed a series of hovercraft PRC-600 with low noise level to minimize the noise generated by the air-cushion. In the course of experiments with B85 biofuel additional noise measurements were prepared in static tests on the land in Institute of Aviation and dynamic tests on the Zegrze lake. Sound Pressure Level (SPL) was measured by SVAN912 Sound Level Meter from distance 30 m at at 10 deg C and the wind below 3 [m/s] versus revolutions of engine (1/min). Experiments indicated characteristic differences between static and dynamic tests of noise of hovercraft fueled by BioE85. Further work on silencing of hovercraft is needed from general point of view.

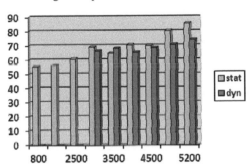

Figure 10. SPL (dB) of hovercraft PRC-600

3 REMARKS

The expansion of hovercraft in the area of commercial transport, platform for moving heavy loads, patrol and rescue services and sport also ecotourism define and promote environmental point of view, which to this moment was practically absent. Actually in Poland we have registered by the Polish Register of Shipings - Polski Rejestr Statków hovercrafts of Institute of Aviation in number of 10. Biofuels are ready to use without serious problem when hovercrafts are powered by modern car engines or turbine aviation engines using renewable Jet A/A-1 fuel compliant with specifications ASTM 7566. Actually very prospective idea is using of hybrid system, where gasoline engine generate electricity for the independent electric motors of the drive system like hovercraft of Mercier-Jones or "zero" emission vehicle like first Polish electric hovercraft "OSA". This experimental light one person hovercraft of Trigger Composites and EVCONVERT.EU is powered by electric motor and Litium battery pack. This battery packs is sufficent for 30-40 minutes of hovercraft flight. The biggest difference between the electrical and gasoline engine is lower noise level. Hovercraft was presented on fair in Warsaw "Wiatr i Woda 2010" (Wind and Water 2010).

Source: http://www.mercier-jones.com/Hovercraft.html
Figure 11. Hybrid hovercraft of Mercier-Jones

Source: Copyright Marek Żygadło
Figure 12. First Polish electric hovercraft "OSA"

REFERENCES

[1] Climate change and the freight industry ,2012 http://www.fta.co.uk/policy_and_compliance/enviroment/logistics_carbon_reduction_scheme/climate_change_and_freight.html
[2] Dąbrowski W. Graffstein J. Masłowski P. Popowski S. 2004. On dynamics of howercraft motion on ice Prace Instytutu Lotnictwa 1/2004 (176) str. 26 – 30, Warsaw, Institute of Aviation
[3] Internal reports of Institute of Aviation: „Hałas poduszkowca PRP-560 cz.1" – Report no. BE1/410/2002 ; „Hałas zewnętrzny poduszkowca PRC-600" - – Report no. LTBA-1/2006, „Hałas wewnętrzny i zewnętrzny poduszkowca PRC-600" – Report no. BE/3/2005
[4] KLIMEK Z.: PODUSZKOWCE → problemy formalno – prawne, http://poduszkowce.net/publikacje_eng.html
[5] Korsak T. Description of the patrol and rescue hovercraft IL prp-560 "Ranger" designed and produced by Institute of Aviation, 2004,Prace Instytutu Lotnictwa 1/2004 (176) str. 9 – 13
[6] Koziarski S. Zmiany w strukturze transportu Unii Europejskiej http://www.cut.nazwa.pl/ptg/konfer/include/wydaw.php?r=2007&ab=1b

[7] Measures to include maritime transport emissions in the EU's greenhouse gas reduction commitment if no international rules agreed http://ec.europa.eu/governance/impact/planned_ia/docs/2012_clima_001_maritime_transport_emissions_en.pdf

[8] Moldenhawer A.: OPIS I WNIOSKI Z PRAC NAD PODUSZKOWCEM „URSYNOW M6"2004, Prace Instytutu Lotnictwa 1/2004 (176) str. 5-8,

[9] Ostrowiecki E. Cheda W. 1975,Pojazdy poduszkowe, Warszawa, Wyd. MON,

[10] Reducing emissions from transport http://ec.europa.eu/clima/policies/transport/index_en.htm

[11] Ślęzak M. Systemy diagnostyki pokładowej (OBD) w warunkach polskich na przykładzie wybranych pojazdów samochodowych , 2010, Archiwum Motoryzacji ,Vol. 7 no3 pp.263-308,

[12] Trigger-Composites "Osa Sport II", 2007 http://www.samolotypolskie.pl/samoloty/1766/126/Trigger-Composites-Osa-Sport-II

[13] Tourist facilities in ports – Growth opportunities for the European maritime economy: economic and environmentally sustainable development of tourist facilities in ports – Study report, 2012,Office for Official Publications of the European Communities, 2009 ISBN 978-92-79-12996-4 European Commission

[14] US Emission,2012, http://www.c2es.org/technology/overview/transportation, Center of Climat and Energy Solutions. Arlington

[15] Zschocke A.2012, Green Operations and Alternative Fuel Testing, Greening and Independence from Fossil Fuel Workshop, Frankfurt, 9. October 2012

Pollution at Sea, Cargo Safety, Environment Protection and Ecology
Maritime Transport & Shipping – Marine Navigation and Safety of Sea Transportation – Weintrit & Neumann (Eds)

Response to Global Environment Education for Disaster Risk Management: Disaster Preparedness of JBLFMU-Molo, Philippines

R.A. Alimen, R.L. Pador & C.D. Ortizo
John B. Lacson Foundation Maritime University-Molo, Iloilo City, Philippines

ABSTRACT: The study determined the disaster preparedness of a maritime university, specifically, John B. Lacson Foundation Maritime University-Molo in the Philippines. This study used quantitative-qualitative modes of data collection. The study determined the different disaster practices, drills, and exercises at the maritime university in its response to Global Environmental Education for Disaster Risk Management. To further reinforce the data generated for this investigation, a document analysis of the written documents of the Disaster Committee, JBLFMU-Molo was utilized. The participants of this study were faculty members and marine officers who had been on-board. Results revealed that majority of the respondents at maritime university preferred to join "Disaster Committee" as part of their social awareness and consciousness on the prevailing imbalance forces of nature. As a maritime university, the participants' first priority is *fire committee* as compared to *earthquake, flood, typhoon, bomb threat, and tsunami committees*. In this study, the researchers found out that JBLFMU-Molo conducted disaster drills to better prepare the students, faculty, and staff to the different uncertainties brought about by natural calamities and fire. The most frequent drill conducted was 'fire drill', few respondents mentioned 'earthquake drill' and 'bomb threat' drills. The results revealed that there were no drills on 'flood,' 'tsunami,' 'volcanic,' 'typhoon' although these natural disasters were considered equally dangerous. This illustrated that the maritime university focused only in combating fire perhaps because of the required STCW competency standard on how to deal with fire on-board. This 'fire drill' increased the participants' awareness of their maritime profession, risk of life on-board, tanker operation-activities, and handling chemical and toxic cargoes and substances. In response to the many challenges faced by many individuals, the maritime university (JBLFMU-Molo) tapped students from the NROTC and CWTS as volunteers to the many activities/approaches to lessen the hazards that brought by "disaster of every life." The volunteers were trained and educated on how to predict incoming disasters. The training helped the volunteers to determine the level of resiliency, capabilities, and susceptibility of certain place and people towards different types of disasters. Furthermore, the study highlighted the qualitative comments, remarks, views of the respondents in relation to "Disaster Preparedness Program" of the maritime university.

1 INTRODUCTION

In the United States of America (USA), ninety percent (90%) of the declared disasters result from natural phenomena, which flooding was considered a major component. According to reports, annually, US averages 100,000 thunderstorms, Glaveston Texas hurricane in 1900 killed more than 6,000. Death figure from Katrina exceeds 2,000 people. Average of 22 "killer tornados" each year, and 13,000 earthquakes each year were reported (Fox, White, Rooney, & Rowland, 2006). These are the nature's occurrences brought about by the climate change and abusive use of natural resources, which need attention of different sectors of the society.

Another study (Peras, Pulhim, Lasco, & Cruz, 2008) cited that drastic changes in climate create some imbalance in the environment and lead to adverse weather condition as stated in the observations below:

seasonal and international variability in climate, extreme weather events such as cyclones, typhoons, prolonged dry spells, and intense rainfall are known to cause adverse effects such as drought and flood in tropical Asia.

Furthermore, another study entitled "Coping with Climate Change in Small Landholdings (2010)", supported the above-stated statements that the advent of climate change and erratic weather conditions have created havoc in cropping schedules, human properties, and activities.

The extreme environmental disturbances in the nature are now considered a significant problem by educators, policy-makers, and government officials. The people from the academe suggest the need to educate the mindset of the students and young ones regarding the severe condition of the Mother Earth. As attested by the statements in the study below:

Environmental education has to be given priority among students as well as community groups and general public. The Earth is in dismal condition due to man's continuous onslaught of its natural resources and callous activities which lead to impending environmental degradation, one of which is global warming. Faced with the gloomy scenario for future generation, school leaders are expected to ensure that the 21st century students develop awareness, skills, and attitudes necessary to pursue successful action for environmental protection and conservation (Orr, 1996, in Cadiao, 2010).

In the Philippines, El Niño brought long drought while La Niña caused heavy floods resulting in death and destruction of properties. This has been the result of an alarming increase in the earth's temperature due to global warming as a result of continuous release of greenhouse gasses into the atmosphere (Peras, Pulhim, Lasco, Cruz, 2008). These extraordinary varieties of environmental and ecological changes were potentially caused by unconcerned and unmindful management of environmental project, scheme, or policy as mentioned in this study (Huerbana, 2009).

However, the assumption that "natural disasters" are inherently and predominantly natural phenomena has tended to exclude the social sciences from consideration in much of the spending that is done in disaster preparedness. Much of the conventional work on disasters has been dominated by "hard science" and has been the product of the "social construction of disasters" as events that demonstrate the human condition as subordinate to "nature". Within such framework, there is the inherent danger that people are perceived as victims rather than being part of socio-economic systems that allocate risk differently to various types of people. People therefore often treated as "clients" in the process of disaster mitigation and preparedness, and as passive onlookers in a process in which science and technology do things to them and for them, rather than with them (Cannon, 2000).

Natural and human-made disasters continue to adversely affect all areas of the world in both predictable and unpredictable situations; therefore, public emergency response to disasters is needed. Public health professionals and others should be involved with disaster management (Logue, 1996).

Legarda (2009) discussed that natural disaster caused by climatic change are among the greatest threats faced by the world, especially the developing countries. She further mentioned that:

Climate change and disaster risks are the defining issues of our time . their increasing trend driven by economic growth brings to fore a human development issue and a human security concern that calls for urgent action.

In connection with the unprecedented climate degradation and deterioration, Orr (1996, in Cadiao, 2010) posited that environmental education is an opportunity to revitalize and enliven the curriculum and pedagogy, an opportunity to create a genuinely interdisciplinary curriculum, to design the campus to reduce costs, lower environmental impacts and help catalyze sustainable economy.

In line with these reports, disaster preparedness and emergency response systems are supposed to be in place. Typically, disaster preparedness and emergency response systems are designed for persons to escape or rescue by walking or running from natural calamities like earthquake, flood, tsunami, fire, and etc.

Cudal (1999, in Cadiao, 2010) stressed that there is a need for the schools to focus their efforts to human resource development strategy to meet the needs of growing population as well as re-structure and re-orient educational system.

2 DISASTER PREPAREDNESS

Disaster preparedness is a process of ensuring that an organization has complied with the prevention measures. It is assumed as a state of readiness to contain the effects of a forecasted disastrous event to minimize loss of life, injury, and damage property. It is extensively defined as a way to provide rescue, relief, rehabilitation, and other services in the aftermath of the disaster. It entails the capability and resources to continue to sustain its essential functions without being overwhelmed by the demand placed on them, first and immediate response—emergency preparedness. Moreover, typically, disaster preparedness and emergency response systems are designed for persons to escape or rescue by walking or running (Fox, White, Rooney, Rowland, 2006).

The advent of climate change and erratic weather conditions is especially crucial to farmers. Erratic conditions have created havoc in cropping schedules (Coping with Climate Change in Small Landholdings, 2010).

Apart from seasonal and international variability in climate, extreme weather events such as cyclones, typhoons, prolonged dry spells, and intense rainfall are known to cause adverse effects such as drought and flood in tropical Asia as mentioned in the studies of Peras, Pulhim, Lasco, Cruz (2008).

An extraordinary variety of environmental and ecological changes can potentially be caused by any project, scheme, or policy (Huerbana, 2009).

3 ENVIRONMENTAL EDUCATION

Environmental education has to be given priority among students as well as community groups and general public. The Earth is in dismal condition due to man's continuous onslaught of its natural resources and callous activities which lead to impending environmental degradation, one of which is global warming. Faced with the gloomy scenario for future generation, school leaders are expected to ensure that the 21st century students develop awareness, skills, and attitudes necessary to pursue successful action for environmental protection and conservation (Orr 1996, in Cadiao, 2010).

According to Orr (1996, in Cadiao, 2010) environmental education is an opportunity to revitalize and enliven the curriculum and pedagogy, an opportunity to create a genuinely interdisciplinary curriculum, to design the campus to reduce costs, lower environmental impacts and help catalyze sustainable economy.

Based on the above-mentioned statements, the researchers of this study attempted to investigate the disaster preparedness of the maritime university in the light of the environmental education drive. It was conceptualized in this investigation how marine officers and maritime instructors view disaster preparedness program to mitigate disaster and environmental problems and issues.

4 THEORETICAL FRAMEWORK

The study employed theoretical framework from the study entitled "Vulnerability Analysis and Disasters" by Cannon (2010) and "Disaster, the Environment, and Public Health: Improving Response" by Logue (2009). This theoretical framework focused on the belief that there is an inherent danger when people are perceived as victims rather than being part of socio-economic systems that allocate risk differently to various types of people. People are often become treated as "clients" in the process of disaster mitigation and preparedness, and as passive onlookers in a process in which science and technology do things to them and for them, rather than with them. It additionally anchored on Logue (2009) that there is a need to highlight the importance of natural disasters, environmental health issues, approaches, and reviews relating to capacity building, training, and collaboration.

Furthermore, Wang (2010) study suggested approaches to disaster emergency relief, these are the following: (a) empowering the communities to prepare for natural disasters and train themselves to be the first responders of calamities, (b) have tools to help the affected people move more quickly into recovery mode. To institutionalize the disaster management, the following are required approaches: (a) creation of capacity, (b) adaptation strategies, and (c) linkages and knowledge. The framework of this study underscored the strengthening of the knowledge and expertise to review relevant risks, calculate probabilities, prepare contingency plans and lay out specific procedures for disasters. This capacity build up may be a great help in responding to emergency situations so that lives can be saved and losses mitigated.

5 OBJECTIVES OF THE STUDY

The objectives of the study were advanced in the following:
1 To identify the disaster preparedness activities, drills, and simulations of the maritime university
2 To present the views and ideas of the marine officers and instructors regarding the disaster preparedness in the JBLFMU-Molo as the only maritime university in the Philippines.
3 To discuss the disaster preparedness plan of the maritime university derived from the qualitative information shared by the participants.

6 CONCEPTUAL FRAMEWORK

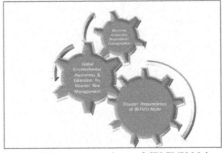

Figure 1. Disaster preparedness of JBLFMU-Molo anchored on global environment education for disaster risk management

The conceptual framework presented in this section illustrated the connectivity of the maritime university disaster preparedness program and the issues on global environment education for disaster risk management. The conceptualization of the maritime university's disaster preparedness was based on the prevalent issues on climate change and social responsibilities of different stakeholders. This leads to global environmental awareness and education towards disaster preparedness of JBLFMU-Molo.

7 METHOD

The present study employed quantitative-qualitative method of gathering data. The respondents were thirty (30) marine officers and faculty members of JBLFMU-Molo, Iloilo City, as the only maritime university in the Philippines. The respondents were classified according to number of years teaching in the university, area of specialization, and the location of residence (Table 1). They were provided by the questionnaire as data-gathering instrument, a rating scale to provide the respondents' profile and information. The open-ended questions were incorporated in the instrument to gather their ideas, views, and perceptions about the disaster preparedness of the only maritime university (JBLFMU-Molo) in the Philippines. This data-gathering instrument was subjected to the panel of experts in research, instrumentation, and environment. Suggestions and recommendations of the validators/jurors and experts were considered in the final draft of the instrument. The instrument was pilot-tested among personnel of selected government organizations to assure the validity and to address some problems arising from the administration and distribution of the research instrument. The researchers consulted experts in the field in relation to the questions on technical and environmental social issues.

8 PARTICIPANTS OF THE STUDY

The participants of the present study were the marine officers and instructors of maritime university in the Philippines (JBLFMU-Molo). The participants were classified according to the different categories such as number of years in teaching at maritime university, area of specialization, and location of the participants. Most of the participants are teaching at the maritime university for 6 years and more (f = 17, 57%), have General education (f = 20, 67%), and residing at the different municipalities of Iloilo, Philippines (f = 18, 60%). The distribution of the participants is shown in Table 1.

Table 1. Distribution of the Respondents

Category	f	%
A. Entire Group	30	100
B. Number of Years in Teaching		
6 years and above	17	57
1-5 years in teaching	13	43
Total	30	100
C. Area		
Professional	6	20
Technical	4	13
Gen Education	20	67
Total	30	100
D. Location of Residence		
District of Iloilo City (IC)	8	27
IC Proper	4	13
Municipalities of Iloilo	18	60
Total	30	100

9 DATA-GATHERING INSTRUMENT

The present study employed the data-gathering instrument entitled "Disaster Preparedness of JBLFMU-Molo" consisted of three (3) parts. The first part contained personal information of the participants and the second part dealt on the disaster preparedness of the maritime university in the Philippines. The instrument has twelve (12) items. Some of which had open-ended questions. The last part asked about the suggestions and comments on disaster preparedness of the maritime university. The information captured by this section was used to generate qualitative descripton of the disaster preparedness of the university.

This data-gathering instrument was subjected to face-content validity by the experts and jurors in the fields of research, instrumentatation, environment, and disaster management. Suggestions and remarks of the experts and jurors were followed by the researchers to improve the final draft of the instrument. Pilot testing was conducted to solicit suggestions for the refinement of the instrument.

10 RESULTS AND DISCUSSION

This section presents the results based on the objectives of the present study. Majority of the respondents (60%) answered that they were members of "Disaster Committee" while the (40%) were not members.

Table 2. Members of the Disaster Committee of a Maritime University in the Philippines

Question	Frequency	Percentage
Are you a member of Disaster Committee?		
YES	18	60
NO	12	40
TOTAL	30	100

The participants of the study preferred to be members of the *fire committee* as the first priority, followed by *earthquake, flood, typhoon*, and the least priority were *fire, bomb threat, and tsunami*. This means that most of the respondents perceived typhoon as the most prevalent disaster the frequently occurrs in the country. Tsunami was the least occuring disaster cited by the respondents or may not be occurring at all.

Data are shown in Table 3.

Table 3. Disaster Committee Participated by the Respondents

Disaster Committee	Frequency	Percentage	Rank
A. Fire	7	39	1.0
B. Earthquake	6	33	2.0
C. Flood	3	17	3.0
D. Typhoon	2	11	4.0
E. Bomb Threat	0	0	5.5
F. Tsunami	0	0	5.5
Total	18	100	

The participants agreed that the maritime university conducts regular drills on disaster (f = 24, 80%) as shown in Table 4. The data confirmed the commitment of the university to address challenges in combating environmental disturbances by conducting regular drills on disaster preparedness.

Table 4. Regular Drill on Disaster Preparedness

Has JBLFMU-Molo conducted regular drill on disaster preparedness?	Frequency	Percentage
YES	24	80
NO	6	20
TOTAL	30	100

The maritime university had the following disaster drills: (1) fire drill (f=27, 90%), (2) earthquake drill (f=2, 6.67%), (3) bomb threat (f=1, 1.33%), (4) flood drill, tsunami, volcanic, typhoon drills (f=0). The results show that the university is well-prepared in combating fire, may be because this particular drill is needed on board and a requirement to become competent officers. This drill helps the respondents to acquire advanced knowledge and awareness of the future accidents that the students might encounter as brought about by overheating of engine, electrical faults, and fire.

Table 5. Emergency Drills conducted by the Maritime University

Which of the emergency drills does your university exercise?	Frequency	Percentage	Rank
A. Fire Drill	27	90.00	1.0
B. Earthquake Drill	2	6.67	2.0
C. Bomb Threat Drill	1	3.33	3.0
D. Flood drill	0	0	5.5
E. Tsunami Drill	0	0	5.5
F. Volcanic Drill	0	0	5.5
G. Typhoon Drill	0	0	5.5
TOTAL	30		

The results in the Table 6 show the number of drill/activities conducted by the university to simulate the procedure needed in order to prepare in case of disasters such as fire, earthquake, flood, typhoon, and others natural environmental disasters caused by human negligence and lack of concern. The results of the study revealed that majority of the participants agreed that disaster drills are conducted "once a year" (f=20, 67%, R=1) followed by "twice a year" (f=6, 20%, R=2) and "three times a year" (f=4, 13%, R=3) according to the participants.

Table 6. Number of Disaster Drills Conducted by the Maritime University

Question	Frequency	Percentage	Rank
How often does the university conduct disaster drills?			
A. Once a year	20	67	1
B. twice a year	6	20	2
C. three times a year	4	13	3
TOTAL	30	100	

The results in Table 7 revealed that most of the participants (f=28, 93%) said that the university had management plan and only few of participants (f=2, 7%) said "no disaster management plan."

Table 7. Disaster Management Plan of Maritime University in the Philippines

Question	Frequency	Percentage
Does JBLFMU-Molo have Disaster Management Plan?		
YES	28	93
NO	2	7
TOTAL	30	100

Table 8. Integration of the Disaster Management Plan

Question	Frequency	Percentage
Is the Disaster Management Plan integrated in the annual development Plan?		
YES	28	93
NO	2	7
TOTAL	30	100

Table 9. Calamity/Disaster Experienced by the Participants

Question	Frequency	Percentage
Have you experienced calamity/disaster?		
YES	30	100
NO	0	0
TOTAL	30	100

The respondents of this study experienced primarily the havoc of flood, followed by typhoon, fire, and earthquake. The flood that affected the properties and lives of the people in Region 6 remained horrible and terrifying even up to this time. This unexplained environmental turbulence/disaster is one of the effects of the lack of concern of the people of their surroundings and Mother Earth.

Table 10. Type of the Disasters Experienced by the Participants

Question	Frequency	Percentage	Rank
Which of the following disasters had you experienced?			
A. Flood	10	33	1.0
B. Typhoon (strong wind)	8	27	2.5
C. Fire	8	27	2.5
D. Earthquake	4	13	3.0
Total	30	100	

Majority (26, 27%) of the respondents agreed that the university provides assistance to the victims of disasters who were faculty members and staff. The assistance was in form of "loans" to alleviate the sufferings caused by faculty members during natural disasters in the region. The university also taps some non-government agencies to help less fortunate people who were victimized by the forces of nature.

Table 11. Disaster Assistance provided by the maritime university

Question	Frequency	Percentage
Is the university providing assistance to the faculty/staff who were victims of disasters/calamity?		
YES	26	27
NO	4	13
TOTAL	30	100

The results in Table 12 and 13 showed that the maritime university (JBLFMU-Molo) utilized students from the NROTC and CWTS as volunteers for any untoward accidents brought about by the natural disasters such as flood, typhoon, earthquake, and fire. These brigades were established by the university as part of social responsibility to the people drastically hit by disasters brought about by environmental pollution and human negligence.

Table 12. Volunteers to Disaster established by the school

Question	Frequency	Percentage
Does the university have volunteers in case of emergency to help those who were victims of disasters?		
YES	25	83
NO	5	17
TOTAL	30	100

11 SUMMARY OF QUANTITATIVE DATA

Majority of the respondents at the maritime university preferred to join "Disaster Committee" as part of their social awareness and consciousness to the prevailing imbalance/forces of nature. As a maritime university, the participants' first priority was *fire committee* as compared to *earthquake, flood, typhoon, fire, bomb threat, and tsunami*.

In this study, the researchers found out that JBLFMU-Molo conducted the drills to better prepare the students, faculty, and staff to different uncertainties brought about by natural disasters. The most frequent drill conducted was 'fire drill', few of the respondents mentioned 'earthquake drill' and 'bomb threat'. The results revealed that there were no drills on 'flood,' 'tsunami,' 'volcanic,' 'typhoon' although these natural disasters are considered as dangerous but are not likely to happen in the region. This illustrates that the maritime university focused only in combating fire perhaps because of the required STCW competency standard in dealing fire on-board. This 'fire drill' increases the participants' awareness of their maritime profession, risks of life on-board, tanker operation-activities especially carrying chemical and toxic cargoes and substances.

The results of the study revealed that majority of the participants agreed that disaster drills are conducted "once a year" to prepare the students in any situation on-board international vessels. The participants also agreed that the university established "disaster management plan" focusing on techniques of fighting fire. Respondents mentioned that they were also victims of flood, typhoon, fire, and earthquake. The most frequent kind of disaster that has devastated properties and lives of the people in Region 6 was typhoon (Typhoon Frank). This typhoon created unforgettable moment in every life of Ilonggos. The impact of this natural disaster is an example of environmental concern that all sectors of society should be aware of and be vigilant with in order to help protect the natural resources of a region.

In response to the many challenges faced by many individuals, the maritime university (JBLFMU-Molo) encouraged students from the NROTC and CWTS as volunteers to activities/approaches that are hazards or "disasters of every life." These volunteers were trained and educated on how to sense or predict the incoming disasters. The training helps them to determine the level of resiliency, capabilities, and susceptibility of a certain place and the people in their readiness toward the different types of disasters.

12 QUALITATIVE DATA

The qualitative data were captured by interviews and open-ended questions. The respondents were requested to answer the two (2) open-ended items such as (1) What are your ideas/comments/views about the Disaster Preparedness of JBLFMU-Molo?, and (2) What can you say about the Disaster Preparedness Plan of JBLFMU-Molo? The selected respondents openly shared some comments and views on Disasater Preparedness of JBLFMU-Molo, Iloilo City, Philippines. The most prevalent comments/views/ideas were transcribed and translated by the researchers and presented in this section of the study.

The most prevalent ideas/comments/views on the Disaster Preparedness of JBLFMU-Molo are stated in the following statements:

1 *Maayo guid kon may ara "Disaster Preparedness Program sa aton school. Kay ano nga oras may preparasyon kita sa tanan nga klamidad kag disaster nga naga kalatabo. Amo gani nga kon may mga volunteer sa disaster training sa tanan nga kalamidad indi kita mabudlayan mag bulig bisan sa gua kag sa sulod sang 'school compound' matabo. Kay indi naton mapaktan ang panahon kon san o ini nga disaster naga katabo*

(It is good to have Disaster Preparedness Program in the school so that whenever disasters happen, we are always prepared. That is why we have volunteers for Disaster Training so that we are always ready to help those outside and inside the school because we do not know when will these disasters occur).

2 *The school has done an excellent job as far as disaster preparedness is concerned. The school already put-up 'escape stairs' in all buildings that will be used as fire or earthquake exits for the teachers, students, and staff. The school installed 'fire alarm device' that can be used by everyone if there is an earthquake or fire. The administration significantly improves the drainage system and acquires some rescue equipment such as ropes, emergency lights, and other equipment needed for the preparation of flood and heavy rains brought about by sudden change of temperature or monsoon.*

3 *Fire drill was conducted annually. But, the members of the Disaster Committee are inactive except for the 'Fire Committee' of the university.*

The following are the prevalent comments/remarks of the respondents on the 'Disaster Preparedness Plan of JBLFMU-Molo,' Iloilo City, Philippines:

1 *The school conducting fire and earthquake drills once or twice a year. These drills are participated by the students, teachers, and personnel. The school also organizes different committees/teams and each team has a direct area of responsibility in case there will be any disaster that would happen. The school board allocates a budget to assist individuals who had been damaged/affected by disasters. Other measures of preparedness were introduced by the school so that those who are badly affected will be supported.*

2 *Kon may plano na gani, sayuron mahimo guid ina tanan. Labi na guid kon may 'well trained" kita nga mga volunteers for 'Disaster Brigade' para sa tanan nga kalamidad nga naga kalatabo sa palibot naton. Maayo guid kon may ara kita nga "Disaster Preparedness Plan" kay bisan ano oras preparado kita tanan sa mga kalamidad nga*
matabo sa sulod man o sa gua sang school compound.

If there is a plan in place, that means that everything can be done as regards disasters. Most especially is there are well trained volunteers that will serve as "Disaster Brigade" on whatever calamity that are going to happen around us. It is better to have a Disaster Preparedness Plan so that we are prepared at any time the calamity happens inside and outside the school premise.

3 *The school has developed preparedness plan for disasters where personnel were assigned to the different committees. Considering that JBLFMU-Molo is a maritime university, the visible one is the 'fire drill.'*

13 CONCLUSIONS

In conclusion, the present study revealed that majority of the respondents at maritime university was members of the "Disaster Committee" as the major issue of the present time. The University's thrust on maritime safety has enhanced the disaster awareness and advocacy of the faculty and staff. Environmental education is an opportunity to revitalize and enliven the curriculum and pedagogy, an opportunity to create awareness and concern to those who are victims of disasters and other calamities.

It was found out that conducting 'drills' better prepare the students, faculty, and staff to different uncertainties brought about by natural disasters. The most frequent drill conducted at maritime university was 'fire drill' among the disaster drills. In line with the University's goal of creating a disaster conscious academe, the academic community makes sure that disaster related drills and activities are in place.

In response to the many challenges facing many individuals, the maritime university (JBLFMU-Molo) utilized students from the NROTC and CWTS as volunteers for any activities/approaches to hazards and other disasters of life. The volunteers were trained and educated on how to sense or predict the incoming disasters. The training helps them to determine the level of resiliency, capabilities, and susceptibility of certain place and the people towards dealing with the different types of disasters. This includes the NROTC and CWTS in responding to disaster preparedness of the school and city.

14 RECOMMENDATIONS

1 It is recommended that 'fire drill' should be continued and more drills be conducted because frequent practice makes the students, faculty, and

staff ready and well-prepared to face and handle disasters.

2 Disaster education and information campaign should be enhanced by inviting more experts in environmental advocacy and disaster preparedness.

3 The administration should devise more disaster programs to stimulate the faculty members to become active participants and volunteers in times of natural and environmental crisis.

4 The political-economic vulnerability of affected individuals from the disasters should be studied by those who are experts in this field. This may generate data that could be used as basis for further discussion on disaster issues and environmental problems.

5 Disaster Brigades of schools should be coordinated with other schools to strengthen more resources and hone competent volunteers.

6 Non-government organizations should join the crusade in dealing with environmental mitigating measures and disaster information-education campaigns.

7 Parallel studies must be conducted to determine other factors that would help in the issues on environment and disaster preparedness programs.

REFERENCES

Alam, J. (1998). Environmental Impact Assessment-Some Issues. Environmental management and Sustainable Development. Vol. 12, No. 2.

Cadiao, L. V. (2010). *Development of Modular Package on Environmental Awareness, Protection and Conservation for College Students*. Compilation of Faculty and Students' Researches. Occidental Mindoro State College, San Jose, Occidental Mindoro, Philippines.

Cannon, T. (2000). Vulnerability Analysis and Disaster.

Coping with Climate Change in Small Landholdings (2010).

Elauria, J.C. (2006). Environmental Studies for Publication and Utilization in the Philippines. The PSERE Journal ISSN 1656-376X. Volume XI, No. 1 (2006).

Fletcher, K. (2001). *Environmental Management and Business Strategy: Leadership Skills for the 21st Century*. New Jersey: Prentice Hall.

Follosco, G. (2001). *Technology Students' Level of Environmental Awareness, Values and Involvement on Environmental Protection and Conservation*. Unpublished Doctoral Dissertation. Technological University, Manila, Philippines.

Fox, White, Rooney, & Rowland (2006). Disaster Preparedness and Response for Persons with Mobility Impairments: Results of the Nobody Left Behind Project. Research and Training Center on Independent Living, University of Kansas.

Huerbana, R.S. (2009). Environmental Impact Assessment of the Rehabilitation, Conservation and Publication of the Cagayan de Oro River. First International Conference on Multidisciplinary Research, Universidad de Zamboanga.

Institutional Research for Internationalization of Philippine Research Journal Publication (2010). Coping with Climate Change in Small Landholdings

Leagarda (2009). A Call on the International Community to focus on Disaster Risk Reduction. LOREN LEGARDA REPORTS. A Special Edition on Climate Change.

Logue, J.N. (1996). *Disaster, the Environment, and public health: Improving our Response. Environmental Science, Working with Earth*. Eleventh Edition. New York: Thomas Book Company.

Orr, D. (1999). *Educating for the Environment: Higher Education Challenge for the Next Century. Journal of Environmental Education. Vol. 27, No. 3.*

Peras, R., Pulhim, J.M., Lasco, R.O., Cruz, R.V. & Pulhim, F.B. (2008). *Climate Variability and extremes in Pantabangan-Carranglan Watershed, Phil: Assessment of Impacts and Adaptation Practices.*

Wals, A. (2000). Caretakers of the Environment, A Global Network of Teachers and Students to Save the Earth. Journal of Environmental education. Vol. 21, No.3.

Wang, Z. (2010). *Geological Environment and Disasters Along Railway Line in the Qinghai-Tibet Plateau.*

www.lorenlegarda.com ph

Students were utilized to employ home-made barricade-structures during the Oil Spill in Guimaras, Philippines: One disaster where the university had helped and actively participated.

IN CASE OF
EMERGENCY
BREAK GLASS

Pollution at Sea, Cargo Safety, Environment Protection and Ecology
Maritime Transport & Shipping – Marine Navigation and Safety of Sea Transportation – Weintrit & Neumann (Eds)

Marine Environment Protection through CleanSeaNet within Black Sea

S. Berescu
Romanian Naval Authority, Constanta, Romania

ABSTRACT: The article presented is intended to highlight the activity and efforts developed by the Romanian Naval Authority (RNA) in order to fulfill the obligations assumed by Romania and required by the IMO Conventions and EU Directives. The organizational structure reveals that RNA is complying with the new requirements and recommendations regarding the pollution prevention and pollution response. It is stressed the good cooperation with the International Maritime Organization (IMO) and the European Maritime Safety Agency (EMSA) and the important achievements concerning the application and the enforcement of the MARPOL requirements in the area of jurisdiction has been brought. The new equipment and advantages of the CleanSeaNet System is described and presented as a new method used to protect the marine environment. A real case of marine pollution shows how MCC is functioning by applying the satellite image information in order to suppress any form of violation in respect with national and international legislation for marine pollution prevention.

1 INTRODUCTION

Cleaner seas represent an objective which involves the maritime states all over the world. The European Union member states are strongly committed to act in a harmonised manner to protect and intervene in case of maritime pollution as per the requirements of specific IMO conventions and European directives and regulations.

To understand how important the struggle against marine pollution is and also the role played by the National Administration, is important to shortly present the Romanian Naval Authority (RNA), specialized technical body, acting as a state authority in the field of the safety of navigation, that represents and fulfils the obligations assumed by Romania with regard to international agreements and conventions such as those connected with environmental protection against marine pollution from ships.

In doing its obligations on pollution protection within the area of jurisdiction, RNA is using the new European service CleanSeaNet since September 2007.

Maintaining a competitive level and a sustainable development are the major objectives even in the context of the world crisis'negative effects. RNA provides high quality level services in accordance with the provisions of the legal and regulatory requirements which are included in the quality system policy and the procedure of the Management System having an essential contribution in the company competitiveness. The Romanian Naval Authority's motto "Safety through Quality" represents the importance given to the highest standards within the company.

2 MAIN TASKS

The main tasks of the Romanian Naval Authority regarding the fight against pollution have been defined including the following:

- Inspection, control and surveillance of navigation in Romanian maritime waters and inland waterways;
- Fulfilment of the obligations assumed from the international agreements and conventions to which Romania is part of;
- Representing the Romanian Government within the international organizations in the field of naval transports;
- Implementation of international rules, regulations and conventions into Romanian legislation;

- Development, endorsement and submission of drafts laws and mandatory norms to the Ministry of Transports for approval;
- Port State Control and Flag State Control;
- Coordination of search and rescue activities in the Romanian navigable waters and of the actions to be taken in case of navigation accidents and casualties;
- Protection of navigable waters against pollution by vessels;
- Sanctioning of the contraventions and investigation of the navigation accidents and casualties;
- Technical surveillance and certification of maritime and inland water ships, offshore drilling units flying the Romanian flag and of naval equipments, as per RNA regulations;
- Supervising the compliance of the Romanian naval transports with the provisions of the ISM Code and ISPS Code.

To meet and apply the requirements set in the international conventions such as SOLAS/1974, SAR/1979, MARPOL 1973/1978 and OPRC/1990, Romanian Naval Authority has been legally appointed as the responsible authority to perform the management and mission co-ordination for SAR and Oil Response activities and also to monitor the vessels' traffic within the area under Romania responsibility, through Maritime Coordination Centre.

3 THE MARITIME COORDINATION CENTRE:

The one roof concept, as per IMO recommendation, which includes VTS and SAR-OPRC under a single umbrella, is one of the strongest points of the related centre which permits to act in a unitary manner for monitoring, controlling, coordinating and intervention in case of maritime incidents, casualties and pollution.

The Maritime Coordination Centre (MCC), through SAR - Pollution Department and VTS Department, performs activities regarding search and rescue, prevention and response to marine pollution, as well as the surveillance and management of the vessels' traffic, 24 h/day. The Maritime Coordination Centre has full responsibility in cooperating between RNA and other international organizations involved in the field of search and rescue of human lives at sea and marine pollution. By attending national and international programs in this field, such as seminars, trainings, conferences and organizing national and international exercises in order to reach the specific standards at highest level has increased the role and importance of the Romanian Marine Coordination Centre within the region.

MCC is involved with the International Maritime Organization in working groups, within the Marine Environment Protection Committee (MEPC) and Sub-Committee on Radio communications and Search and Rescue (COMSAR), providing technical assistance, issuing international legal regulation regarding SAR and oil response activities. MCC has been also designated as national operational contact point (NOCP), according to the National Contingency Plan for Oil and HNS pollutions.

For the application of the stipulations of art.6 of the International Convention (1990) regarding the preparation, response and cooperation in case of hydrocarbons pollution, there is established a harmonized system for action in case of pollution with hydrocarbons named the National System which includes the measures for preparation and response in case of pollution with hydrocarbons.

4 THE NATIONAL PLAN FOR PREPARATION, RESPONSE AND COOPERATION

The National Plan for preparation, response and cooperation in case of marine pollution with hydrocarbons is part of National System.

The authorities involved with this Plan are:
- Ministry of Water and Environmental Protection (Designated Department) - for coordination of activities connected with the function of National System, elaboration and updating of National Plan, Contact National Point with international authorities;
- Civil Protection Commandment - for terrestrial intervention;
- Romanian Naval Authority of the **Romanian Ministry of Transport** – for maritime operations;
- Ministry of Defense - Navy Forces – for military navies.

The Plan consists in application of the provisions of International Convention for preparation, response and cooperation in case of pollution with Hydrocarbons, adopted at London in 30 November 1990 (OPRC 1990). The application of this Plan covers the Romanian Black Sea coast, territorial sea and Romanian exclusive economic area, with the aim to mitigate promptly and efficiently the incidents with hydrocarbon pollution, for protecting the sea environment, coastal area and human life and health. The Plan includes also the requests for organizing at national level preparations, cooperation and intervention in case of pollution, the methods of action after receiving the report regarding the hydrocarbon pollution, promoting the international cooperation and research for mitigation of marine pollution with hydrocarbons.

At national level the coordination for the application of the Plan is performed by the Ministry of Waters and Environmental Protection, in cooperation with the Ministry of Transports, the Ministry of Internal Affairs, the Ministry of Public Administration and the Ministry of Defense, which have responsibilities in coordination of actions for preparation, cooperation and response, through their designated departments and subordinated units, as set in the Plan.

For interpretation of the Plan, regarding the juridical status of interior maritime waters, territorial sea, contiguous area and Romanian exclusive economical area, the terms are defined as per art.2 of OPRC 1990 and as per those specified in Law 17/1990 as amended.

Commanders of maritime navies under Romanian flag, commanders of marine drillings units, of ports and installations for hydrocarbons, have the obligation to immediately report at RNA all incidents of hydrocarbons pollution by them or by others or the presence of hydrocarbons in water.

Pilots of civil or military airplanes must report to the Center for air traffic all incidents of hydrocarbons pollution observed in water.

The Center for control of air traffic will transmit immediately the information to RNA or if the case, to National Agency "Romanian Waters", DADL.

The Operative Commandment for Marine De-pollution (CODM) – is the coordinator of activities in case of marine pollution with hydrocarbons and is conducted by a General Coordinator (Prefect of Constanta County) and by a Deputy General Coordinator (President of Constanta County Council).

CODM is the National Center for Intervention in case of Emergency, the operational base of all actions for mitigation of marine pollution, in case of entrance in action of Plan or in case of regional cooperation.

CODM organizes and leads annually for all units included in the National System a general exercise for intervention in case of major marine pollution with hydrocarbons. List of intervention forces, the equipments for response and communication of the units which are parts of National System is permanent updated and kept by the Permanent Secretariat of CODM.

Rules of the Plan are stipulated by the Governmental Decision nr.1232/2000 for approving the Methodological norms for implementation of provisions of International Convention regarding the civil responsibility for damages produced by hydrocarbons pollution, 1992 (CLC, 1992).

In case of pollutions other than those stipulated in 1992 Protocol, according by Governmental Ordinance nr.15/2000, compensation of spending for de-pollution and restoration is made under the principle *polluter pays*.

5 ACTIVITIES AND RESPONSIBILITY

As an important tool within MCC, Constanta VTS is a modern and integrated system for maritime traffic management and it has the responsibility of monitoring, conducting, surveillance and coordinating vessel traffic, in order to improve the safety and efficiency of navigation and to protect the environment in the VTS area fully connected with AIS and LRIT systems.

As an obligation for an EU costal state, Romania, performs the application of the CleanSeaNet service, which is a satellite based monitoring system for marine oil spill detection, tracing and surveillance by checking on scene the satellite images.

MCC is connected also to SafeSeaNet (SSN) System for port and alert notifications. This system consists essentially of setting up an electronic network between the maritime administrations of the member states in order to facilitate the implementation of maritime safety aspects.

The connection with the Global Integrated Shipping Information System (GISIS) database of the International Maritime Organization, allows introducing all information regarding port reception facilities, SAR and marine pollution incidents.

To protect the marine environment by coordinated prevention response and limitation of consequences of pollution from ships as well as by monitoring and managing the vessel traffic, RNA is committed to minimize loss of life, injury, property damage and risk to the environment by maintaining the highest professional standards.

These objectives are permitting to provide an effective SAR services for all risks, to protect the marine environment and to improve the safety and efficiency of navigation within maritime responsibility area.

The legislative frame has been implemented containing all the relevant provisions with regard to EU Directives and IMO Conventions.

Romania has ratified important conventions, protocols and agreements concerning the protection of the marine environment such as MARPOL 73/78 with all annexes; OPRC1990; CLC 1992; Bunkers 2001, as well the regional agreements: Bucharest Convention, 1992, Odessa Ministerial Statement, 1993, Regional Contingency Plan.

The following EU Directives were transposed, implemented and enforced:

- Directive 2000/59/EC on port reception facilities for ship-generated waste and cargo residues;
- Decision 2850/2000/EC setting up a Community framework for cooperation in the field of accidental or deliberate marine pollution;
- Directive 2005/33/EC amending Directive 1999/32/EC as regards the sulphur content of marine fuels;

– Directive 2002/59/EC of the European Parliament and of the Council of 27 June 2002.

At EU level, Constanta MCC is part of the Consultative Technical Group for Marine Pollution Preparedness and Response.

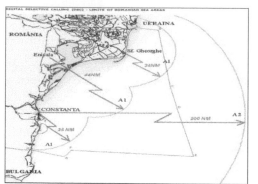

Figure 1. Area of responsibility for national intervention in case of pollution and SAR operations

All staff of the Maritime Coordination Centre is duly qualified and trained by authorized body of the International Maritime Organization to act as coordinators for SAR missions and Oil response incidents. The MCC's entire personal are trained to operate all the modern equipments, being able to perform missions in a close cooperation with the other appropriate organizations from the Black Sea region. In order to improve the efficiency of personnel and for maintaining a high level of response, trainings and the exercises are carried out at regular intervals of time.

With regard to oil preparedness, prevention, response and cooperation, MCC:
– Coordinates the prevention and response of pollution activities within RNA system including Harbour Master Offices (elaborating operational procedures for these activities);
– Participates to the investigation of oil incidents and evaluation of oil spill effects, in accordance with national legislation;
– Is connected to the CleanSeaNet service delivered by European Commission through EMSA, receives analyzed satellite images, for routine monitoring of illegal discharges from sea going ships and coordinates the activities of checking by aircrafts and surface units available;
– Introduces alerts in SSN system and updates system database;
– Monitors the sulphur content in the fuels of the vessels on Romanian territory;
– Elaborates pollution statistics for IMO, Black Sea Commission, EMSA, annually and on request;
– Coordinates and participates to all types of national and regional pollution response exercises.
In case of a major pollution, MCC:

– Receives alerts for oil pollution incidents and ensures off-shore response communications;
– Participates to the evaluation of the effects and establishes the causes of major marine pollution incidents, in conformity with its competences;
– In case of marine heavy pollution, asks partial or total activation of the National Contingency Plan (NCP), through General Coordinator of Operative Commandment for Marine Pollution (OCDM);
– Sends alerts and keeps the contact for the emergency situations with relevant national and international authorities (including IMO, EMSA and Black Sea Commission).
– According to NCP, MCC has been designated as Maritime National Operational Contact Point (M-NOCP), 24 hours capability. The main tasks of MCC as M-NOCP are to receive alerts for oil pollution incidents and to ensure off-shore response communications, directly or through RADIONAV SA. In accordance with the Regional Contingency Plan, M-NOCP exchanges information with Black Sea Commission and all Black Sea MCC's regarding the major pollution incidents in Black Sea and keeps informed national competent authorities on related situations.

The CleanSeaNet system was developed for the detection of oil slicks at sea using satellite surveillance and was offered by EMSA to all EU member States, according Directive 2005/35/EC. The system is based on marine oil spill detection by checking on scene the satellite images.

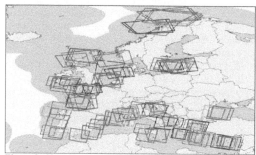

Figure 2. CleanSeaNet EU planned images

The service integrated into the national and regional response chain, aims to strengthen operational pollution response for accidental and deliberate discharges from ships and assist Coastal States to locate and identify polluters in areas under their jurisdiction. CleanSeaNet is delivering oil spill alerts in near real time (30 minutes) to both the Coastal State(s) and EMSA for detected slicks as well as giving access to the satellite image(s) and associated information over the web (and via email for low resolution images). In case of a detected oil slick, an alert message is delivered to the operational contact point.

6 REACTION IN CASE OF OIL POLLUTION:

In order to better understand how the company is acting in case of pollution incident it is important to present a real case which occurred on 3rd of November 2008 at 08.12 utc., when Constanta MCC has received from Italian Monitoring Satellite Image Centre an alert message regarding the existence of a large oil sleek of 340 metres width and 20, 33 km long oriented direction SW - NE. Located position Lat. 44°53'100"N and Long.029°49'350"E.

Following the cooperation with the ECDIS system, analysing the satellite image received in accordance with the vessel traffics' monitoring has been revealed that the polluter was m/v GUZIDE S, Turkish flag. The said vessel was navigating from Turkish port Martas to Romanian port Galati located on the Danube River. On 4th of November 2008 at 14.00 lt m/v GUZIDE S arrived in the port of Galati and after the completion of the arrival formalities the representatives of pollution department within RNA commenced the investigation of the reported pollution incident.

The investigation consisted in analysing and verification of the ship's certificates, navigation log book, engine log book and the log book for oil and bunkering evidence. The notifications transmitted by the ship to Galati Harbour Master prior arrival and the documents submitted upon arrival were analysed as well. An expended control regarding the ship's fulfilment with MARPOL requirements including the inquiry of the crew has been performed.

As evidence it was found that the quantity of bilge water from the port side bilge tanks was smaller than the quantity of bilge water mentioned in the Oil Record Book.

Figure 3. Satellite image received showing the existence of an oil sleek

The figures from the Oil Record Book did not correspond with the reality revealed by tanks soundings. According the ship's Navigation Log Book, on the 3rd of November 2008 at 08.12 UTC the vessel was in the position reported by Italian Monitoring Satellite Image Centre in fully accordance with the pollution moment.

After such evidences the captain of the vessel has recognized the violation of the MARPOL requirements and that marine pollution was related to his ship due to negligent transfer of the bunker. According with the provisions of the Government Decision 876/2007, the Captain of the ship was punished with a substantial contravention fine.

The results of the satellite imagistic system were appreciated as valuable and functionality of the CleanSeaNet has been demonstrated. This presentation demonstrates that the results on environment protection obtained by implementing the new technology are positive encouraging further development.

7 CONCLUSIONS:

The pollution response in Romania through RNA by presenting the organizational framework and the using of new CleanSeaNet EU service, followed by the proactive discussions within Gdynia conference will conduct to new results regarding the improvement of preparedness and response to marine pollution in general. A scientific analysis of complex improvement of information transmitted by satellite can be considered an important issue for future.

The organizational system described in this article and the case report presented having access to new technology imposed in EU could be an example for other national systems and an incentive for better developments and cooperation in the benefit of all mankind. Ecological disasters due to major pollution from ships can be avoided only through a legal frame, good organization, preparedness for response and cooperation between countries.

REFERENCES

[1] Gucma L., Juszkiewicz W., Łazuga K.: The Method of Optimal Allocation of Oil Spill Response in the Region of Baltic Sea.TransNav - International Journal on Marine Navigation and Safety of Sea Transportation, Vol. 6, No. 4, pp. 489-493, 2012

[2] Bąk A., Ludwiczak K.: The Profile of Polish Oil Spill Fighting System. TransNav - International Journal on Marine Navigation and Safety of Sea Transportation, Vol. 6, No. 4, pp. 501-503, 2012

[3] Arsenie P., Hanzu-Pazara R.: Human Errors and Oil Pollution from Tankers. TransNav - International Journal on Marine Navigation and Safety of Sea Transportation, Vol. 2, No. 4, pp. 409-413, 2008

[4] E. Oanţă "Virtual Reality Original Instrument Employed in Crises Management" Proceedings of the 12th international Congress of the International Maritime Association of the Mediterranean (IMAM 2007), Varna, Bulgaria, 2-6 September 2007, Maritime Industry, Ocean Engineering and Costal Resources – Editors: Guedes Soares & Kolev, 2008 Taylor & Francis Group, London, ISBN 978-0-415-45523-7, pag.1095-1102.

Pollution at Sea, Cargo Safety, Environment Protection and Ecology
Maritime Transport & Shipping – Marine Navigation and Safety of Sea Transportation – Weintrit & Neumann (Eds)

Phytoplankton Diversity in Offshore, Port and Ballast Water of a Foreign Vessel in Negros Occidental, Philippines

B.G.S. Sarinas, M.L.L. Arcelo & L.D. Gellada
John B. Lacson Foundation Maritime University, Iloilo City, Philippines

ABSTRACT: Introduction of harmful aquatic organisms and pathogens in our ocean is one of the greatest threats according to the IMO (International Maritime Organization). Alien or invasive species travel from one ocean to the other through ballast water from the international shipping industry which is very inevitable. In the Philippines, few existing studies were established on phytoplankton composition in ballast tanks of a foreign vessel; thus this study is conducted. This study aimed to identify the phytoplankton diversity of offshore, port and ballast water from a foreign vessel docking in Negros Occidental, Philippines. Furthermore, this study aimed to determine the cell density, generic diversity and evenness and physicochemical characteristics such as pH, temperature and salinity. A total of 39 liters were taken from the middle column of the offshore, port and ballast tanks through sounding pipe and siphon technique. Temperature, pH and salinity were measured, in situ. All water samples were preserved with Lugol's solution and transported in the Phycology laboratory at Southeast Asian Fisheries Development Center-AQD. This study provides baseline information on phytoplankton diversity present in offshore, port and ballast water from a foreign-going vessel in the Philippines.

1 INTRODUCTION

Shipping industry transports 80% to 90% of commodities around the world (Popa, 2009; Marrero and Rodriguez, 2004). Nowadays, thousands of ships navigate the Philippine waters both inter-island and international route. Ship that is actively part of shipping industry is called merchant vessel. These breakthroughs of man can pose several hazards and threats to mankind as well as in the marine environment.

Ships use ballast water since 1850 (Hallegraeff and Bolch, 1992). This contains suspended particles which are primarily used by ships to provide trim, list, draught, stability, or even stresses created by the ship (IMO, 2005). Approximately 150, 000 metric tons of ballast water and sediment are carried by an individual vessel. This could be a mixture of waters from many ports (Deacutis and Ribb, 2002). It was noted that one of the contributory factors to marine pollution is the process of ballasting and deballasting (Popa, 2009).

Ballast water is taken on board usually at the source port if there is less cargo, a process called ballasting. On the other hand, if seawater is pumped out in the ocean after loading of cargo at the destination port, this process is called deballasting (Deacutis and Ribb, 2002). It is estimated that about 12 million tones of ballast water are transferred around the globe annually (Marrero and Rodriguez, 2004). While Popa (2009) said that every year, shipping transfers 21 billion gallons of ballast water in the ocean yearly. Consequently, about 3, 000 to 4, 500 marine species are being carried by the ships through ballasting and deballasting processes around the globe. This led to the transfer of non-indigenous, alien, non-native, exotic, invasive marine species across the globe making it as the one of the four greatest threats in the world's ocean (IMO, 2005) and the most pressing environmental issue for biodiversity loss (Manaaki Whenua Landcare Research, 1994). It was further believed that the ships ballast water and sediment served as the major vector of transferring non-indigenous, alien, non-native, exotic, invasive marine species across the globe (Chu, Tam, Fung and Chen, 2006; Hallegraeff, 1998; MacDonald, 1995; Williams, Griffiths, Van der Wal and Kelly, 1988). Meanwhile, the ships ballast tanks serve as incubators during the voyage for certain species such as diatom and dinoflagellate (Hallegraeff and Bolch, 1992).

Now that we have more equipped and high speed ships, the survival of these non-indigenous, alien, non-native, exotic, invasive marine species survive in new habitats with no natural predators and compete for native species (Popa, 2009). However, for slower speed of ships, species richness, species abundance and diversity decrease or do not survive during the voyage which may be attributed to hostility of ballast tanks (Popa, 2009; Klein, MacIntosh, Kaczmarska and Ehrman, 2009; Chu et al., 2006; Gollasch, Lenz, Dammer and Andres, 2000; Williams et al., 1988). Marrero and Rodriguez (2004) said that there is no present method that can completely eradicate alien species from the ocean. However, mid-ocean ballast water exchange can mitigate the process of transferring these unwanted stowaways (Klein et al., 2009a; Popa, 2009; Hallegraeff and Bolch, 1992; Williams et al., 1988).

Phytoplankton serves as the major food of zooplankton and other marine animals. They are microscopic plants floating in the ocean with size ranging from 2 μm to more than 20 mm (Zaiko, Olenina and Olenin, 2010). Other species are harmful such as they produce toxins affecting fish, shellfish and humans. These toxic phytoplankton form dense color of red, green and brown in the ocean (Carver and Mallet, 2003). The origins of phytoplankton are difficult to determine and in most cases are impossible to determine (Zaiko et al., 2010). Minchin and Gollasch (2002) and Hallegraeff and Bolch (1992) said that the factors affecting phytoplankton biogeography is the ocean currents. Other environmental conditions that regulate phytoplankton composition are temperature (Badylak and Phlips, 2004; Laamanen, 1997), salinity (Badylak and Phlips, 2004), nutrient concentrations or nutrient availability (Badylak and Phlips, 2004; Marshall, Burchardt and Lacouture, 2005), weight ratio of inorganic nitrogen to inorganic phosphorus (Laamanen, 1997), depends on grazing rates (Badylak and Phlips, 2004), successional patterns and seasonal variations (Marshall et al., 2005), hydrography, winds, hull fouling, movement of live shellfish and lastly, by shipping (Minchin and Gollasch, 2002).

Numerous research papers were conducted on phytoplankton composition of ships' ballast water across the globe. In the Philippines, due to the absence of baseline data, Sarinas et al., (2010) conducted an initial survey on the plankton composition of the ballast tanks of the inter-island passenger-cargo vessel with the route Iloilo to Manila, Philippines and vice-versa. The species composition of phytoplankton found was noninvasive but difficult to conclude whether the phytoplankton was native or nonnative. Nevertheless, the study serves as the baseline information.

Marshall et al. (2005) found out 1, 454 phytoplankton taxa. Meanwhile, Klein et al. (2009a) studied the diatom survivors in ballast water during trans-Pacific crossings and found out 41 diatom taxa having 29 species. They also found out 86, 429 live diatom cells per liter that were contained in the ballast tanks. These authors pointed out that the invasive biology was poorly understood. Martin and LeGresley (2008) examined the phytoplankton species in the Bay of Fundy since 1995 and found eight new dinoflagellate species, 14 new diatom species and additional of five species of flagellates, small zooplankton, cyanonobacteria and haptophytes. The new species were speculated to be transported by ballast water. Meanwhile, the Manaaki Whenua Landcare Research (1994) and Hallegraeff and Bolch (1992) pointed out that dinoflagellates were a threat to the marine ecosystem because they could survive in a long journey due to their ability to form cysts in the ballast water tanks.

On the other hand, Gollasch et al. (2000) studied the survival of tropical ballast water organisms during a cruise from the Indian Ocean to the North Sea and found out that the ballast water coming from Singapore contained 30 species of diatoms and 24 taxa of juvenile and adult copepod. While in the ballast water of Colombo, they found out 16 species of phytoplankton and 21 taxa of zooplankton were dominated by 11 copepod species and three calanoids. Hallegraeff and Bolch (1992) and Hallegraeff (1998) studied comprehensively the resting spores and cysts of diatom and dinoflagellates in a ship's ballast water and concluded that there were certain species were resistant resting spores present in ballast water. They had found the following toxic dinoflagellates in their laboratory: *Alexandrium catenella, A. tamarense, and Gymnodinium catinatum* wherein these species could contaminate shellfish causing paralytic shellfish poisoning. They further added that, during dinoflagellate blooms, seafarers should not take in ballast water to avoid contamination to other ports.

Impacts of invasive stowaway species are irreversible (Popa, 2009). Deacutis and Ribb (2002) pointed out impacts of invasive stowaway species such as ecological, human health and economic impacts. Ecological changes or disruption of these invasive unwanted stowaways maybe due to lack of natural predators, abundance of food sources, tolerance to pollution, diseases and other stressors and out competing less aggressive species (Deacutis and Ribb, 2002). They also pointed out that 400 aquatic nonnative species had been found since 1990. Europe, Australia, New Zealand, Russia (Black Sea) and United States were being affected by toxic algal blooms and pathogenic organisms. On the other hand, for human health, the *Vibrio cholera* and the toxic phytoplankton caused diseases and paralytic shellfish poisoning (PSP). In terms of economic impacts, one important invasive species were the zebra mussel that clogged the cooling tanks of power plants and drinking water reservoir species. In Great Lakes, the economic impact of zebra mussel cost up to tens of millions of dollars just to control this invasive species. The sea lampreys, ruffle and round gobby harmed the native fish community and considered as a threat to ports and fishing industry. On the other hand, Zaiko et al. (2010) enumerated the impacts of invasive plankton and these were eutrophication, bloom of harmful algae, change in native phytoplankton community and water quality (hydrochemistry, transparency, nutrients). Olenina et al. (2010) cited *Prorocentrum minimum* as invasive species and caused impacts. On the other hand, Zaiko et al. (2010) revealed that the impacts of invasive zooplankton species were: they compete with the native species for food and space, transfer parasites and diseases, they changed the normal zooplankton community and invasive species acting as predators to the native species.

Various studies had been conducted in different countries on species composition in the ship's ballast water, but only few studies existed in the Philippines. It is recommended that there should also be an assessment of ballast water in the foreign-going vessels docking in Negros Occidental, Philippines hence this study was conducted. In addition, since a lot of vessels are docking in Negros Occidental, an assessment should be made to find out if there was the presence of alien or invasive species in Negros Occidental. Lastly, this study will establish a baseline data on the phytoplankton diversity in offshore, port and ballast tanks from a foreign-going vessel docking in Negros Occidental, Philippines.

1.1 Objectives of the Study

This study aimed to determine the phytoplankton diversity of the offshore, port and ballast tanks of a foreign-going vessel docking at Negros Occidental, Philippines. Moreover, this study sought to determine the phytoplankton density profile (cells/ml) using the Sedgwick Rafter counter cell, diversity, evenness and physicochemical characteristics such as pH, temperature and salinity.

2 METHOD

2.1 Research Design

This study utilized the survey design. This is for the fact that phytoplankton was determined in various sources such as port, offshore and ballast water of a foreign vessel docking in Negros Occidental, Philippines through siphon technique and by the use

of Sedgwick-Rafter colony counter cell at the SEAFDEC-AQD, Tigbauan, Iloilo, Philippines.

2.2 Procedure

Sampling was done in May 2012. The sampling points were: offshore, port and ballast tanks of a foreign vessel. Three areas from the port were sampled (left, middle and right sides) with a total length of 250 m and 10 m away from the port. While a 1-inch wide hose was used to collect seawater through a siphon technique in the middle portion of the water column. A sounding tape was used to determine the depth of the water column. Then, three liters of seawater were collected for each area of the port. The same procedure was applied in offshore with 111.29 m parallel to the ship's length over-all (LOA).

In ballast water tank sampling, the researchers asked permission to conduct the study from the Philippine Ports Authority (PPA), Bureau of Customs, ship agent and the master of the ship. One foreign-going vessel docking in Negros Occidental, Philippines was selected as the source for ballast water samples that came from the waters of China. Seven different ballast tanks were chosen purposely: (1) forepeak tank (FPT), (2) double bottom tank (DBT) 2P, (3) DBT 3C, (4) DBT 2S, (5) DBT 1, (6) DBT 4P and (7) DBT 4S. The siphon technique was also applied in collecting ballast water among the seven ballast tanks that was at the middle layer of the ballast tanks.

Each liter was preserved with 10 ml Lugol's solution and was screened in a 20 µm mesh size screen. All remaining particles were collected in aliquot bottles (20 ml) through a wash bottle containing filtered (20 µm filter) sea water. Water physicochemical characteristics such as pH, temperature and salinity were also measured, *in situ*.

All samples were transported at the Phycology laboratory-Southeast Asian Fisheries Development Center-AQD for the intensive identification and counting of cells/ml (density profile) of phytoplankton diversity using the Sedgwick-Rafter counter cell.

The keys used in the identification of genera were Matsuoka and Fukuyo (2000), Botes (2003) and other keys and monographs available in the literature.

2.3 Data Analyses

The following environmental indexes (Atlas and Bartha, 1998) were used in this study:

Shannon Weaver-Index (H) where,

$$\left(\overline{H}\right) = \frac{C}{N}\left(N \log N - \sum ni \log ni\right)$$

where:

C = 2.3

N = number of individuals

ni= number of individuals in the ith species ($i1$, $i2$, $i3$, $i4$,…ix)

Shannon Evenness (e) where,

$$e = \frac{\overline{H}}{\log S}$$

where:

$\overline{H}$ = Shannon-Weaver diversity index

S = number of species

Generic Richness on the other hand, could be measured by counting the number of genus present in offshore, port and in every ballast tank.

Kruskal-Wallis test set at .05 level of significance was employed to see if there was a significant difference in the cells/ml of plankton diversity in offshore, port and ballast water tanks.

3 RESULTS AND DISCUSSION

3.1 *Plankton diversity, density (cells/ml), generic richness and evenness in offshore, port and ballast water tanks*

Table 1 shows the phytoplankton diversity and density in offshore, port and ballast tanks, generic richness, generic diversity and generic evenness. Most of the phytoplankton found in this study were diatoms similar to the study of Gollasch et al. (2000), Klein et al. (2009a) and Klein, Kaczmarska and Ehrman (2009b). For plankton diversity in offshore sample, the most dominant genus was the *Melosira* (88 cells/ml). It was followed by *Coscinodiscus* (16 cells/ml) and *Ditylum* (14 cells/ml). *Melosira* had a great range of taxa and a common genus in freshwater and marine epibenthic habitats (Guiry and Guiry, 2012). A total of 10 genera of plankton were found in offshore with 4.9 as its generic diversity and generic evenness, respectively (Table 1). *Coscinodiscus* are found to be a free-living marine and often abundant phytoplankton (Guiry and Guiry, 2012) and were also found in the study of Klein et al. (2009a). On the other hand, the genus *Ditylum* was a cosmopolitan genus excluding the polar waters (CIMT, 2012a).

For the port area sample, the highest number of cells was still *Melosira* (43 cells/ml), followed by *Ditylum* and *Thalassionema* (24 cells/ml), then *Asterionella, Lauderia* and *Synedra* (23 cells/ml), respectively. A total of 12 genera of plankton was found in the port area with the generic diversity and generic evenness of five, respectively. *Synedra* could be found throughout the world-in freshwater and saltwater environments (Ashley, 2012). *Synedra* was also present in the study conducted by Sarinas et al. (2010) in an inter-island vessel in Iloilo port with route Manila to Iloilo to Bacolod, Philippines.

Table 1. Phytoplankton diversity showing the genera and density (cells/ml) in offshore, port and ballast tanks of a foreign vessel (in mean values) and generic richness and evenness.

Genera	Offshore	Port	FPT	DBT 2P	DBT 3C	DBT 2S	DBT 1	DBT 4P	DBT 4S
Arachnoidiscus H. Deane ex G.Shadbolt	-	18	-	-	-	-	-	23	-
Asterionella Hassall	-	23	-	-	-	-	-	-	11
Chaetoceros Ehrenberg	12	16	-	14	13	-	23	12	11
Coscinodiscus Ehrenberg	16	18	24	22	14	14	-	21	-
Ditylum J. W. Bailey	14	24	16	-	15	-	-	24	13
Eucampia Cleve	11	21	22	-	-	-	-	25	12
Grammatophora Ehrenberg	12	-	-	-	-	-	-	-	-
Isthmia C. Agardh	11	22	18	24	-	13	19	17	13
Lauderia Cleve	13	23	-	-	-	-	-	26	14
Leptocylindricus Cleve	-	-	7	-	-	-	-	-	-
Licmophora C. Agardh	-	-	12	-	-	-	-	-	-
Melosira C. Agardh	88	43	-	-	-	11	-	26	-
Odontella C. Agardh	-	21	-	17	11	11	-	25	15
Pseudo-nitzschia Peragallo	-	-	-	26	-	-	-	-	-
Rhabdonema Kutzing	-	-	13	-	-	-	-	-	-
Rhizosolenia Brightwell	-	-	-	13	-	-	24	-	-
Skeletonema Greville	-	-	-	18	-	-	-	-	-
Striatella C. Agardh	-	-	-	-	-	-	19	-	-
Synedra Ehrenberg	12	23	15	23	17	15	15	-	13
Thalassionema Grunow ex Mereschkowsky	12	24	13	-	11	-	22	-	13
Summary									
Generic Richness	10	12	9	8	6	5	6	9	9
Generic Diversity ($\overline{H}$)	4.9	5	5	4.9	4	4	4	5	4.2
Generic Evenness (e)	4.9	5	5	5	5.5	5	5.5	5	5

Table 1 shows the genera of phytoplankton found in offshore, port and ballast water tanks such as FPT, DBT 2P, DBT 3S, DBT 2S, DBT 1, DBT 4P and DBT 4S with corresponding density (cells/ml).

For ballast tank 1 (FPT), a total of nine genera of plankton were recorded with *Coscinodiscus* (24 cells/ml) as the dominant plankton, followed by *Eucampia* (22 cells/ml) and *Isthmia* (18 cells/ml). Both generic diversity and generic evenness had a value of five. On the other hand, *Eucampia* was common marine planktonic genus while *Isthmia* spp. are epiphytic on seaweed (Guiry and Guiry, 2012). *Eucampia* was also found in the study of Klein et al. (2009a) from the ballast water during the trans-Pacific crossings in Canada.

For ballast tank 2 (DBT 2P), eight genera of plankton were recorded with 4.9 and 5 as generic diversity and generic evenness, respectively. The most dominant plankton was *Pseudo-nitzschia* (26 cells/ml), followed by *Isthmia* (24 cells/ml) and *Synedra* (23 cells/ml). *Pseudo-nitzschia* is an alarming diatom because it produces the neurotoxin domoic acid that can cause amnesic shellfish poisoning and domoic acid poisoning (Thessen, 2010). It was evident that toxic or harmful diatoms and dinoflagellates were indeed transported through ballast water similar to the study of Klein et al. (2009a), Doblin et al. (2004), Hallegraeff (1998) and Hallegraeff and Bolch (1992).

For ballast tank 3 (DBT 3C), six genera of plankton were identified with 4 and 5.5 generic diversity and generic evenness were recorded. This was dominated by *Synedra* (17 cells/ml), followed by *Ditylum* (15 cells/ml) and *Coscinodiscus* (14 cells/ml).

For ballast tank 4 (DBT 2S), the plankton genera went down to five genera with four and five generic diversity and generic evenness, respectively. It was dominated by *Synedra* (15 cells/ml), followed by *Coscinodiscus* (14 cells/ml) and *Isthmia* (13 cells/ml).

While in ballast tank 5 (DBT 1), a total of six genera of plankton were identified with four and 5.5 as generic diversity and generic richness, respectively. The most dominant genera of plankton was *Rhizosolenia* (24 cells/ml), *Chaetoceros* (23 cells/ml) and *Thalassionema* (22 cells/ml). The distribution of *Rhizosolenia* and *Chaetoceros* was widespread throughout the world's ocean (CIMT, 2012 b, c) while *Thalassionema* was cosmopolitan genus in temperate to tropical waters (CIMT, 2012d). *Rhizosolenia* sp. was also found in the study of Gollasch et al. (2000) from the ballast water of ships from the Indian Ocean to the North Sea. Meanwhile, the *Chaetoceros* spp. was also found in the study of Klein et al. (2009b) found in the ballast water of ships arriving at Canadian ports on the West Coast, East Coast and Great Lakes and in trans-Pacific crossings, Canada.

In ballast tank 6 (DBT 4P), nine genera of plankton were recorded as the highest genera of plankton with five as generic diversity and generic evenness, respectively. The most dominant genera were *Lauderia* and *Melosira* (26 cells/ml), followed by *Eucampia* and *Odontella* (25 cells/ml) and *Ditylum* (24 cells/ml). *Lauderia* is common and widespread in the ocean (Guiry and Guiry, 2012). *Odontella* on the other hand, was very abundant throughout the oceans (Guiry and Guiry, 2012).

Finally, for ballast tank 7 (DBT 4S), nine genera of plankton were identified with 4.2 and five as the generic diversity and generic evenness were recorded. The most dominant was the *Odontella* (15 cells/ml), followed by *Lauderia* (14 cells/ml), then *Ditylum, Isthmia, Synedra* and *Thalassionema* (13 cells/ml).

It was noted that genus *Grammatophora* in offshore, *Leptocylindricus* in the FPT, *Rhizosolenia* in DBT 2P and DBT 1 of this study were also shown in the study of Klein et al. (2009a) from the ballast water of ships during the trans-Pacific crossings in Canada.

Statistical analysis from this study showed that the cells/ml of plankton genera, showed no significant difference, $H = 14.224$, $p = .076$ in offshore, port and ballast water tanks.

3.2 *Physicochemical characteristics of offshore, port and ballast water tanks*

Determination of physicochemical characteristics such as temperature, pH and salinity were recorded, *in situ* with three replications for each area and tank. Temperature was measured through a digital thermometer, pH was recorded through a pH meter and salinity was measured through a salinometer. In the offshore area, the highest temperature was 30.2 °C at the middle and at the right side facing the ocean where the ship was situated during anchorage. The pH was highest at the right side at 8.1 while salinity was also the same with port area which was 11 ppm at three areas of the offshore (Fig. 1).

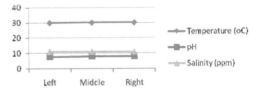

Figure 1. Temperature, pH and salinity in port area (left, middle and right sides).

In the port area, the highest temperature and pH recorded was 30.9 °C and 8.08 at the middle of the port facing the ocean, respectively. While salinity was consistent at three areas (left, middle and right sides) of the port at 11 ppm (Fig. 2).

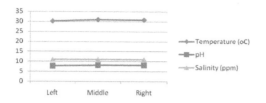

Figure 2. Temperature, pH and salinity in port area (left, middle and right sides).

In ballast water tanks, the highest temperature recorded was 39.1 °C in DBT 4S while lowest at DBT 2S (31.1 °C). The highest pH recorded was 8.05 at DBT 2P and lowest in DBT 1 (6.97). On the other hand, salinity varies from a range of 2-7 ppm (Fig. 3).

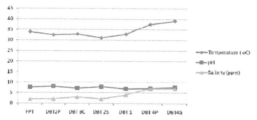

Figure 3. Temperature, pH and salinity in ballast water tanks.

4 CONCLUSIONS

A total of 20 genera of diatoms was identified in this study. Some of the diatoms could be found in offshore and port samples, but not in ballast water and vice versa. Generic richness is highest at port area with 12 genera present. This might be due to the favorable environment which allowed the growth of the diatoms most especially in the port area. Moreover, most of the plankton found in this study could be found throughout the world's ocean. However, one genus of diatom, *Pseudo-nitzschia* found in this study was toxic or known as harmful algae. Indeed, this study presents that harmful or toxic algae can be carried through international shipping through ballast water and thus provides the baseline information of plankton composition in the offshore, port and in the ballast tanks from a foreign-going vessel. It is recommended that foreign ships entering the ports of the Philippines should follow the rules implemented by the IMO to conduct ballast water exchange in the open ocean to minimize the transfer of alien and harmful organisms. A yearly sampling should also be conducted in the ballast water in foreign-going vessels and analysis of cysts should also be given attention so as to determine the toxic or nontoxic plankton.

ACKNOWLEDGMENT

We would like to thank Dr. Ronald Raymond Lacson Sebastian, Chief Executive Officer, John B. Lacson Foundation Maritime University for the full support extended in this study; the Southeast Asian Fisheries Development Center (SEAFDEC)-AQD for the laboratory equipment; Dr. Melchor M. Magramo, Research Coordinator of John B. Lacson Foundation Maritime University-Arevalo and Dr. Resurreccion B. Sadaba of University of the Philippines Visayas (UPV) for the words of encouragement; Dr. Wilfredo Campos of UPV and Mr. Peter Paolo Rivera, Research Assistant of the Marine Science Institute, College of Science, University of the Philippines Diliman (UPD) for the knowledge imparted to the researchers.

REFERENCES

Ashley, B. 2012. *Synedra* alga facts. eHow. Retrieved July 30, 2012 from http://www.ehow.com/facts_7663450_synedra-algae.html

Atlas, R. & Bartha, R. 1998. *Microbial Ecology Fundamentals and Applications*. 4th ed. California: Benjamin/Cummings Pub. Comp., Inc.

Badylak, S. & Phlips, E. J. 2004. Spatial and temporal patterns of phytoplankton composition in a subtropical coastal lagoon, the Indian river lagoon, Florida, USA. *Journal of Plankton Research* 26(10): 1229-1247.

Botes, L. 2003. *Phytoplankton identification catalogue-Saldanha Bay, South Africa, April 2001*. Globallast Monograph Series No. 7. Global Ballast Water Management Programme. IMO London.

Carver, C.E. & Mallet, A. L. 2003. A preliminary assessment of the risks of introducing non-indigenous phytoplankton, zooplankton species or pathogens/parasites from South Amboy, New Jersey (Raritan Bay) into White Paint, Digby Neck, Nova Scotia. Mallet Research Services. Global Quary Products.

Center for Integrated Marine Technologies (CIMT) 2012a. Phytoplankton identification. Retrieved July 30, 2012 from http://oceandatacenter.ucsc.edu/PhytoGallery/Diatoms/dityl um.html

Center for Integrated Marine Technologies (CIMT) 2012b. Phytoplankton identification. Retrieved July 30, 2012 from http://cimt.ucsc.edu/habid/phytolist_diatoms/24rhizosolenia .html

Center for Integrated Marine Technologies (CIMT) 2012c. Phytoplankton identification. Retrieved July 30, 2012 from http://cimt.ucsc.edu/habid/phytolist_diatoms/13chaetoceros speices.html

Center for Integrated Marine Technologies (CIMT) 2012d. Phytoplankton identification. Retrieved July 30, 2012 from http://cimt.ucsc.edu/habid/phytolist_diatoms/31thalassione ma%20species.html

Chu, K. H., Tam, P. F., Fung, C. H. & Chen, Q. C. 2006. A biological survey of ballast water in container ships entering Hong Kong. *Hydrobiologia* 352(1-3): 201-206.

Deacutis, C. F & Ribb, R. C. 2002. Ballast water and introduced species: management options for Narragansett bay and Rhode island. Narragansett Bay Estuary Program. R.I. Department of Environmental Management. Retrieved December 5, 2010 from http://www.nbep.org/publications/other/ballast/BallWaterIntroSpeciesRpt.pdf

Doblin, M. A., Popels, L. C., Coyne, K. J., Hutchins, D. A., Cary, S. C. & Dobbs, F. C. 2004. Transport of the harmful alga *Aureococcus anophageffferens* by oceangoing ships and coastal boats. *Applied and Environmental Microbiology* 70 (11): 6495-6500.

Gollasch, S., Lenz, J., Dammer, M. & Andres, H-G. 2000. Survival of tropical ballast water organisms during a cruise from the Indian Ocean to the North Sea. *Journal of Plankton. Research* 22(5):923-937.

Guiry, M.D. & Guiry, G.M. 2012. **AlgaeBase**. World-wide electronic publication, National University of Ireland, Galway. Retrieved July 31, 2012 from http://www.algaebase.org

Hallegraeff, G. M. 1998. Transport of toxic dinoflagellates via ships' ballast water: bioeconomic risk assessment and efficiency of possible ballast water management strategies. *Marine Ecology Press Series* 168:297-309.

Hallegraeff, G. M. & Bolch, C. J. 1992. Transport of diatom and dinoflagellate resting spores in ships' ballast water: implications for plankton biogeography and aquaculture. *Journal of Plankton Research* 14(8):1067-1084.

International Maritime Organization (IMO). 2005. *Ballast water management convention*. UK: CPI Books Limited, Reading RG1 8EX.

Klein, G., MacIntosh, K., Kaczmarska, I. & Ehrman, J. M. 2009a. Diatom survivorship in ballast water during trans-pacific crossings. *Biology Invasion*. Springer. DOI 10.1007/s10530-009-9520-6.

Klein, G., Kaczmarska, I. & Ehrman, J. M. 2009b. The diatom *Chaetoceros* in ships' ballast waters – survivorship of stowaways. *Acta Botanica Croatia* 68(2): 325-338.

Laamanen, M. J. 1997. Environmental factors affecting the occurrence of different morphological forms of cyanoprokaryotes in the northern Baltic sea. *Journal of Plankton Research* 19(10):1385-1403.

MacDonald, E. M. 1995. Dinoflagellate resting cysts and ballast water discharges in Scottish ports. *ICES* Document CM. O:10-17.

Manaaki Whenua Landcare Research. 1994. International Union for the Conservation of Nature (IUCN) guidelines for the prevention of biodiversity loss caused by alien invasive species. Invasive Species Specialist Group of the IUCN Species Survival Commission.

Marrero, J. P. & Rodriguez, E. M. 2004. Marine pollution from ships' ballast water. *Journal of Maritime Research* 1(1):35-42.

Marshall, H. G., Burchardt, L. & Lacouture, R. 2005. A review of phytoplankton composition within Chesapeake Bay and its tidal estuaries. *Journal of Plankton Research* 27(11):1083-1102.

Martin, J. L. & LeGresley, M. M. 2008. New phytoplankton species in the Bay of Fundy since 1995. *ICES Journal of Marine Science* 65:759–764.

Matsuoka, K. & Fukuyo, Y. 2000. *Technical guide for modern dinoflagellate cyst study*. WESTPAC-HAB: Japan.

Minchin, D. & Gollasch, S. 2002. Vectors-how exotics get around. In *Invasive Aquatic Species of Europe. Distribution, Impacts, and Management*. pp.183-192. Ed. E. Leppakoski, S. Gollasch and S. Olenin. Kluwer Academic, Dordrecht. 583 pp.

Olenina, I., Hajdu, S., Wasmund, N. Jurgensone, I., Gromisz, S., Kownacka, J., Toming, K. & Olenin, S. 2010. Assessing impacts of invasive phytoplankton: The Baltic Sea case. *Marine Pollution Bulletin* 60(10):1691-1700.

Popa, L. 2009. Marine bio pollution through ballast water. *Marine Transportation and Navigation Journal* 1(1):105-109.

Sarinas, B. G., Gellada, L., Garingalao, A., Baria, L., D. B. Tirazona & Sorio, L. R. 2010. "Initial assessment of phytoplankton and zooplankton composition in ballast water tanks of an inter-island passenger-cargo vessel in the Philippines." In *Proceedings of 12th Annual General Assembly by the International Association of Maritime University (IAMU) Held in Gdynia Maritime University, Gdynia, Poland 2011*, edited by B. Laczynski, 277-284. Gdynia Maritime University: Gdynia, Poland, 2011.

Thessen, A. 2010. Diatom *Pseudo-nitzschia*. Retrieved July 31, 2012 from http://diatoms.lifedesks.org/pages/990

Williams, R. J., Griffiths, F. B., Van der Wal, E. J. & Kelly, J. 1988. Cargo vessel ballast water as a vector for the transport of non-indigenous marine species. *Estuarine, Coastal and Shelf Science* 26(4): 409-420.

Zaiko, A., Olenina, I. & Olenin, S. 2010. Overview of impacts of alien invasive plankton species. Klaipeda University, Coastal Research and Planning Institute Heraklion, Crete. Retrieved December 2, 2010 from: http://www.meece.eu/meetings/Crete/presentations/tuesday/Zaiko_tues.pdf.

Study of Trawling Impacts on Diversity and Distribution of Gastropods Communities in North of Persian Gulf Fishing Area

M. Shirmohammadi, B. Doustshenas, A. Savari & N. Sakhaei
Khorramshahr University of Marine Science and Technology, Iran

S. Dehghan Mediseh
South Aquaculture Research Center, Iran

ABSTRACT: This study took place to survey the changes of diversity and distribution of the gastropods in an important fishing area in the northwest coasts of the Persian Gulf due to bottom trawling. Sampling carried out before the shrimp trawling season then repeated two weeks and three months after the end of the shrimp-fishing season. Two set of nine sampling stations selected in area with less than 6m and more than 10m depths respectively. In both depths, the abundance of the gastropods, diversity, richness and evenness indices excluding Simpson dominant index decreased in two weeks and three months after trawling (P<0.05). The abundance of the gastropods increased in size class of < 2mm in after trawling. In both depths, the most abundance species was *Acteocina involuta* in two weeks after trawling. The grain size of the sediment can be categorized mostly silt-clay in area. Increasing pattern of trawling intensity due to trawling were found in regions with less than 6m depth comparison of deeper areas with more than 10m depth.

1 INTRODUCTION

The impacts of fishing gears on the marine environment have been a matter of great concern to the sustainable management of oceanic resources (Smith et al., 2000). The Bahrakan is one of the most important sites for fishing activities in the Persian Gulf (ROPME, 2004). Many studies have focused on benthic community responses in northwest of Persian Gulf (Nabavi SB, 1992; Dehghan MS, 2007; Hovizavi Sh., 2009; Roozbahani et al., 2010; Shokat et al., 2010). However, this the first study on trawling affects on benthos communities in the area. Despite concerns, the intensity and extent of bottom trawling have continued to increase throughout the world, particularly over the last few decades (Hannah et al., 2010).Trawling is believed to affect stock abundances directly by removing or killing individuals, and indirectly by affecting structures and organisms that serve as habitat and food for demersal fish species (Kumar and Deepthi, 2006). Trawling intensity depended on size and weight of trawl, rapidity trawling, type of seabed, power of water flow and tide and natural confusion in the zone (Dellapenna et al., 2006). Previous studies suggested that gastropods assemblages were good indicators of impacts trawling because of their sensitivity to habitat alterations (Morton, 1996).

2 MATERIALS AND METHODS

Impacts of bottom trawling were investigated at 18 stations in the northwest of the Persian Gulf in the south Khuzestan province inshore waters in along the coast Bahrakan (30° 03' - 30° 06' N, 49° 43' – 49° 46' E)(Fig. 1). Two groups of station selected in the 6m and 10m bathy metric lines because of different fishing impact. The trawling efforts occurred from the end of July until the end of August (30 days). Sampling were taken before the start of the trawling season (15 May) and two weeks (5 September) and three months (14 November) after the end of the trawling season in 2010 repeated. Triplicate samples collected by van Veen grab (0/025 m^2) for studying the macrobenthos of each station. Samples of the macrofauna were sieved through 0.5 mm mesh then preserved in 4 % neutralized formalin (Joice et al., 2006) and then transferred into Rose Bengal solution (1 g.l^{-1}) . In addition, biomass was estimated using Ash-free dry weight method.

The ignition method (Buchanan, 1984) used to estimation organic contents of sediments. Grain size studied using wet sieve method (Buchanan, 1984). Furthermore, Shannon diversity, Margalef species richness, Simpson dominant and Pielou evenness indices calculated for every sampling period (Ludwig and Reynolds, 1988). Normality data examined with Shapiro-Wilk test. Sediment analysis investigated by ANOVA. Slight different done by Tukey test, using 95% confidence limits. Differences in density, mean abundance of species and indices between sampling periods were examined using nonparametric Kruskal-Wallis H test and slight different used by Mann-Whitney U test. Statistical computation performed using SPSS 11.5 and PRIMER 5.0 softwares.

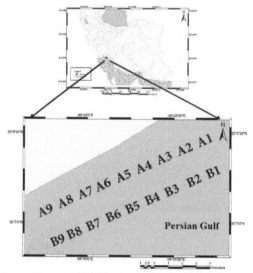

Figure 1. Map of the investigation area, including the sampling stations in the coasts of the Bahrakan, 6m depth (A1 to A9) and 10m depth (B1 to B9)

3 RESULTS

The results showed that more than 93% of the grain sizes of the sediment in all sampling periods were <0.063 mm thereupon, the sediment of in the Bahrakan zone can be categorized mostly silt-clay (table 1). There were significant differences between the silt-clay portions of sediment before trawling with the silt-clay percentage of sediment in other sampling period (P < 0.05).

A totaly 62 species of gastropods identified that only, the temporal change in average number of species by dominant species for each depth is shown in Table 2. The results showed that, species *Acteocina involuta* (family Cylichnidae) displayed a higher abundance than other species in both depths in two weeks after trawling (Table 2).

Table 1. Mean of sediments parameters analysis in the Bahrakan (2010)

Time sampling	Depth (m)	Organic matter (%)	Silt-Clay (%)
Before trawling	6	21.28 ± 0.34a	95.50 ± 0.34b
	10	21.10 ± 0.36a	93.49 ± 0.55a
Two weeks after Trawling	6	26.33 ± 0.20c	98.21 ± 0.03cd
	10	26.30 ± 0.19c	98.30 ± 0.08cd
Three months after trawling	6	23.82 ± 0.54b	99.08 ± 0.06d
	10	23.36 ± 0.46b	97.55 ± 0.28c

Dissimilar words within each column are significantly different (P<0.05) (Mean ± Standard error).

Table 2. Average abundance (no.m^{-2}) of dominant organisms that have found in the Bahrakan (2010)

	Before trawling	Two weeks after trawling	Three months after trawling
6m depth			
Acteocina involute	1.48 ±1.48a	78.54±10.35b	109.62±37.63b
Scaliola arenosa	88.89±27.20c	0a	2.96±2.05b
Truncatella sp.3	271.11±86.47c	0a	25.18±8.84b
10m depth			
Acteocina involute	0a	80±7.70c	1.48±4.91b
Gibberula bensoni	19.25±6.17b	0a	19.25±5.80b
Nassarius ephamillus	17.77±5.78b	2.05±2.05a	17.77±5.78b
Scaliola arenosa	66.66±30.34c	0a	1.48±1.48b
Zafra comistea	17.77±4.93b	1.48±1.48a	14.81±4.84b

Dissimilar words within each row are significantly different (P<0.05).

The results showed that the total of mean abundance was related 273.29 (no.m^{-2}) in 6m depth and 154.37(no.m^{-2}) in 10m depth during studies period. The changes over time in mean abundance of gastropods is shown in Figure 2. In both depths, there were significant differences between the abundance of gastropod communities before trawling and after it (P < 0.05). On the other hand, there were no significant differences between the abundance in two weeks and three months after trawling (P>0.05).

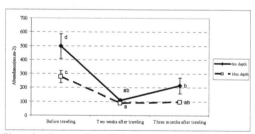

Figure 2. Temporal variations of mean abundance of gastropods (no.m^{-2}) in the Bahrakan (2010), dissimilar words are significantly different (P<0.05)

In both depths, the highest mean of abundance obtained before trawling and the lowest was in two weeks after trawling (Fig. 2).

Figure 3 shows dissimilar data for the mean biomass in different sampling phases. The total of mean biomass was reported 0.16 (g.m^{-2}) in 6m depth and 0.13 (g.m^{-2}) in 10m depth during study period. In both depths, the highest and the lowest biomasses of gastropods were recorded before trawling and two weeks after trawling, respectively (Fig. 3).

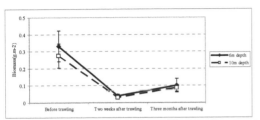

Figure 3. Temporal variations of mean biomass of gastropods (g.m-2) in the Bahrakan (2010)

The results showed that, the highest percentage abundance obtained in size classes 2 to 4 mm and smaller than 2 mm in before trawling and after trawling, respectively While, size class >4mm wasn't observed in after trawling (Fig. 4&5).

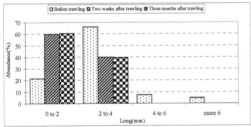

Figure 4. Variations of percentage abundance with different size classes (mm) in 6m depth in the Bahrakan zone in different sampling time (2010)

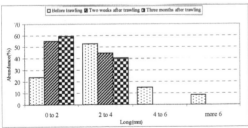

Figure 5. Variations of percentage abundance with different size classes (mm) in 10m depth in the Bahrakan zone in different sampling periods (2010)

4 DISCUSSION

The results of the study revealed that bottom trawling impacted gastropod communities. Morton, 1996 also recorded such effect. In two selected sampling depth, abundance of the gastropods was decreased significantly after trawling (P <0.05). This result is according with Pearson and Rosenberg

(1978) study. Trawling gears may destroys shelters of benthos and expose them to prey. Some previous studies record same results (Kaiser and Spencer, 1996; Morton, 1996; Engel and Kvitek, 1998; Jenning et al., 2001; Tanner, 2003; Kumar and Deepthi, 2006). In this study, no significant difference observed between the abundance of the gastropods in samples of two weeks and three months after the trawling because of slow recovery of gastropod communities. Thereupon, impacts of trawling on them were too much, and they need more time for recovery. Result consistent with previous studies (Engel and Kvitek, 1998; Rumohr and Kujawski, 2000). Ball et al., 2000 was recorded muddy bottoms need more time for recovery than sandy one and trawling made more turbidity in muddy beds with higher suspended particle that preventing larva settlement in bottom substrate. Results consistent with others studies (Tuc et al., 1998; Smith et al., 2000; Dellapeana et al., 2006). Bahrakan sediments are mostly muddy also reported in previous study (Shokat et al., 2010). Only abundance of A. involuta (Cylichnidae) increased and became dominant species two weeks after trawling. Other authors found increasing of burrower and scavenger gastropods after trawling. They suggested removing sediment of upper layers by trawling induce emerging of burrowers and scavenger from adjacent zone for nourishment from the damaged benthos in trawling in the fishing zone (Rumohr and Kujawski, 2000; Lokkeborg, 2005). Most species of family Cylichnidae are burrower and scavenger (Faucci et al., 2007). Shannon diversity, Margalef species richness, and Pielou evenness indices decreased two weeks after trawling season too. Others expressed that increasing organic matter and diminution abundance of macrobenthos decreased diversity (Tuck et al., 1998; Thrush et al., 2001; Sparks-McConkey and Watling, 2001). Three months after trawling period, all indices increased except Simpson dominant index. In both depths, decreasing the biomass of the gastropods after trawling observed similar as the abundance pattern. Similar to this study, Prena et al (1999) that concluded direct contact of trawling gears, burial benthos and invasion scavengers caused decreasing the biomass of the benthos. In this study, in both depths, the relative abundance of the gastropods increased in size class < 2mm and decreased in size class Upper 2mm after trawling. Larger individuals and species could find easier by prayer and may affect by trawl force more. Howevere gastropod communities tend to decrease in individual size. Most gastropods had the small size, which showed continual stress on the benthos in the area. Jennings et al (2001) supported this thought. Engel and Kvitek (1998) noted that, small organisms could pass from trawl net and trawling affected them less. Results showed that impacts of trawling were very intense in 6m depth, because, abundance and biomass of the scavenger gastropods increased and their diversity declined in 6m depth more than 10m depth. It seems that the area needs to more researches to find better way for sustainable yield.

5 CONCLUSION

Fishing areas are vulnerable sites due to increasingly fishing efforts. Decreasing abundance and biomass of benthic community with trawling could affect fishing industries in long time. Because of long time recovery of benthic community it's necessary to design better fishing gears with less destructive effects on sea floor.

REFERENCES

Ball, B.J., Fox, G. and Munday, B.W., 2000. Long- and short-term consequences of a Nephrops trawl fishery on the benthos and environment of the Irish Sea. ICES Journal of Marine Science, 57, 1315–1320.

Buchanan, J.B, 1984. Sediment analysis. In. Holme A. and McIntyre A.D. (eds), Methods for the study of marine benthose. NBlackwell, Oxford. pp.41–64.

Collie, J.S., Hall, S.J., Kaiser, M.J. and Poiner, I.R., 2000. A quantitative analysis of fishing impacts on shelf-sea benthos. Journal of Animal Ecology, 69, 785–798.

Dellapenna, T. M., Allison, M. A., Gill, G. A., Lehman, R. D. and Warnken, K. W., 2006. The impact of shrimp trawling and associated sediment resuspension in mud dominated, shallow estuaries. Estuarine, Coastal and Shelf Science, 69, 519–530.

Duplisea, D. E., Jennings, S., Malcolm, S. J., Parker, R. and Sivyer, D. B., 2001. Modelling potential impacts of bottom trawl fisheries on soft sediment biogeochemistry in the North Sea. Geochemical Transactions, 2(1), 112–117.

Eleftheriou, A. and McIntyre A., 2005. Methods for the study of marine benthos. Third Edition. Blackwell Science Ltd a Blackwell. 418P.

Engel, J. and Kvitek, R., 1998. Effects of otter trawling on a benthic community in Monterey Bay National Marine Sanctuary. Conservation Biology, 12, 1204–1214.

Faucci, A., Toonen, R.J. and Hadfield, M.G., 2007. Host shift and speciation in a coral-feeding nudibranch. Proceedings of the Royal Society, 274,111–119.

Hall, S. J., 1999. The effects of fishing on marine ecosystems and communities. Oxford: Blackwell. 274p.

Hannah, R. W., Jones, S. A., Miller, W. and Knight, J. S., 2010. Effects of trawling for ocean shrimp (Pandalus jordani) on macroinvertebrate abundance and diversity at four sites near Nehalem Bank. Oregon. Fishery Bulletin, 108, 30–38.

Jenning, S., Dinmore, T. A., Duplisea, D. E., Warr, K. J. and Lancaster, J. E., 2001. Trawling disturbance can modify benthic production processes. Journal of Animal Ecology, 70, 459–475.

Joice, V. T., Sreedevi, C. and Madhusoodana Kurup, B., 2006. Variations on the infaunal polychaetes due to bottom trawling along the inshore waters of Kerala (south-west coast of India). Indian Journal of Marine Sciences, 35(3), 249–256.

Jones, J. B., 1992. Environmental impact of trawling on the seabed: a review. New Zealand Journal of Marine and Freshwater Research, 26, 59–67.

Kaiser, M.J. and Spencer, B., 1996. The effects of beam trawl-disturbance on infaunal communities in different habitats. Journal of Animal Ecology, 65, 348–358.

Kumar, A. B. and Deepthi, G. R., 2006. Trawling and by-catch: Implications on marine ecosystem: Review. Current Science, 90, 922–931.

Lokkeborg, S., 2005. Impacts of trawling and scallop dredging on benthic habitats and communities. FAO Fisheries Technical. pp.1–58.

Ludwig, J. A. and Reynolds, J. F., 1988. Statistical Ecology: A Primer on Methods and Computing. John Wiley and Sons, New York. 337P.

Morton, B., 1996. The Subsidiary Impacts of Dredging (and Trawling) on a Subtidal Benthic Molluscan Community in the Southern. Marine Pollution Bulletin, 32(10), 701–710.

Pearson, T.H. and Rosenberg, J., 1978. Macrobenthic succession in relation to organic enrichment and pollution of the marine environment. Oceanography and Marine Biology Annual Review, 16,229-311.

Prena, J., Schwinghamer, P., Rowell, T.W., Gordon Jr., D.C., Gilkinson, K.D., Vass, W.P. and McKeown, D.L. 1999. Experimental otter trawling on a sandy bottom ecosystem of the Grand Banks of Newfoundland: analysis of trawl bycatch and effects on epifauna. Marine Ecology Progress Series, 181, 107–124.

Queiros, A.M., Hiddink, J.G., Kaiser, M.J. and Hinz, H., 2006. Effects of chronic bottom trawling disturbance on benthic biomass, production and size spectra in different habitats. Experimental Marine Biology and Ecology, 335, 91–103.

ROPME, 2004. State of the Marine Environment 2003. Regional Organization for the Protection of the Marine Environment (ROPME), Kuwait. 253P.

Rumohr, H. and Kujawski, T., 2000. The impact of trawl fishery on the epifauna of the southern North Sea. ICES Journal of Marine Science, 57, 1389–1394.

Shokat, P., Nabavi, S.M.B., Savari, A. and Kochanian, P., 2010. Application of Biotic Indices in Assessing the Ecological Quality Status of Bahrekan Estuary. Pakistan Journal of Biological Sciences, 13(22), 1085–1091.

Simpson, A. W. and Watling, L., 2006. An investigation of the cumulative impacts of shrimp trawling on mud-bottom fishing grounds in the Gulf of Maine: effects on habitat and macrofaunal community structure. ICES Journal of Marine Science, 63: 1616-1630.

Smith, C. J., Papadopoulou, K. N. and Diliberto, S., 2000. Impact of otter trawling on an eastern Mediterranean commercial trawl fishing ground. ICES Journal of Marine Science, 57, 1340–1351.

Sparks-McConkey, P. J. and Watling, L., 2001. Effects on the ecological integrity of a soft-bottom habitat from a trawling disturbance. Hydrobiologia, 456, 73–85.

Tanner. J.E., 2003. The influence of prawn trawling on sessile benthic assemblages in Gulf St. Vincent South Australia. Canadian Journal of Fisheries and Aquatic Sciences, 60, 517–526.

Thrush, S. F., Hewitt, J. E., Funnell, G. A., Cummings, V. J., Ellis, J., Schultz, D., Talley, D. and Norkko, A., 2001. Fishing disturbance and marine biodiversity: role of habitat structure in simple soft-sediment systems. Marine Ecology Progress Series, 221, 255–264.

Tuck, I. D., Hall, S. J., Robertson, M. R., Armstrong, E. and Basford, D. J., 1998. Effects of physical trawling disturbance in a previously unfished sheltered Scottish sea loch. Marine Ecology Progress Series, 162, 227–242.

Chapter 2

Gas and Oil Transportation

Gas and Oil Transportation
Maritime Transport & Shipping – Marine Navigation and Safety of Sea Transportation – Weintrit & Neumann (Eds)

Future Development of Oil Transportation in the Gulf of Finland

O-.P. Brunila & J. Storgard
University of Turku, Centre for Maritime Studies, Kotka, Finland

ABSTRACT: Approximately 290 million tonnes of oil and oil products were transported in the Baltic Sea in 2009, of which 55% (160 million tonnes) via Gulf of Finland. Oil transportation volumes in the Gulf of Finland have increased from 40 million to almost 160 million tonnes over the last ten years. To forecast oil transportation in the Gulf of Finland three alternative scenarios were produced for years 2020 and 2030. The future oil volumes are based on the expert estimates. The study showed that no dramatic increase is to be expected in oil transportation volumes in the Gulf of Finland. Most of the scenarios forecasted only a moderate growth in maritime oil transportation compared to the current levels.

1 INTRODUCTION

Maritime traffic in the Baltic Sea is very dense, especially if the traffic volumes are compared to the size of the sea area. More than 2,000 vessels continuously transport different kinds of cargoes on the Baltic Sea. Approximately 25% of the vessels are loaded with oil, oil products or chemicals. (Baltic Sea Action Group 2008; HELCOM 2009) The Baltic Sea is 392,000 square kilometres wide with an average depth of only 54 meters. As a comparison, the average depth in the Mediterranean Sea is 1,550 meters and in the Atlantic and Pacific Oceans 4,000 meters. The Baltic Sea is connected to the North Sea and the Atlantic Ocean only by the narrow Danish straits. (Myrberg et al. 2011)

The Gulf of Finland is the easternmost part of the Baltic Sea and it is approximately 400 km long. Width varies between 65 km and 135 km. The Gulf of Finland is surrounded by three countries: Finland, Estonia and Russia. The narrowest point is found between Helsinki and Tallinn. The shores and the archipelago are mainly rocky, and there are hundreds of islands of various sizes. The total length of the coastline including the islands in the Gulf of Finland is up to 6,500 km. (Knuutila 2011) The Gulf of Finland is very shallow, the average depth only being 37 meters. The Gulf and the archipelago are very sensitive and vulnerable to pollution, due to for example the low volumes and slow turnover of water, low temperatures and ice cover during winter,

and stratification of water into layers with different temperatures. (Knuutila et al. 2009)

The purpose of this study is to produce maritime oil transportation scenarios for the years 2020 and 2030 in the Gulf of Finland. An analysis of the current state and factors affecting future development of oil transportation in the Gulf of Finland is also provided. Transported oil volumes in 2020 and 2030 were estimated by the group of experts.

2 MATERIALS AND METHODS

The study was carried out in two stages. First, a statistical analysis and literature review on oil transportation in the Baltic Sea and on the development of oil transportation in the Gulf of Finland was made. The current state analysis and development of oil transportation are mainly based on information from ports, operators, terminals, previous studies, governmental and EU databases. The future scenarios are based on international and national energy strategies and on other future scenarios (e.g. the Baltic Transport Outlook 2030, the Technology Outlook 2020, Hänninen et al. and the Baltic transport system – Finnish perspective). The purpose of literature review was to find out the current state and the main future trends in Russia and in EU which affect the amount of maritime oil transportation.

Second, *expert panel* gave an estimation of transported oil volumes in 2020 and 2030 in the Gulf of Finland. An expert workshop was arranged in May 2012 and anyone who was interested was able to participate. The workshop was arranged as part of the larger seminar "Security threats of maritime oil transport in the Baltic Sea". In addition, some experts who were not able to participate the workshop were asked to give their evaluations by e-mail. Total 9 experts gave their evaluations of future oil transportation scenarios.

3 OIL TRANSPORTATION DEVELOPMENT DURING LAST DECADE IN THE GULF OF FINLAND

In the last ten years the volumes of oil and oil products transportation has tripled in the whole Baltic Sea area. The main reason for the increase in maritime oil transportation volumes in the Baltic Sea is Russia`s new oil terminals in the eastern part of the Gulf of Finland. The Port of Primorsk started operating in 2002, and other ports have increased their capacity as well (Hänninen & Rytkönen 2004; Leningrad Region 2011). Even the economic recession in 2008-2009 didn't decrease the amount of transported oil. Oil terminals in the port of Ust-Luga started operating in March 2012. This will increase the transportation of oil in the Gulf of Finland in the future. (UK & Trade Investment 2010; Ust-Luga Company 2012)

In 2010, almost 290 million tonnes of oil and oil products were transported in the Baltic Sea, of which more than 55% via the Gulf of Finland (Holma et al. 2011). The relatively small sea areas, crossing traffic between Helsinki and Tallinn and oil tankers going to the west from the eastern part of the Gulf of Finland are a combination which can cause a huge environmental disaster. Maritime oil transportation is also vulnerable to security threats, an issue that has attracted less attention in the Baltic Sea.

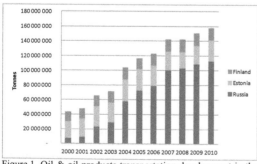

Figure 1. Oil & oil products transportation development in the Gulf of Finland 2000-2010.

As it can be seen in Figure 1, the oil transportation has quadrupled in the past ten years in the Gulf of Finland, and in the Baltic Sea oil transportation has tripled at the same time period. In the year 2000 transported oil volume in the Gulf of Finland was 43 million tonnes and in 2010 almost 160 million tonnes. The increase in volumes is mainly due to the growing Russian export while the volumes in Estonia and in Finland have been quite stable during the last decade.

In the Gulf of Finland totally 17 ports handle oil and oil products: six ports in Finland, six in Estonia and five in Russia. (Holma et al. 2011) Finland`s largest oil port is Sköldvik and port handles 96 % of the Finnish volumes. In Estonia the largest port is Muuga and ports share of Estonian oil volumes is 82 %. In Estonia the majority of oil transportation is transit traffic from Russia via Estonian ports to Europe. (Port of Tallinn 2011& Holma et al.2011) Estonia has no crude oil reserves or oil production plants or refineries of its own.

In 2000, the handled oil volume in Russia was only 7.3 million tonnes, while in 2009, the volume was almost 109 million tonnes. Russia opened the oil port of Primorsk in 2002. Since that time, the volumes in Primorsk have increased six-fold from 12.4 million tonnes to 79.2 million tonnes. The port of Vysotsk started to handle oil in 2007, and its oil volumes are currently about 14 million tonnes annually. In the port of St. Petersburg, the annual oil volumes have increased from 7.4 million tonnes to 16 million tonnes. (Holma et al. 2011 & Pasp Russia 2012) Besides these ports, the port of Ust-Luga started operating in 2012 (World maritime news 2012). The Baltic Pipeline System 2 was connected to the port of Ust-Luga and in future the oil volumes in Ust-Luga are expected to increase.

4 PERSPECTIVES ON THE FUTURE DEVELOPMENT OF MARITIME OIL TRANSPORTATION

This section will begin by going through Russia`s energy strategy and oil production plans as well as the EU's climate and energy strategies and the national energy strategies of Estonia and Finland. The scenario outlines concerning future transportation oil volumes in the Baltic Sea and in the Gulf of Finland were formulated on the basis of these trends.

4.1 *Russia`s Energy Strategies 2020 and 2030*

Russia has one of the world's largest energy resources and it controls one third of the global natural gas reserves, one tenth of the oil resources and one fifth of the coal reserves. These resources are the basis of the economic growth, external trade and policy of Russia. Many countries in Europe are dependent on Russia´s energy resources. The key elements of Russian long-term national energy policy are energy safety, energy efficiency, budget effectiveness and ecological energy security. In 2009 it was reported that Russia will invest 1.8-2.1 trillion US dollars in oil and gas projects by 2030. These projects will increase oil production for European and Asian demand and also for Russia´s own demand (Kupchinsky 2009). The realization of a social and economic long-term energy state policy has been divided into two phases. The first phase includes the basis for Russia´s progressive development with different scenarios of social and economic development. The key scenarios for the first phase are a normative and legislative base without barriers, energy efficiency, and exports of oil and gas in the internal and external fuel and energy markets. The second phase comprises the formation of a new fuel and energy complex. The policy aims for growth, transparency and competition in the energy markets. Existing energy modes will also be developed. (Ministry of Energy of Russian Federation 2003)

Oil production will continue in the traditional oil production regions in Russia, but also new oil and gas fields will be broached in North-eastern Russia (the Timan-Pechora region). In the future, Russia´s main oil base will be in Western Siberia. In oil production, the main priority will be the rationalization and increase of productivity. Russia's Energy Strategy also contains objectives related to renewable and local energy resources, although these are not very clearly defined. (Ministry of Energy of Russian Federation 2003)

4.2 *EU and national future policies in Finland and in Estonia*

The aims of the European Union`s Energy and Climate strategy are to reduce emissions of greenhouse gases by 20% (compared to the 1990 levels), to increase the share of renewable energies to 20 % of the EU's final energy consumption and to increase energy efficiency by 20 %. The aims also include increasing the share of biofuels in the transport sector to 10%. (Finland`s Ministry of the Environment 2011) The EU climate and energy package includes four directives: revision and strengthening of The Emission Trading System (ETS), An "Effort Sharing Decision", Binding national targets for renewable energy and carbon capture and storage. (European Commission 2010)

The European Union has also created the concept of Short Sea Shipping, which refers to a maritime transport space without barriers. The maritime mode of transport is highly efficient in terms of environmental performance and energy efficiency.

The EU`s maritime transport policy until 2018 includes the main strategic goals for the European maritime transport system until 2018 and identifies key areas for action which will strengthen the competitiveness of the sector while enhancing its environmental performance in the EU. (Commission of the European Communities 2009; European Commission 2011)

Finland`s and Estonian national energy strategies follow the EU`s policy. The challenge faced by Estonia is that 90% of Estonia`s energy consumption is produced with oil shales. In addition, the country's connection to Russia's energy network causes carbon leakage. Estonia has already invested in cutting down on emissions from the oil shale power plants. (Hamburg 2007; Kisel 2011)

In Finland, the main goal is to decrease the consumption of energy and to cut down on emissions. Emissions that are outside Emission Trading must be reduced by 16% by 2020. In Estonia and in Finland, national energy strategies go hand in hand with the EU`s climate and energy package. (Turunen et al. 2008)

5 SCENARIOS AND FORECASTS FOR MARITIME OIL TRANSPORTATION IN THE GULF OF FINLAND IN 2020 AND 2030

5.1 *About the scenarios*

The volumes of oil and oil product transportation in the Gulf of Finland in 2020 and 2030 were formulated by using expert assessments. The workshop was attended by seven persons. In addition, an e-mail inquiry was sent to seven experts of whom two gave their estimations by e-mail. In the workshop, the basic outlines for each scenario were presented and then experts were asked to give their estimations about oil volumes in each scenario. Estimations were given as probability distributions, in other words experts gave for each scenario a most probable and minimum and maximum values of transported oil volumes.

The reason for three different scenarios for the years 2020 and 2030 is that the future development of maritime oil transportation in the Gulf of Finland depends on many factors, and it is not realistic to give only one estimation. By giving three scenarios we can analyze how the volumes of oil transportation would differ in different situations, taking into consideration for example economic, political or energy issues. For each scenario, three different figures of oil transportation volumes were formulated in order to take probability distributions into account.

5.2 Scenarios for 2020

5.2.1 The "Slow development" scenario
Basic assumptions:
- Stagnant economic growth in the EU
- Recession prevails
- Heavy industries move to Asia, South America and other continents
- Demand for oil decreases
- Oil production in Russia fails to increase
- No investments in oil production technology

The "Slow development" scenario is based on the assumption that the European countries and Russia will suffer from a long-term economic slowdown. It is expected that the economic crisis prevails especially in Southern Europe. The demand for consumer goods will decrease or remain at the current level. Also it is assumed that heavy industries, such as the metal and forest industry, will continue to move to Asia, South America or other continents in order to cut production costs and because of the growing demand in developing countries.

The demand for oil and oil products will decrease because of high oil prices and that is why the logistics costs will become more expensive. Russia has no interest in investing in new oil production technology as the demand for oil is decreasing. Despite the low economic situation, in this scenario the Baltic Pipeline System 2 is connected to port of Ust-Luga, in other words 15-20 million tonnes of oil will be transported via Ust-Luga, starting in late 2012 or 2013. The oil transportation will increase in Russia, but it will cut the transit transportation via Estonian ports.

5.2.2 The "Average development" scenario
- Economic growth continues in the EU and Russia
- Heavy industries develop new products and production capacity is maintained in Europe
- Demand for oil increases despite of the increase in the price of oil and oil products
- Russia produces oil at full capacity with Soviet era equipment
- Some investments in oil production/technology are made

This scenario depicts a "business as usual" situation. The population, economy, technology and society continue to develop in a similar manner to the past decades in Europe. It is believed that economic growth will be more rapid in Russia than in the rest of Europe. New oil and gas pipelines connect Russian ports and gas lines to Europe. In Russia oil ports, Ust-Luga and Primorsk will operate at full power and full capacity. In Europe, fossil fuels and oil will remain the main energy sources for transport. Present investments plans for efficient energy technologies will be realised by 2020. Also growing demand for oil will lead to investments in

new and more efficient oil products technologies. In this scenario, heavy industries will continue their operations in Europe and new green products and innovations will be made.

In Finland and Estonia, the demand for oil products will only increase a little, because the share of bio fuels and other alternative energy sources will increase. Oil transportation in Estonia will decrease, because Russia will concentrate its oil transportation to its own ports and only small amounts of Russian oil will be transported via Estonian ports.

5.2.3 The "strong development" scenario
- Fast economic development in Europe and Russia
- Despite new innovations in transportation, oil remains the main energy resource despite its high price
- Russia speeds up its investments in production and refinement technology to increase its oil producing capacity
- Russia starts to explore new oil deposits in Arctic areas and seas
- The EU invests more and more in green technology and renewable energy resources
- Large investments in ports, vessels and tankers

The "Strong development" scenario is the most optimistic vision of economic development and transport in Europe and Russia. The overall economic situation and trends are very positive at the global level.

The demand for oil will remain high all over the world, regardless of its high price. Russia will invest in oil production and refinement technologies for the current oil production areas and expand oil drilling in the Arctic areas, but the production will not start yet. Oil production in the Arctic areas is more expensive, but the high price of oil products will make drilling economically viable.

The EU will invest more and more in green technologies and renewable energy resources. The targets of the climate and energy package for 2020 will be achieved as planned in the strategies. This scenario assumes that the physical size of ports will not grow much, meaning that there will be no new pier areas for larger tankers.

5.3 Oil transportation volumes in 2020

In the Slow development 2020 scenario, the expected volume of oil transported by ships in the Gulf of Finland will be 170 million tonnes. The minimum volume was estimated to be 151 million tonnes and the maximum volume 182 million tonnes. The minimum figure almost equals the volume of oil transported in 2009.

In the *Average development 2020* scenario, the expected volume of oil transported is 187 million tonnes, the minimum volume 169 million tonnes and the maximum volume 207 million tonnes. In the

Strong development 2020 scenario, the expected volume of oil transported in the Gulf of Finland is 201 million tonnes, the minimum volume 177 million tonnes and the maximum volume 218 million tonnes. In the following table the expected volumes of oil transportation in each scenario for 2020 are presented. The table also shows the change in percentage points in comparison with the 2009 volumes.

Table 1. Estimated oil and oil products transportation in the Gulf of Finland in 2020

	Million tonnes	Change (%)
Year 2009	150.6	
Slow development 2020	170.6	13.3
Average development 2020	187.6	24.2
Strong development 2020	201.5	33.8

5.4 Scenarios for 2030

5.4.1 The "Stagnating development" scenario
– Targets of the EU's climate and energy strategy will not be achieved
– Lack of investments in green technologies
– Transportation costs are high
– Heavy industries relocate production to Asia, South America or to other continents
– No new oil or gas production areas have been taken into use in Russia because of a lack of investments
– No new investments in tankers or new maritime technology
– The Arctic sea route has been opened

The forecast for 2030 is more negative than that for 2020 because of the longer time span. More radical changes may take place, for example in the economy, politics and the climate. The *Stagnating development* scenario assumes that the implementation of the EU's climate and energy package has failed and there is no interest in greener technology. It is believed that despite of the recession, some development must have happened over the next two decades. In this scenario the main energy resource for transportation will still be oil. The demand for oil will increase the price of oil and transportation costs. Because of the poor economic situation all over the world, there are no investments in new energy saving transportation technology.

Heavy industries have relocated their production from Europe to Asia, South America or other continents because of lower labour and production costs. Russia has not been able to increase its oil or gas supply. Oil production in the Arctic area is so expensive that only test drillings have been made, but production has not started. No new investments are being made in ports or vessels, except for the compulsory investments to replace outdated tanker fleet. The opening of the Arctic Route will change transportation routes to some extent by 2030.

5.4.2 The "Towards a greener society" scenario
– The EU's climate and energy package will be implemented as planned
– The EU will become a more and more carbon neutral society following the visions of the energy strategy 2050
– Demand for oil products has decreased
– In the industrial sector, new, successful openings will be made in Europe
– New innovations in transportation are made, which helps to handle increasing volumes more efficiently
– Russia will have expanded its oil production activities to Arctic areas
– The Arctic sea route will be opened

In the *Towards a greener society* scenario, growth continues in a similar way as in the *Average development* scenario for 2020. The economic situation will be mainly positive. In 2030, the EU's climate and energy package objectives will be achieved and new, more ambitious strategies will be formulated for the following decades. The EU is becoming more and more carbon neutral society. Renewable energy is increasingly replacing fossil fuel energy recourses.

Despite of economic growth it is believed that the demand for oil will decrease. The reason for the decreasing oil demand is the increasing amount of renewable energy sources and bio fuels. In the transportation sector, there will also be a shift towards railways and multimodal transportation modes. Heavy industries have relocated their production to countries in what are now called developing countries, but the green wave will have brought new innovative industries to Europe.

Russia will have been building up its oil production capacity especially in the Arctic. The growing demand for oil will be in Asia, so the majority of the oil from new oil production areas will go to China and India, where private and public transportation will increase strongly.

5.4.3 The "Decarbonised society" scenario
– The EU will implement new stricter environmental policy
– Many vehicles use electricity
– Demand for oil products is decreasing in EU
– Russia will build new gas lines to Europe and oil pipelines to the Arctic Sea and Asia
– Oil volumes in Russia will increase, but most of the new oil production will be consumed in Asia
– The Arctic sea route will be opened

In this scenario, the EU will implement very strict environmental policies, which all EU member states follow. This trend will also be apparent in other western countries. Green technology will be one of the major export products in Europe, and less wealthy countries in particular are attempting to solve their energy problems with green technology.

Still it is believed that oil and fossil fuels will remain the main energy recourse in poorer countries and also in Russia. Russia will still use oil and fossil fuels because of its national production capacity. New battery technologies will enable the use of electricity as a power supply in cars. The remainder of the world will still be mainly using combustion engines in cars.

Russia´s oil production will expand to the Arctic areas, and the country will produce increasing amounts of oil. As in the *Towards a greener society* scenario, in this scenario, too, almost all Arctic oil will travel through pipelines to Asia. Russia`s domestic demand for oil will decrease and Russia will begin to use greener technologies in transportation and energy production. New gas pipelines will be built to Europe via the Baltic Sea. Europe will prefer using gas because the carbon capture is one of the EU´s key tools in the fight against global warming and climate change.

5.5 *Volumes in 2030*

In the *Stagnating development 2030* scenario, the expected volume of oil transportation is 165 million tonnes, and the minimum and maximum volumes are 148 and 177 million tonnes respectively. Compared to the oil volumes transported in 2009, oil transportation will increase by 9.5%.

In the *Towards a greener society 2030* scenario, oil transportation in the Gulf of Finland will increase by almost 18% compared to 2009. The expected volume is 177.5 million tonnes. The minimum volume is 156 million tonnes and the maximum volume 192 million tonnes. The demand for oil in Europe will probably decrease or remain stable, while Russia will continue transporting oil and oil products overseas via the Gulf of Finland.

In the *Decarbonised society 2030* scenario, the transportation of oil and oil products will be almost at the same level as in the *Stagnating development* scenario. It is expected that oil transportation volumes will increase by 12.5 % from the 2009 level. The expected volume is 165.5 million tonnes and the minimum and maximum volumes 153 and 190 million tonnes respectively.

In the following table the expected volumes of oil transportation in each scenario for 2030 are presented. The table also shows the change in percentage in comparison with the 2009 volumes.

Table 2. Expected volumes in oil transportation scenarios for the year 2030 and change in percentage.

	Million tonnes	Change (%)
Year 2009	150.6	
Stagnating development 2030	165	9.5
Towards greener society 2030	177.5	17.8
Decarbonised society 2030	169.5	12.5

6 SUMMARY AND CONCLUSIONS

Oil transportation in the Gulf of Finland has almost quadrupled in the past ten years. In 2000, little over 43 million tonnes of oil and oil products were transported and handled in the Gulf of Finland. In 2009, this amount was 150.6 million tonnes and in 2010 almost 160 million tonnes. The economic recession, which started in late 2008 had not much effect on the volumes of maritime oil transportation. The increase in oil volumes is due to the increasing oil production and exports in Russia. Russia´s oil volumes are the largest in the Baltic Sea. They have increased exponentially from the year 2000.

According to the experts, there will only be a moderate increase in the oil transportation volumes compared to the statistics for 2009. Variations in the scenarios are not major, but the minimum and maximum volumes of oil and oil products transported in each scenario have quite a wide range. This variation between the minimum and maximum scenarios varies around 30 to 40 million tonnes. Transported oil volumes in the *Slow development* scenario in 2020 and the *Stagnating development* scenario for 2030 are almost the same. We can see that the estimated volumes of Ust-Luga increase the transportation volumes, but depending on the scenario, the volumes for Estonia and Finland might decrease a little. In the *average and strong* development scenarios for 2020, the transported oil volumes will increase. The reason for the increasing oil volumes is that new technologies and the share of renewable fuels will not develop rapidly in the next seven years. If new technologies, for example electric vehicles, are in a wide use, the demand for oil products will decrease. However, especially in transportation of passengers and goods, fossil fuels will remain the main energy source.

ACKNOWLEDGEMENTS

This study has been made as a part of the MIMIC (Minimizing risks of maritime oil transport by holistic safety strategies) project. The project is funded by the European Union and it has been approved as an EU flagship project. The financing comes from the European Regional Development Fund, the Central Baltic INTERREG IV A Programme 2007-2013; the Centre for Economic Development, Transport and the Environment of Southwest Finland (VARELY); the City of Kotka; Kotka-Hamina Regional Development Company (Cursor Oy); Kymenlaakso University of Applied Sciences; the Finnish Environment Institute; the University of Tartu; Tallinn University of Technology and the Swedish Meteorological and Hydrological Institute. We express our gratitude to the financers of the project.

REFERENCES

Baltic Sea Action Group 2008. Clean and Safe Maritime Activities. Available at URL:http://www.bsag.fi/focus-areas/clean-and-safe-maritime-activities

Commission of the European Communities (2009). Communication from the commission to the European Parliamnet, the Council, the European Economic and Social Committee and the Committee of the Regions. Strategic goals and recommendations for the EU`s maritime transport policy until 2018. Available at URL:http://eur-lex.europa.eu/LexUriServ/LexUriServ.do?uri=COM:2009: 0008:FIN:EN:PDF

European Commission 2010. Climate Action. The EU climate and energy package. Available at URL: http://ec.europa.eu/clima/policies/package/index_en.htm

European Commission 2011. Maritime transport. Motorways of the Sea. Available at URL: http://ec.europa.eu/transport/maritime/motorways_sea/motorways_sea_en.htm

Finland`s Ministery of the Environment 2011. EU´s climate- and energy packet. Available at URL: http://www.ymparisto.fi/default.asp?node=22013&lan=fi

Hamburg, A. 2007. Estonian National Energy Strategy. Oil Shale, 2007, Vol 24, No.2 Special. Pages 332-336. Estonian Academy Publisher. Available at URL: http://www.eap.ee/public/oilshale/oil-2007-2s-13.pdf

Hänninen M., Kujala P., Ylitalo J.,Kuronen J.2012. Estimating the Number of Tanker Collisions in the Gulf of Finland in 2015. TransNav - International Journal on Marine Navigation and Safety of Sea Transportation. Volume 6. Number 3, pp. 367-373, 2012.

HELCOM 2009. Overview of the shipping in the Baltic Sea. Vailable at URL: http://www.helcom.fi/stc/files/shipping/Overview%20of%2 0ships%20traffic_updateApril2009.pdf

Holma, H., Heikkilä, A., Helminen, R. & Kajander S. 2011. Baltic Port List 2011. Market review of cargo development in the Baltic Sea ports. University of Turku. ISBN 978-951-29-4785-0.

Hänninen, S. & Rytkönen, J. 2004. Oil transportation and terminal development in the Gulf of Finland. VTT Publications 547. Espoo 2004. Available at URL: http://www.vtt.fi/inf/pdf/publications/2004/P547.pdf

Kisel, E. 2008. Developing Estonian energy policy hand in hand with EU energy packages. Available at URL: http://web-static.vm.ee/static/failid/122/Einari_Kisel.pdf

Knuutila, S. 2011. The easternmost waters of the Baltic. Available at URL: http://www.ymparisto.fi/default.asp?node=16516&lan=en

Knuutila, S., Jolma, K. and Asanti, T. 2009. The Imapcts of oil on the marine environment. Available: at URL http://www.ymparisto.fi/default.asp?contentid=334186&lan=fi&clan=en

Kupchinsky, R. 2009. Russian Oil and Gas in 2030. Publication: Eurasia Daily Monitor Volume: 6, Issue: 164. Available at URL: http://www.jamestown.org/single/?no_cache=1&tx_ttnews%5Btt_news%5D=35460

Leningrad Region 2011. The Baltic pipeline System – the key federal project in the Leningrad Region. Available at URL: http://eng.lenobl.ru/economics/investment/principlefederalprojects/balticoilpipeline?PHPSESSID=640241724ce1bc428 39f3b18b9e71f1a

Ministry of Energy of the Russian Federation 2003. The Summary of The Energy Strategy of Russia for the period of up to 2020. Moscow. Available at URL:http://ec.europa.eu/energy/russia/events/doc/2003_strategy_2020_en.pdf

Myrberg, K., Raateoja, M. and Lumiaro, R. 2011. Itämeren peruskuvaus. Suomen ympäristökeskus and Edita. Available at URL: http://www.itameriportaali.fi/fi/tietoa/yleiskuvaus/peruskuvaus/fi_FI/peruskuvaus/

Pasp Russia 2012. Большой порт Санкт-Петербург (Big Port of St. Peterburg). Available at URL: http://www.pasp.ru/bolshoy_port_sankt-peterburg1

Port of Tallinn 2011. Muuga Harbour. Available at URL: http://www.portoftallinn.com/muuga-harbour

Turunen, T., Lepistö, A. and Ritonummi, T. (2008). Pitkän aikavälin ilmasto- ja energiastrategia (A Longterm climate and energy strategy). Available at URL: http://www.tem.fi/files/20585/Selontekoehdotus_311008.pdf

UK Trade & Investment 2010. Russia – Opportunities for UK-based companies in the ports sector. Available at URL: http://www.beckettrankine.com/downloads/UKTI_Russia_Ports_Report.pdf

Ust-Luga Company 2012. The Internet site of Ust-Luga Company. Available at URL: http://www.ust-luga.ru/info/?lang=en

World maritime news 2012. Russia: First Shipment from Ust-Luga Crude Oil Terminal to Rotterdam. Posted on March 28th.2012. Available at URL: http://worldmaritimenews.com/archives/50691

Possibilities for the Use of LNG as a Fuel on the Baltic Sea

S. Jankowski
Maritime University Szczecin, Poland

ABSTRACT: The environment still is exposed on degradation caused by industries activities. Fortunately for the environment, the international community (IMO, EU, HELCOM) agreed to reduce emissions of some pollutants. These requirements are being implemented gradually and will have full force in 2015 and 2016. On the other hand that situation is a big challenge for a shipping industry to meet these requirements, especially on ECA (emission control area). One option is to use LNG as a fuel. This paper describe the LNG situation on Baltic Sea, and aims of project "MarTech LNG" as well. "MarTech LNG" is one of projects, which support the LNG solutions for south Baltic Sea region.

1 INTRODUCTION

During the next few years, according to IMO regulations, all vessels must decrease air pollutant in the exhaust gases especially inside emission control areas (ECA).

In 1997 a new annex was added to the International Convention for the Prevention of Pollution from Ships (MARPOL). The main aim of the Annex VI "Regulations for the Prevention of Air Pollution from Ships" is finding a solution to minimize emissions from ships oxides of sulfur (SOx – Table 1), particulate matter (PM), nitrogen oxides (NOx – Table 2), ozone depleting substances (ODS), volatile organic compounds (VOC) and their contribution to local and global air pollution and environmental problems. Annex VI entered into force in 2005, but in 2008 was revised. The significant tighten emissions limits adopted in 2008, are gradually introduced from 2010.

The tables below present the limits of emission of air pollution caused by engines of vessels: the regulation 14 (SOx) and the regulation 13 (NOx).

Table 1. Emission limit for SOx and particulate matter (IMO 2008)

Outside an ECA	Inside an ECA
4.50% m/m prior to 1 January 2012	1.50% m/m prior to 1 July 2010
3.50% m/m on and after 1 January 2012	1.00% m/m on and after 1 July 2010
0.50% m/m on and after 1 January 2020	0.10% m/m on and after 1 January 2015

Table 2. Emission limit for NOx (IMO 2008)

		Tier I	Tier II	Tier III (only for NOx ECA)
Ship construction date on or after		1 January 2000	1 January 2011	1 January 2016
Emission limit (g/kWh) as a function of engine rated speed n [rpm]	n < 130	17.0	14.4	3.4
	n = [130÷1999)	$45 \cdot n^{-0.2}$	$44 \cdot n^{-0.23}$	$9 \cdot n^{-0.2}$
	n ≥ 2000	9.8	7.7	2.0

In addition IMO has adopted mandatory technical and operational energy efficiency measures which will significantly reduce the amount of CO_2 emissions from international shipping.

Currently Baltic Sea and North Sea are established as an ECA only for SOx, but everybody engaged in sea transport business should think perspectively. North America and from 1 January 2013 United States Caribbean Sea are SOx, NOx and PM ECA.

There is a high probability that new ECAs will be established (Figure 1) or that the existing ones will be more restrictive.

Existing
Possible future ECA

Figure 1. DNV's map of current and possible ECAs in the future (DNV 2011)

2 TECHNOLOGICAL SOLUTIONS

The review of existing engine technology and its development indicates that currently only three solutions are in accordance with SOx regulations. If shipowners wish to continue sailing on Baltic Sea after 2015 they have to choose (DMA 2012):
– low sulphur fuel,
– an exhaust gas scrubber,
– LNG fuel (liquefied natural gas).
The first solution require only minor modifications on vessel fuel systems. The content of sulphur in a fuel like MDO (marine diesel oil) and MGO (marine gas oil) can be below 0.1%. The main disadvantage such a choice is limited availability of low sulphur fuel is that rising demand is expected to increase its price uncertainty.

The second solution requires installation of an exhaust gas scrubber to remove sulphur from the engine exhaust gas by using chemicals or seawater. This technology require significant modifications on ship systems. Additional tanks, pipes, pumps, and a water treatment system. The sulphur-rich sludge produced is categorized as special waste, to be disposed of at dedicated facilities. Moreover, scrubbers increase the power consumption, thereby increasing its CO_2 emissions.

The third solution is using LNG (liquid natural gas) as a fuel. Natural gas is the cleanest form of fossil fuels available, and when fuelling a ship with LNG no additional abatement measures are required in order to meet the ECA requirements. However, an LNG-fuelled ship requires purpose-built or modified engines and a sophisticated system of special fuel tanks, a vapouriser, and double insulated piping. Available space for cylindrical LNG fuel tanks on board ships has been a key challenge, but new hull integrated tanks are expected to simplify this issue.

For new ships delivered after 1 January 2016, exhaust gas purification by Selective Catalytic Reduction (SCR) or LNG fuel are the only two currently available abatement measures to meet Tier III requirements.

3 BENEFITS OF USING LNG

LNG means liquefied natural gas. The natural gas is temporarily converted to liquid form at -163 Celsius, under atmospheric pressure. It takes up 600 times less space than as a gas, therefore it is more efficient for storage and transport. LNG is currently tested as a fuel on more than 20 vessels sailing on Norwegian waters.

In addition LNG is clean not only in aspect of exhaust gases, but also in case of spill. LNG does not cause environmental disaster because in such a case it will evaporate quite fast. The main hazard in case of LNG spill, are frostbites due to extremely low temperature (Starosta 2007).

Taking account above mentioned three solution it should be said, that LNG is the best alternative in aspect of economic and environmental impact to Baltic Sea.

3.1 Environmental impact

LNG as a fuel has the lowest emission of all three pollutants NOx, SOx, and particles, as well as the greenhouse gas CO_2 (GHG). SOx and particles are reduced by close to 100% (Figures 2 and 4), NOx emissions close to 85–90% (Figure 3), and net GHG emissions by 15–20% (Figure 5).

Pollutant emissions for a typical, cargo Baltic Sea vessel is shown below (DNV). The typical cargo vessel was determined as follows:
– Gross tonnage: 2700,
– Power of main engine: 3300 kW,
– Yearly sailing hours: 5250.
Low sulphur fuel means a fuel contains maximum 0.1% sulphur, and conventional fuel contains 1% sulphur (according to 01.07.2010 status).

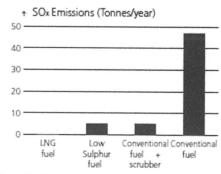

Figure 2. SOx emission of different fuel solutions for typical Baltic Sea cargo vessel (DNV 2010)

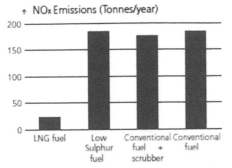

Figure 3. NOx emission of different fuel solutions for typical Baltic Sea cargo vessel (DNV 2010)

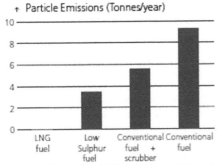

Figure 4. PM emission of different fuel solutions for typical Baltic Sea cargo vessel (DNV 2010)

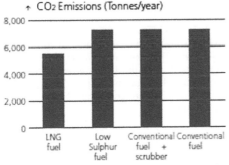

Figure 5. CO_2 emission of different fuel solutions for typical Baltic Sea cargo vessel (DNV 2010)

3.2 Economic benefits

Nowadays the LNG trade market is large and flexible. The forecast developed by U.S. Energy Information Administration (EIA) in 2008 are optimistic and indicates an expanding gap between conventional fuel and LNG (Figure 6).

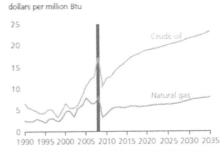

Figure 6. LNG price forecast by EIA (DNV 2010)

Lloyd's Register's forecasts of the fuel market indicate that demand of LNG as a fuel will depends on number of vessels fueled by LNG and its price (figure 7).

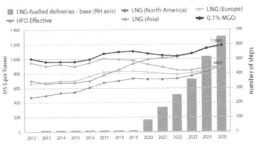

Figure 7. LNG price forecast by Lloyd's Register, 2012 (Lloyd's Register 2012)

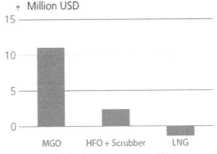

Figure 8. Exploitation costs over 20 years related to conventional fuel (DNV 2010)

The cost of a new vessel equipped with LNG propulsion is higher about 10-20% than conventional vessel with similar gross tonnage. The additional cost is mainly due to the sophisticated LNG storage tanks, the fuel piping system and in some cases a slightly larger ship. Based on experience from ships built, the additional investment cost for the LNG fuelled typical Baltic Sea cargo vessel has been estimated to about 4 million USD. Estimated cost of scrubber installation should be around 1 million USD. Taking these assumptions into account and forecasting price of marine gas oil (MGO) in 20 years perspective the lowest exploitation cost are in case of LNG vessel (Figure 8).

The exploitation costs analysis indicates that fueling LNG is cheaper even in comparison to HFO, and differences between MGO option is up to 12 million USD.

4 STATUS OF LNG INFRASTRUCTURE ON BALTIC SEA

In order to enable navigation of vessels using LNG as a fuel, a grid of bunker stations is required. An average period between bunkering for the LNG vessels today is about one week, and vessels should have possibilities to obtain LNG in one of the ports during their trips. Currently there is no LNG infrastructure on Baltic Sea (figure 9).

Figure 9. LNG terminals on Baltic Sea (DNV 2012)

The number of import terminals is not enough to provide a supply of LNG for every route on Baltic Sea. They should operate rather as a hub of LNG and distribute it to small scale bunker stations.

In case of decision about building new import terminal, it belongs to government in order to securing energy independence of given country, but decisions about building small scale LNG terminals or bunker stations, depend on market. Currently there is no LNG bunker stations on Baltic because there are a small number of LNG powered vessels, and lack such vessels is a result of lack of bunker stations. It seems correct that at least at the beginning, the bunker stations should also have a political support.

MarTech LNG – "Marine Competence, Technology and Knowledge Transfer for LNG in the South Baltic Sea Region (SBSR) is one of the projects which aims are dissemination of LNG technology by exchanging experiences, knowledge and competencies within SBSR. The project

supports the activities related to LNG technology, promotes LNG as a green energy and the cleanest marine fuel. Main idea of the project is to create a better access to technology and knowledge on LNG related business activities to build up a better competences and specialization among the SBSR maritime business supply chain. The main idea will be achieved by realize following aims:
– Develop the LNG related competences for the Maritime industries in SBSR
– Foster LNG targeted scientific research
– Create LNG supply/value chain in SBSR
– Support LNG development and operation processes in SBSR

One of the first task of the project was region study in terms of existing education, research, training and consulting institutions providing activities related to LNG technology. The region was also analyzed in aspect of stakeholders dealing with LNG technology. The region LNG profile or joint study "Mapping LNG knowledge and competence in the SBSR" indicates on lack or too small amounts of LNG that could enable development of the LNG business. Cooperation between stakeholders and institutions is weak and really hard to find LNG supply chain. But LNG enterprises have a big potential which can be activated when LNG as a fuel will be available (MarTech LNG).

5 SUMMARY

LNG is one of the best solution for Baltic region to protect environment against pollution caused by conventional fuels. Now is the time for owners to decide which solution to choose to be in compliance with the MARPOL Convention. They will choose LNG, if on Baltic Sea the LNG infrastructure will exist. Unfortunately it seems that without political support, building infrastructure may be difficult or even impossible.

REFERENCES

Danish Maritime Authority, North European LNG Infrastructure Project, 2012
Det Norske Veritas, Greener shipping in the Baltic Sea, 2010
IMO, The Regulations for the Prevention of Air Pollution from Ships (Annex VI), 2008
Jon Rysst, DNV report on the potential of LNG shipping in the Baltic, 2011
Lloyd's Register, LNG-fuelled deep sea shipping, 2012
Project MarTech LNG, Mapping LNG knowledge and competence in the SBSR - Joint Study, 2013
Starosta A.: Safety of Cargo Handling and Transport Liquefied Natural Gas by Sea. Dangerous Properties of LNG and Actual Situation of LNG Fleet. TransNav - International Journal on Marine Navigation and Safety of Sea Transportation, Vol. 1, No. 4, pp. 427-431, 2007

Identification of Hazards that Affect the Safety of LNG Carrier During Port Entry

P. Gackowski & A. Gackowska
Gdynia Maritime University, Gdynia, Poland

ABSTRACT: This paper presents the general division of ports for the identification of hazards that affect the safety of LNG carrier for port and LNG terminal in Świnoujście located on Pomeranian Bay. Examining in detail each of the steps of the vessel's approach to the port, naming common threats to ships safety and identifying wind as the specific factor that have a direct impact on the safety of this port's inbound ship. This paper presents final results of studies and simulations run for various wind conditions that can be met on the area of the Pemerianian Bay and LNG Terminal in Świnoujście.

1 INTRODUCTION

Ships adopted to transport of liquefied natural gas – the LNG tankers are a group of most dangerous types of ships used nowadays in sea transportation of cargoes. It is so due to nature of their cargo, the way of its transport and all the consequences in case of accident which they can participate in.

Methane, with a chemical symbol CH_4 is a dangerous cargo according to IMO and UN classification. In Code of Dangerous Cargoes – IMDG Code it is located in class 2.1 and with UN number UN 1971/1972. Although methane itself is not toxic, it is highly suffocating gas, especially in enclosed compartments. Furthermore, it is dangerous because of the low temperature under which is being transported – below -160°C. When spilled methane instantly changes state from liquid into vapour creating visible vapour cloud which can easily dissolve in air. There is no spontaneous explosion, but it is extremely vulnerable to any external source of ignition. The largest LNG carriers have the capacity of liquefied methane of around 260 thousand cbm and more. During the simulations carried out in various research centers around the world it was discovered that the effect of leakage and ignition of cargo of an LNG tanker of this size would burst with such strength that exceeds a thousand atomic bomb used during World War II.

These vessels carrying liquefied gas in such form - LNG, are narrowly specialized group of ships carrying liquid-tankers. As each type of commercial ships, LNG tankers may be divided into several groups, depending on the distribution of the indicator used for classification. The first division is the division of these units due to the arrangement of cargo tanks and their construction. In this division, we can distinguish two classes of ships:
- spherical tank vessels
- membrane tank vessels.

However, the most tangible indicator is the capacity of vessels, usually measured in cubic meters - cbm. It is the clearest division of these vessels, which gives an idea of the amount of cargo they can carry. This division is as follows:
- 70000-80000 cbm
- Conventional - 125000-150000 cbm
- Q-Flex - up to 217 000 cbm
- Q-Max - up to 266 000 cbm

Figure 1. Comparison of LNG carriers[1]

[1]From web page: *http://www.webmar.com/2008/03/q-flex-y-q-max.php*

This provides the clear division into classes, which will be used later in this article.

2 DIVISION OF PORTS

Each LNG tanker is designed to carry cargo, which is natural gas on a certain route between the port of loading and port or ports of discharging. When identifying hazards that affect the safety of LNG vessels there must be introduced the overall division of commercial ports, which can be points of LNG terminals being targeted trade routes of LNG tankers. Introduction of ports' division (Figure 2.) will help to distinguish groups of factors and indicate the precise security risks in the specific conditions of the port.

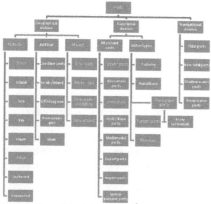

Figure 2. General division of ports[2]

From the point of view of navigators, of utmost importance is the navigational division, along with geographical distribution. It is for reasons of geographical and natural situation, that there is a vast majority of potential dangers for ships entering and leaving the port.

3 IDENTIFICATION OF HAZARDS

In order to correctly detect and then identify threats and potentially dangerous situations for a ship calling at a port it is necessary first to identify groups factors affecting them. Information such as data on the maneuverability of the trade and the port of other ships, current weather and its impact on all ships and port facilities must be taken into account in determining the LNG ship safety parameters. These informations are collected in equal measure by the ship's staff and port services in equal way.

Depending on the date of entry of the vessel, the risk may be due to the different variables that affect the same ship or port of the same characteristics of the reservoir port traffic of other vessels. The last group of factors is one containing so-called human error.

The entrance to the port of LNG carrier can be divided into inextricably linked with each other stages in which different groups of factors will have a different impact on the safety of individuals served by the port. The first stage occurs when ship's entering to the fairway and goes up to the port's heads. Passing major Heads of port (port entrance) to the turning circle is counted as second stage. The third stage is swing by LNG tanker and its final mooring to the quay. At each of these stages factors from different groups will have a dominant role, "(...) especially hydro meteorological factors and phenomena affecting the maneuver input:

- reservoir bathymetry,
- breadth, depth and orientation of the approach fairway,
- wind speed and direction,
- The strength and direction of the current,
- Other impediments to movement (shoals, wrecks). "[2]

For outer harbor and LNG terminal in Świnoujście main threat will be wind and bathymetry of the basin.[3]

In many studies it is assumed that the passing by a ship fairway buoys designating the traffic lane leading to the port. It can be, however, recognized that effective cooperation starts after the final declaration of the ship in the reporting point. The newly established terminal in Świnoujście uses the reporting system, in which the vessel wishing to call the LNG terminal is required to check-in service port for a long time before the actual entrance to the fairway.[4]

LNG tanker maneuvering on the fairway has limited space for her maneuvers. This is due to the fact that, by definition, all the fairways and approaches to merchant ports are considered as restricted waters. This limitation is due not only to the lack of horizontal space that sets the fairway, the existing shoals and shallow water, the deployment of underwater wrecks and rocks. Also important is the fact that there are vertical movements of water masses, which are called tides. They are particularly important for the ports located on the shores with considerable fluctuations in water level, and thus, where the availability of the port for

[2] Based on: Gackowski P. (2011). Warunki zewnętrzne wpływające na bezpieczeństwo manewrów statku podczas wejścia do portu. Zeszyty Naukowe Akademii Morskiej w Gdyni 69

[3] Based on: Gackowski P. (2011). Analiza wybranych warunków zewnętrznych wpływających na bezpieczeństwo wejścia gazowca LNG do terminalu gazowego LNG w Świnoujściu. PPEEm conference, Wisła
[4] Based on: Śniegocki H. and Gackowska A. (2011). Vessel traffic system VTS for newbuild LNG terminal in Świnoujście. PPEEm conference, Wisła

the vessels of bigger size is only temporary (as described by expression Tidal window period). In the Pomeranian Bay, where the External Port and LNG terminal in Świnoujście is located, like the area around the Baltic sea, level changes are mainly caused by hydro-meteorological conditions. The most important is wind - its direction and speed. It mainly causes stacking or pushing water. Tides occurring in the Baltic are so small that they can be omitted. As has been shown in the development of state of the environment IMiGW PIB Polish Baltic coastal zone between 1986-2005[5], the average long-term (1986-2005.) In Świnoujście sea level stood at 502cm. The information from the website of the Maritime Office in Szczecin[6], gives state of changes in water level caused by southern storms. It cause a decrease by up to 130cm in relation to the average water level adopted at the level of 500cm. Because of such sea level fluctuations, for vessels (including a change in the UKC, maneuverability) even a minimal reduction in the level of water on the track to Świnoujście is extremely negative, on entering or leaving the ship.

The most important hydro meteorological factor forming potentially dangerous obstacle for ships is wind. It happens that the port itself is naturally protected from it, but the approach itself is marked by a huge risk of sudden and strong gusts of wind. It is extremely dangerous for ships with a large lateral surface. Analysis of the winds over the area of Pomeranian Bay and drawn up on the basis rose plotted for the whole year shows a clear advantage winds from SW, NW and SE. The least frequently reported winds from the N-and NNW direction. In each month of the year the situation is generally similar. There is a dominance of winds from the south-western sector. In each of the dominant directions of their occurrence in a given month, it is less than 10%. If we take into account wind speed, this can be noted that the strongest winds are generally blowing from the NW, N and NE and NE and NW. In each month, the situation is similar. Posted in the following figure (Figure 3.) box-diagram is showing the occurrence of a specified force winds (speed) in Świnoujście.

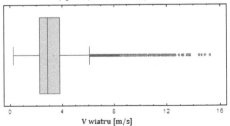

Figure 3. Graph box-of wind speeds for station Świnoujście

---

[5] Based on: Average state of polish shoreline environment in Baltic sea during *1986-2005*. Chosen topics. IMiGW, Warsaw 2011
[6] *http://ms.ums.gov.pl/pl/Karty/Świnoujście/podej.php*

3.1 *Simulation of passing through the fairway*

Simulation studies were carried out using the navigational-maneuvering simulator Navi Trainer Professional 5000, simulator of electronic charts systems and Navi-Sailor ECDIS 4000 and Model Application Wizard (v. 5.0) application. Used ship's model was a model of a loaded membrane type LNG carrier, with parameters L = 315m, B = 50 m, T = 12 m. Ship was on an even keel and in upright condition.

Figure 4. Model of LNG carrier used in the simulations

Traffic lane which was used in simulations was 14.3 m in depth, headed in the direction of 170.2° - 350.2 °. The initial (nominal) speed of the vessel was the value of 6 and 8 kts. Wind adopted in simulations changed every 45 degrees from the direction of N, thus blowing in the eight cardinal directions. In the simulation scenario, the behavior of the model in motion along the approach assumes that the initial position of the model is on the axis of the existing track in the most northerly part and in the line of symmetry of the model and the bottom of the course coincides with the center of the track, in example 170.2 °. Adopted at a constant depth equal to the entire track 14.3 m. A loop was built of a continuous timeline, which for a given range of wave, wind speed and wave height (caused by wind) are continuous-decreasing functions. Wind speed is limited to a value of below 6 m /s. Simulation results obtained for winds (without squalls) with speed for 6, 9, 12, 12,2-15,0 m/s at 0.2 m/s and 15,1-18,0 m/s in 0.1 m/sec. As the end of the simulation were taken when adopted model ran out from the track with a width of 300m. In the scenario simulation assumed the conditions of good visibility during the day to offset the adverse effects for this particular study, the external conditions, such as the movement of other factors. The main objective of the simulation was to investigate the conduct of the volatility function of distance from the track model with given parameters of noise – hydro meteorological conditions. Parameters of conducted simulation – as examples are given in table Tab.1. These simulations did not include the additional impact of current.

Table 1. Simulation parameters for the main directions of existing winds

Simulation	Ship's speed Vs [kt]	Wind direction [points]	Wind wave direction [points]	Swell			Run out of lane		Remarks
				Direction [points]	Wave height Hf [m]	Swell period T [s]	Wind speed Vw [m/s]	Side of lane	
1	8	N	N	-	-	-	15,3	E	
2	8	NE	NE	-	-	-	15,4	E	
3	8	E	E	-	-	-	14,8	W	
4	8	SE	SE	-	-	-	14,8	W	
5	8	S	S	-	-	-	-	-	Ship stayed on lane of breadth 300 m
6	8	SW	SW	-	-	-	14,2	E	
7	8	W	W	-	-	-	15,3	E	
8	8	NW	NW	-	-	-	15,6	W	
1a	6	N	N	-	-	-	-	-	Ship stayed on lane of breadth 300 m
2a	6	NE	NE	-	-	-	14	W	
3a	6	E	E	-	-	-	12,2	W	
4a	6	SE	SE	-	-	-	12,4	W	
5a	6	S	S	-	-	-	12,8	E	
6a	6	SW	SW	-	-	-	12	E	
7a	6	W	W	-	-	-	12,6	E	
8a	6	NW	NW	-	-	-	15,1	W	

In the accompanying graphs below, as examples, are shown graphically as a function of time wind speed (v), change the under keel clearance in the bow and stern, the speed of the model over ground (SOG) and the angle of drift (drift angle) obtained from measurements of the table (Table 1.) representing group of only few test runs (simulations) carried out during this study.

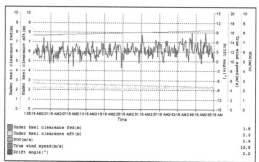

Figure 5. Graph of changes in UKC, v_w, SOG, drift angle (simulation no. 5)

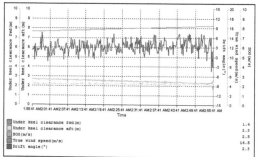

Figure 6. Graph of changes in UKC, v_w, SOG, drift angle (simulation no. 6)

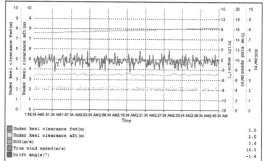

Figure 7. Graph of changes in UKC, v_w, SOG, drift angle (simulation no. 1a)

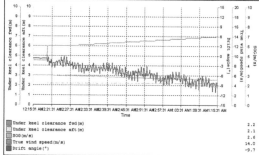

Figure 8. Graph of changes in UKC, v_w, SOG, drift angle (simulation no. 2a)

4 CONCLUSIONS

As a result of the study it can be concluded that the greatest impact on the safety of a vessel in the waters adjacent to the port and along the approach to the LNG Terminal in Świnoujście are weather conditions, especially wind, are the most affecting ones. Big ships, such as the LNG tankers class Q-Max and Q-Flex envisaged as targets of these factors, calling at the port and LNG Gas Terminal in Świnoujście located in southern part of Pomeranian bay are very sensitive to strong winds. Wind, his strength and direction is the most important factor and is more important for ships than for port itself, especially when space for maneuvering is limited, as in the case of the track in Świnoujście. The study shows that in vast majority of cases, the vessel is able to stay on track with a width of 300m, even when the strength of the prevailing winds is 14 m/s. According to the recommendations of Classification Societies it is a value equal to 6 widths of largest vessels. This article is just a one of a series dedicated to discover and study this vast problem.

REFERENCES

[1] IMiGW (2011). Average state of polish shoreline environment in Baltic sea during 1986-2005. Chosen topics., Warsaw

[2] Gackowski P. (2011). Warunki zewnętrzne wpływające na bezpieczeństwo manewrów statku podczas wejścia do portu. Zeszyty Naukowe Akademii Morskiej w Gdyni 69

[3] Gackowski P. (2011). Analiza wybranych warunków zewnętrznych wpływających na bezpieczeństwo wejścia gazowca LNG do terminalu gazowego LNG w Świnoujściu. PPEEm conference, Wisła

[4] OCIMF (2010). ISGINTT- International Safety Guide for Inland Navigation Tank- barges and Terminal.

[5] SIGTTO (1997). Site selection and Design for LNG Ports and Jetties, Information Paper No. 14

[6] Śniegocki H. and Gackowska A. (2011). Vessel traffic system VTS for newbuild LNG terminal in Świnoujście. PPEEm conference, Wisła

[7] Śniegocki H. and Gackowska A. (2012). Bezpieczeństwo tankowców LNG podczas podejścia do portu ze szczególnym uwzględnieniem zmian istniejącego oznakowania nawigacyjnego na przykładzie Portu Zewnętrznego i Terminala Gazowego LNG w Świnoujściu. PPEEm conference, Radom

Gas and Oil Transportation
Maritime Transport & Shipping – Marine Navigation and Safety of Sea Transportation – Weintrit & Neumann (Eds)

The Mooring Pattern Study for Q-Flex Type LNG Carriers Scheduled for Berthing at Ege Gaz Aliaga LNG Terminal

S. Nas, Y. Zorba & E. Ucan
Dokuz Eylul University Maritime Faculty, Turkey

ABSTRACT: Ever growing energy industry requires larger quantities of LNG to be transported by bigger ships between terminals. Every day, new kind of large vessels created by new technologies, and these are used to trade around the globe. This is the dynamic change in shipping industry. But on the other hand these new vessels need to safely berth to existing terminals which we may accept as more static part of the trade. Thus this study born by the request of Ege Gaz Aliaga LNG Terminal management to determine if it is safe to berth to the terminal by a new breed of large LNG carrier type named as Q-Flex and Q-Max. Transas Bridge Simulator NTPRO 5000 series was used in this study for extensive experiments which had been simulated by the use of hook function. During the study, every force applied to mooring hooks and dolphins by the ship lines were divided into 3 dimensions and then measured by simulated experiments. With analysis of the data, required hook and dolphins strengths were determined for the safe mooring arrangements. Upon the completion of the study Ege Gaz Aliaga LNG Terminal became the first safe berth for Q-Flex type vessels in the Mediterranean and the Black Sea. And finally all experiments were confirmed with real life experience when the first Q-Flex type LNG carrier berthed to the Ege Gaz Aliaga LNG Terminal.

1 INTRODUCTION

Due to ever energy hunger economies of the world the natural gas trade market has seen a rapid growth in the last 20 years. (David A. Wood 2012) Also due to geographical and political reasons natural gas industry has been forced to provide other ways than the pipe lines to deliver the commodity to required markets.

Thus Liquefied Natural Gas Carrier ships have born. First LNG carrier "Methane Pioneer" has left Calcaseieu River on 25 January 1959 and she was a tiny one compared to today's standards only having 5034 tons deadweight. Growing market and advantage of cost in bulk carriage also boost the size and change the characteristic of LNG carriers (Starosta, 2007). Today's the biggest LNG carrier is a giant of a ship "Q-Max" carrier with 345 meters length over all and has a cargo capacity of a 266.000 cubic meters which equals to 161.994.000 cubic meters of natural gas. This is the dynamic change in shipping.

But on the other hand as this trade is done via seaway since 1959 there are lots of already constructed coastal facilities which are used to accept this highly specialized cargo but from smaller ships. Due to the great initial building costs of such facilities, it is extremely hard to construct new terminals from scratch for every new breed of LNG carrier borns. Thus it is been accepted that this part of the trade is much more static.

Nevertheless with careful planning, study and with the calculation of risks an existing terminal might be evaluated that if it has correct arrangements and qualities to receive the new class of ships or the new requirements for the terminal to safely accept the ships which might not even been discovered when the terminal was first planned.

Mostly above risk analyses for LNG terminals and shipping depend on computer models (Er, 2007)

2 MOTIVATION OF THE STUDY

This study is born by the request of Ege Gaz Aliaga LNG Terminal Management to determine if it is safe to berth to their existing terminal by a new breed of large LNG carrier type vessel named as Q-Flex.

Aliaga LNG Terminal which one of the two existing LNG Terminals in Turkey is constructed between 1998 and 2002 and has started its operation in 2006. But as described below first Q-Flex vessel was not even built in that year.

In detail, Q-Flex type LNG carriers are the state of the art vessels which are designed to be more efficient and clean than the regular LNG carriers and first delivered to service in 2007. They were designed as membrane type carriers and have a capacity between 210.000 and 216.000 cubic meters which makes them the world's largest carriers until the entry into service of the Q-Max type LNG carriers.

In the completion of this study it is revealed that if the terminal is compatible with the Q-Flex type vessels without any new upgrade in terms of equipment or construction or not.

3 METHODOLOGY

3.1 *The aim of the study*

The main aim of this study is to find a reliable and verifiable and risk free solution to determine the required mooring pattern arrangements of a specific terminal for safe maneuvers of a specific ship type before such an event happens in reality.

The term "mooring pattern" refers to the geometric arrangement of mooring lines between the ship and the berth (OCIMF, 2008). Manifolds connection and other interfaces are not in the scope of the study.

It should be noted that the industry has previously standardized on the concept of a generic mooring layout, taking into account Standard environmental criteria. The generic mooring layout is mainly applicable to a "multi-directional" environment and to the design of ship's mooring equipment. For general applications, the mooring pattern must be able to cope with environmental forces from any direction. This can best be approached by splitting these forces into a longitudinal, lateral and a vertical component and then calculating how to most effectively resist them.

The ship's mooring lines should be able to hold the ship in position with wind speeds of 60 knots (31m/sec) (OCIMF. 2008; 18). Q Flex type vessel have 20 sets of mooring lines with ø44 mm. ultra high molecular weight polyethylene. Each mooring line has MBL of 137 tons. However, the SWL value of terminal mooring arrangements should be not less than MBL of the ship mooring line (OCIMF. 2008; 26).

3.2 *Preparation for Simulation Experiment*

Studies were prepared according to OCIMF (2008) Mooring Arrangements Guide 3rd Edition, SIGTTO "Prediction of Wind Loads on Large Liquefied Gas Carriers" and various other sources.

Study was begun with the preparation of model simulation of Ege Gaz Aliaga LNG Terminal and development of area simulation. Q-Flex Type LNG carrier's loaded and ballasted models were supplied from Transas Company.

Supplied model LNG 10, had been installed to bridge simulator. During the verification test on the LNG 10 model, mooring arrangement was found inconsistent with real Q-Flex Type LNG carrier.

In this context, simulation model's mooring arrangement was requested to be re-developed by Transas Company according to real Q-Flex LNG carrier's mooring arrangements blueprint. Re-developed and corrected LNG 10 model had been tested for re-validation.

In the process, simulation model's different characteristics were compared with the real ship and aerodynamic coefficients of LNG 10 Model have been found inappropriate. Simulation model's aerodynamic coefficients were requested to be re-developed by Transas Company.

Lastly re-developed and corrected model and area simulation were installed to bridge simulator to be used in project experiments. Also experience of RasGas Company and Aliaga LNG Terminal Management's experts are consulted to confirm reliability and validity of simulation models.

Then the Transas Bridge Simulator NTPRO 5000 series was used for extensive experiments which had been simulated by the use of "Hook" function. Every force applied to mooring hooks and dolphins are calculated and recorded.

3.3 *Reliability of Data collection method*

To analysis the data obtained from simulation experiment hook function, the mathematical model of the simulation is have to be examined according to the "mathematical models technical description" of Transas;

The ship itself is a controlled object. The steering gears are propellers, rudders, anchor and mooring systems etc. These mean the forces on ship hull. The value of forces directly depends on the control value changes.

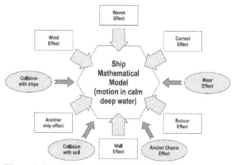

Figure 1. External Forces on the Ship Model

Though actual ship motion in real world can only be described with consideration of additional external forces. External forces deriving from wind, current, waves, channel geometry (the influence of water depth, walls, bottom inclination, etc.) and the presence of other objects (moving or immovable) as shown in Figure 1.

Thus, it is extremely important to consider the environmental conditions used for mooring model in the study.

For the environmental conditions used in simulation experiments, OCIMF's Mooring Equipment Guidelines Third Edition (2008) is taken as reference for Mooring Model Study in compliance with the local meteorological data gathered and verified from various sources.

Wind   : 30m/sec
Current   : 0,1m/sec – 004^0
Wave   : 0 meters

As can be seen from above values the most dominant and important environmental condition is determined as wind. Even though the prevailing wind in the region is from NNE – NE, according to multi-directional environmental forces mentioned in Mooring Equipment Guidelines, wind was taken from every angle in 10 degree steps for 180 degree.

Relative and true directions of the winds used in simulation experiments are shown in Figure 2. In relation to the wind velocity, 30m/sec is the highest possible wind velocity which can be applied in Bridge Simulator, Thus 60 knots of wind velocity couldn't be reached in simulation experiments and the highest value, 58,3 knots (30m/sec) was applied.

Figure 2. Wind Directions Used in Simulation Experiments

Simulation's mathematical wind model can be expressed as follows: The air flow around the ship is considered as uniform flow of constant direction and velocity. The true wind velocity at given Beaufort number is obtained as average value of wind velocity at 6 meters height above the sea level.

Formulas for calculating the aerodynamic forces and moments are as follows; aerodynamic longitudinal force (F_{xA}), aerodynamic lateral force (F_{yA}), aerodynamic vertical force (F_{zA}), vessel course (φ), relative wind direction (φ_w), and speed (V_w) as shown in Figure 3.

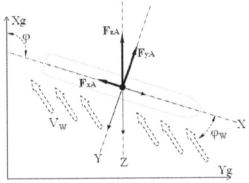

Figure 3. Wind Forces

Structural formulas for complete aerodynamic characteristics calculating are defined by functions represented by partial sums of Fourier series.

Coefficients are depended on the superstructure area and above-water hull area and are provided in the database of the software.

Another important factor for mooring pattern study is of course the mooring arrangement of ship and the terminal.

For the reliability and validity of the study, a real Q-Flex Type vessel, M/T Al Huwaila's data was taken as a reference in simulation experiments and data is listed below. They are used with same values in experiments.

Number of Mooring Ropes : 10 at bow 10 at aft total 20 pieces mooring ropes.
Type of Mooring Ropes   : Ultra High Molecular Weight Polyethylene
MBL of Mooring Ropes   : 137 tons (Max. BL)

Terminals mooring arrangements are also shown in the Figure 4 in relation to the Q-Flex type LNG carrier. Each mooring line fastened to a hook on the dolphins. Angles and distances of the mooring lines were automatically calculated by simulation software which accepted the mooring pattern of the vessel as generic mooring layout.

Figure 4. Mooring Model

Transas software's mooring model can be described as follows: The mooring gear consists of the mooring lines and docking winches. Mooring line diameter and type, winch type, choke positions are taken into account.

A mooring line is modeled as a weightless stretchable thread without considering its special configuration. The model describes the influence of mooring lines from the winch to the another ship's choke, bollard or mooring ring. Two winch operation conditions are examined: constant length conditions ("stop" conditions) and constant tension condition. The last condition supposed to have constant value of the force at ship's end of tow-line. The force is considered to be directed along the tow-line. Both conditions are used in this study.

3.4 *Mooring pattern simulation Experiments and Analysis*

It was determined that the ballasted ship with bigger wind effect area is always giving higher value results. Thus it was considered that the ballasted ship model was better suited to measure the required strength of mooring arrangements of the terminal.

Totally 20 ropes had been moored to hooks on dolphins for LNG model 10 and then 16 tons of force was applied to each one automatically. For start of LNG 10 model simulation experiment, equilibrium of forces on the hooks and equalization of rope lengths were waited.

Whenever an environmental condition was changed, to acquire the correct data, equilibrium of forces on the hooks should be waited. To reduce the waiting times "Fast Time Simulation" practice was used.

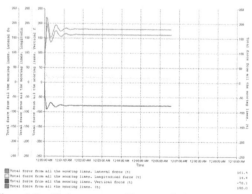

Figure 5. Equilibrium process of forces on the hooks

First installed environmental condition was current in the experiment. Current's effects on the hooks had been observed but found negligible for recording. Thereupon wind was installed to experiment from 000^0 relative direction with the velocity of 10m/sec. When forces had been equalized wind velocity was increased to 20m/sec. After the new equilibrium had been achieved, wind velocity was increased to upper limit of Bridge Simulator, to 30m/sec. Equilibrium of forces on hooks was waited when 30m/sec velocity was reached. Data recorded after forces were equalized. Equilibrium process of forces on the hooks is shown in Figure 5.

Afterward wind direction had been started to change by 10^0 steps in clock wise rotation. Each time oscillations had been waited to reach equilibrium and then data of longitudinal, lateral and vertical forces on the each hook was recorded.

During simulation experiments the force applied to hooks separately identified as 3 different components longitudinal, lateral and vertical. Diagram of these force components are shown in Figure 6. As can be seen in Figure 6 longitudinal forces' direction towards to the bow of ship was taken as (-) and direction towards to the stern of the ship was accepted as (+). Lateral forces' direction to the terminal was taken as (-) and direction away from terminal was accepted as (+). Vertical forces' upwards direction was taken as (-) and downwards direction was accepted as (+).

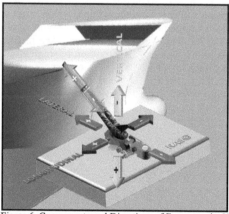

Figure 6. Components and Directions of Forces on the Hooks

In simulation experiments, each force component occurred on each hook from every wind direction was measured and recorded separately. Furthermore, obtained data is shown in graphics. An example graphic of the #1 hook is given in Figure 7.

3.5 *Mooring Model Simulation Experiment Results*

Detailed data of each longitudinal, lateral and vertical force component on every mooring hook is measured. Based on this data, detailed total longitudinal, lateral and vertical forces occurring on each dolphin were calculated.

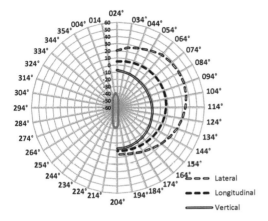

Figure 7. Force Components on the # 1 Hook

From this data because of environmental conditions, the maximum occurred longitudinal, lateral and vertical forces, on dolphins were calculated and shown in Table 1. As the lateral forces, the maximum force of 139,3 tons occurred on M7 dolphin. As the longitudinal forces, the maximum force of 109,1 tons occurred on B1 dolphin. As vertical forces, the maximum force of -52,5 tons occurred on M7 dolphin.

Table 1. Total Forces Occurred on Dolphins

Dolphin Number	Lateral (tons)	Longitudinal (tons)	Vertical (tons)	Total SWL (tons)
M1	125,8	-33,4	-33,4	3x150=450
M2	117,3	17,9	-34,4	3x150=450
M3	49,0	33,4	-16,8	2x150=300
B1	14,6	109,1	-34,3	2x150=300
B4	15,4	-100,1	-33,2	2x150=300
M6	85,1	-30,6	-36,7	2x150=300
M7	139,3	14,5	-52,5	3x150=450
M8	115,3	23,7	-38,4	3x150=450

The maximum calculated force values that caused by Q-Flex Ballasted LNG Carrier which is berthed at Ege Gaz Aliaga LNG Terminal and under environmental conditions of 30m/sec wind from various directions and with a stable current of 0.1m/sec are compared with the SWL value of terminal mooring arrangements.

4 CONCLUSIONS

According to the comparison between the existing hooks' SWLs and the forces which will be occurred on the hooks when Q-Flex vessel berthed, it is determined that mooring equipment are well sufficient for such forces.

This study revealed that Ege Gaz Aliaga LNG Terminal is well suited and a safe berth for Q-Flex Type LNG carriers. Upon the completion of the study Ege Gaz Aliaga LNG Terminal became the first safe berth for Q-Flex type vessels in the Mediterranean and the Black Sea. And finally all experiments were confirmed with real life experience when the first Q-Flex type LNG carrier berthed to the Ege Gaz Aliaga LNG Terminal (25.11.2011).

Also this study revealed that adequate bridge simulation systems can be used to evaluate the compatibility of the more static part of shipping bussiness like terminals to more dynamic part of shipping as ships reliably, efficiently and without taking any unnecessary risks in terms of mooring pattern and arrangement studies.

ACKNOWLEDGEMENT

Authors gratefully acknowledge to Mr. Ibrahim Akbal, General Director of Ege Gaz.

Authors also would like to give a special thank to Mr. Masum Guven, Manager of Ege Gaz Aliaga LNG Terminal and Capt. Khaled Djebbar, Marine Support & Projects Manager of Ras Gas.

REFERENCES

Er, İ. D. (2007) "Safety and Environmental Concern Analysis for LNG Carriers" *International Journal on Marine Navigation and Safety of Sea Transportation*. Vol 1 (4), pp. 421-426.

Hensen, H. (2003) Tug Use in Port A Practical Guide, 2nd Ed. The Nautical Institute. London.

Nas, S. (2008) "Enhancement of Safety Culture in Harbor Pilotage and Towage Organizations", IMLA 16th Conference on MET. Pp.385 - 292. Izmir /Turkey.

Nas, S., Y. Zorba and E. Uçan (2012) The Project of Ship Maneuver Risk Assessment Report for the Aliaga Petkim Container Terminal.

OCIMF (1997) "Prediction of Wind and Current Loads on VLCCs", Witherby & Co. Ltd.

OCIMF (2008) Mooring Equipment Guidelines 3rd Ed. Oil Companies International Marine Forum, Witherby Seamanship International.

SIGTTO (2007) Prediction of Wind Loads on Large Liquefied Gas Carriers, Hyperion Books.

Starosta, A. (2007) "Safety of Cargo Handling and Transport Liquefied Natural Gas by Sea. Dangerous Properties of LNG and Actual Situation of LNG Fleet" *International Journal on Marine Navigation and Safety of Sea Transportation*. Vol 1 (4), pp. 427-431.

Tension Technology International Ltd (2010) "Optimoor Seakeeping Analysis Q-Flex Class at Ege Gaz Aliaga Izmir for Ege Gaz".

Thoresen, C. A (2010) Port Designer's Handbook, 2nd Ed., Thomas Telford Ltd. London.

Zorba, Y. and S. Nas (2008) The Project Report of Determining Required Tug Bollard Pull for the Design Ship at Botas Terminal.

Transas (2003) Mathematical Models Technical Description

Gas and Oil Transportation
Maritime Transport & Shipping – Marine Navigation and Safety of Sea Transportation – Weintrit & Neumann (Eds)

Natural Gas as Alternative Fuel for Vessels Sailing in European Waters

J. Pawelski

Faculty of Navigation, Maritime University of Gdynia, Poland

ABSTRACT: Revised Annex IV to MARPOL 73/78 sets maximum sulphur content for marine fuels in order to cut on air pollution. Fuels compliant with limits applicable for time being in ECA (Emission Control Areas) are expensive and are seriously affecting profits of ship operators. After 1 January 2015 only fuel with sulphur content not higher than 0.10% will be allowed for vessels sailing within ECA. Such restriction will leave low sulphur diesel oil as only available but costly option unless another alternative fuel is found which can comply with future regulations.

1 INTRODUCTION

Shipping is being considered as major contributor to world-wide emission of SO_x (sulphur oxides) due to poor quality of cheap fuels used onboard vessels. Annual amount of shipping related to air pollution at Baltic Sea alone is given in figure 1.

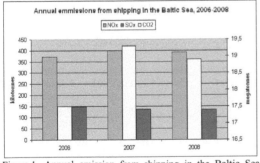

Figure 1. Annual emission from shipping in the Baltic Sea 2006-2008 [HELCOM, 2008].

Currently used marine fuels are 2700 dirtier than fuel used in road transport sector which adopted strict limits many years ago. Estimations based on "business-as-usual scenario" predicted that in year 2020 emission of SO_x from vessels would be higher than from all EU land-based sources, including power plants. IMO and EU are trying to curb this emission by imposing maximum sulphur content in marine fuels. Another significant air pollutant NO_x (nitrogen oxides) is currently not limited in European ECA but IMO aims to cut down on its emission by 80% on 1st January 2016. Lowering content of sulphur in fuel not only reduces SO_x emission but also linearly affects engine PM (particulate matter) emission [Kalli, Kravonen, Makkonen, 2009]. For time being heavy fuels with sulphur content less than 1% are made from crude oil, which naturally contains less sulphur or are blended from high sulphur and low sulphur fuels. Introduction in ECA fuels with sulphur content 0.1% probably will leave vessels with only low sulphur MGO (Marine Gas Oil) which is much more expensive. Increased demand for such fuel will cause sharp rise in its price. Due to manufacturing process is extremely difficult to remove sulphur from heavy fuel oils below certain level [Kalli, et al.,2009]. Other ways of limiting SO_x emission like: seawater and fresh water exhaust scrubbers could be used for time being and certainly in future. So far they are not too popular. In case of problem with scrubber and filtering system, pollution from air will be transferred to the sea. It is serious drawback in areas of slow water exchange as Baltic Sea. This method also requires collection onboard and disposal on shore of toxic sludge.

2 CURRENT LEGISLATION

IMO have set in MARPOL regulation 14 sulphur limits in fuels and time schedule of implementation of particular requirements. Details of it are given in figure 2.

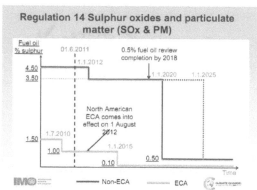

Figure 2. Sulphur limits and dates of introduction [Hughes, 2011].

Table 1. Typical raw natural gas composition [ABB, 2009].

Gas		Composition Range
Methane	CH_4	70-90%
Ethane	C_2H_6	
Propane	C_3H_8	0-20%
Butane	C_4H_{10}	
Pentane and higher hydrocarbons	C_5H_{12}	0-10%
Carbon dioxide	CO_2	0-8%
Oxygen	O_2	0-0.2%
Nitrogen	N_2	0-5%
Hydrogen sulfide, carbonyl sulfide	H_2S, COS	0-5%
Rare gases: Argon, Hellum, Neon, Xenon	A, He, Ne, Xe	trace

For North Atlantic and Baltic Sea ECA's were established for SOx and particulate matter emission control with boundaries given in MARPOL. Graphic representation of European ECA is shown in figure 3.

Gas processing removes NGL's (natural gas liquids) and acid gases like H_2S and CO_2 with purpose to meet certain commercial standards. For example, EASEE-gas (European Association for Streamlining of Energy Exchange-gas) CBP's (Common Business Practices) regarding commercial natural gas parameters are listed in table 2.

Table 2. EASEE-gas Gas quality CBP specifications [Cagnon & Sweitzer, 2011].

Parameter	Unit	Min	Max
Wobbe index[1]	MJ/m^3	46.44	54
Relative density		0.555	0.70
Total Sulphur[2]	mgS/m^3		30
H_2S + COS Sulphur	mgS/m^3		5
Mercaptan Sulphur	mgS/m^3		6
Oxygen	Mol%		0.001
CO_2	Mol%		2.5
Water dew point	^0C at 70 bar		-8
HC dew point	^0C from 1 to 70 bar		-2

Figure 3. Boundaries of North Atlantic and Baltic Sea ECA [ECG, 2011].

Listed above natural gas standard parameters are common in EU however some EU countries have their national standards which are slightly different.

3 NATURAL GAS AS FUEL

3.1 Natural gas basic properties and standards.

Typical composition of natural gas taken directly from well is given in table 1.

3.2 LNG fuel properties.

Average net calorific value or LHV (Lower Heating Value) of natural gas at standard conditions is 35.22 MJ/m^3 and is similar to diesel oil LHV equal 35.94 MJ/L. [Staffel 2011]. Heating value of gas depends on source of origin and method of transportation. Generally LNG (liquefied natural gas) is better quality than pipeline gas due to liquefaction process resulting in cleaner gas, with higher methane percentage. Taking into account natural gas average density 0.768 kg/m^3 [Staffel, 2011] and maximum allowed total sulphur content 30 mg/m3, the highest sulphur content by weight is 39 mg/kg, which is well below future MARPOL limits for fuel allowed to use in ECA. Burning natural gas instead of diesel oil allows reduction of CO_2 emission by 20-30%, thus reducing total emission of GHG (greenhouse gas) if natural gas upstream losses are kept to the minimum. Gaseous fuel also solves problem of PM emission which in this case is very low. Tests carried out on existing vessels have shown NO_x emission reduction from 89% to 75% for engines running on gas, which well corresponds with future NO_x limits [Osberg, 2008]. For time being price of natural gas is much lower than price of diesel oil. Running vessels on gas may bring considerable economic savings despite initial investments in purchasing new types of engines or conversion of existing equipment. It is difficult for now to forecast price of natural gas in future but each year more gas becomes available on international market and price of it remains fairly stable.

3.3 Transportation and storage of natural gas.

Most common and basic method of natural gas transportation is pipeline where gas is pumped under pressure 80-150 Bar. Pipelines are expensive to build and for long distances is cheaper to transport liquefied gas by LNG tankers at temperature -163°C. Pipeline gas is stored in spacious underground caverns or in case of large pipeline networks by packing lines (increasing pressure inside network pipelines). Smaller amounts of gas can be stored as CNG (compressed natural gas) inside tanks at pressure 200-220 Bar. LNG is stored ashore in cryogenic tanks before re-gassing and delivering to client. The last method of gas storage offers best volume ratio where 600 m^3 of gas takes only 1 m^3 as LNG. For this very reason LNG is being taken under consideration as the most viable solution for accommodating on board of vessel reasonable amount of fuel.

4 LNG STORAGE ON BOARD

Most common tanks used today onboard for storage LNG as fuel are cylindrical tanks IMO type C offering high degree of safety due to their high pressure design. They can withstand high loading rates and pressure increase due to LNG boil-off. Tanks are installed either on main deck or well hidden inside hull. On deck installation prevents accumulation of gas and keeps tank away from danger in case of grounding or minor collision. Picture 1 shows vessel with fuel tanks on deck.

Photo 1. Bit Viking oil product tanker Wartsila dual fuel engine [Einang, 2011].

Another typical fuel tank installation is gas tank deep inside hull, where is well protected against grounding or collision by leaving lateral space B/5 to ship's side and vertical distance B/15 to ship's bottom (B – hull width) as shown on figure 4.

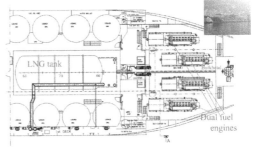

Figure 4. LNG tank inside hull of supply vessel [Osberg, 2008].

Despite their very good safety record cylindrical tanks type C have one drawback. Large size tanks of such type are difficult to fit inside hull or even on deck due to their shape. To increase fuel storage capacity tanks with better fit to vessel hull are needed. DNV (Det Norske Veritas) is now working on approval of LNG tank IMO type A for gas fuel storage. Such tank is presented on figure 5.

Figure 5. LNG fuel tank type A under approval [Graugaard 2010].

This type of tank offers increased LNG fuel storage capacity.

5 NATURAL GAS SAFETY ISSUES

5.1 LNG safety related problems

Introducing natural gas into engine room increases risk of serious accident. Natural gas ignited in open space burns without explosion. Situation is dramatically changing when gas is released in confined space and resulting mixture with air is within 5-15% of gas per weight, which is explosive range for methane. Ignition of such atmosphere in closed compartment has catastrophic consequences. Gas escaping in liquid form has boiling temperature -162°C. It is dangerous not only for personnel but also for ship's hull. Steel used for hull construction is not designed to withstand such extremely low temperature and becomes brittle and liable to cracking. Methane itself is not considered a toxic gas but may displace oxygen and lead to asphyxia of crew in affected compartment.

5.2 Mitigation of hazards.

First method of safety improvement while dealing with gas is creation of double barriers for fuel system. In practice it means that fuel mains are double walled with space between walls filled with inert gas under pressure or purged continuously with dry, clean air and monitored by gas sensors. It works well on larger diameter piping but is not feasible for various engine fuel system components which have to remain single walled. Second method of protection is fitting engine room with ESD system (emergency shut down) activated by gas detecting sensors. ESD should shut off fuel supply and switch off all possible ignition sources. Engine room ventilation system needs to be designed to provide 30 charges of air per hour to prevent accumulation of gas or to remove gas in short time. Sufficient drip trays below LNG tanks are required to contain possible leaks of liquid gas and to prevent contact of extremely cold fluid with a hull. Bunkering operations involve handling of liquid with extremely low temperature, which in case of loss of containment, is harmful for crew carrying out operation. After evaporation LNG poses risk of fire. Personnel involved in bunkering operation should have proper training and detailed procedures to follow.

6 GAS ENGINES

Gas driven engines have been used in offshore petroleum industry for a long time due to abundance of very cheap fuel. It eliminates necessity to provide large quantities of expensive diesel oil as well as fuel oil. Offshore petroleum fields use either gas turbines when amount of used fuel is not a problem or internal combustion engines (Otto cycle) with much better fuel efficiency if availability of gas is limited. This latter solution has been widely adopted for vessels propulsion to ensure longer cruising range for given amount of fuel.

Leading engine manufacturers like: MAN, Rolls-Royce,Wartsila, Mitsubishi are providing now shipping industry with LNG fuelled engines generally divided in two types. First of them are mono-fuel, lean LNG burning engines. Second type are more flexible dual-fuel (LNG+Diesel) engines. Existing diesel engines can be rebuilt for dual-fuel. Such refurbishment is available for both low-speed and medium-speed Diesel engines [Eingang, 2011].

7 POWER GENERATION WITHOUT INTERNAL COMBUSTION ENGINES

Adopting natural gas as fuel on board vessels opens new way to electric power generation. Fuel cells can use natural gas for power generation and such attempts have been made on board of real vessel as on figure 6 below:

"A Battery – without need for recharging"

**Silent
Clean
Effective**

- 340 kW fuel cell was delivered to OSV Viking Lady March 2009.
- Replaces one Auxiliary engine

Figure 6. A 340 kW fuel cell replacing auxiliary engine onboard vessel [Graugaard, 2010].

Elimination of internal combustion engine from power generation cycle removes mechanical energy losses and provides cleaner fuel–energy conversion. Technology of fuel cells is still in stage of development and each year progress is made to improve their efficiency and elimination of expensive components like platinum.

8 LNG AVAILABILITY

European waters are well covered by LNG terminals for importing of gas with new buildings planned in near future. Such terminals may offer bunkering points for vessels when such need arises.

Figure 7. European LNG terminals [Graugaard, 2010].

Figure 7 depicts location of European LNG terminals, including import and export facilities as well as small production and storage terminals with vessel refueling capabilities.

For time being there are only few small terminals with vessel refueling capabilities, located in Norway and Sweden where is demand for such fuel. Introducing LNG as alternative fuel for shipping in European water may change this situation very quickly as small refueling terminals are not too expensive to build.

9 CONCLUSIONS

Impending changes regarding maximum sulphur content in fuel for vessels trading within European ECA may prove difficult to comply with for many shipping companies. Low sulphur diesel oil is an expensive option and its price will go up with increased demand. Adopting natural gas as fuel can solve problem of high fuel price and helps to reduce significantly emission of such air pollutans like: SO_x, NO_x, CO_2 and PM. LNG is considered as well suited for marine applications due to its good volume ratio 1/600 ($1m^3$ LNG/600m^3 of natural gas). Storage on board of LNG fuel is based on well proven technologies developed for LNG carriers. Existing marine engines can be refurbished to use LNG as a fuel and dual fuel (LNG + Diesel) engines are available. Introduction of natural gas into engine room requires secondary barriers for fuel mains, gas detection system with ESD capability and augmented engine room ventilation. Fuel cells fed by natural gas can provide with electric power and can replace power generators driven by internal combustion engines.

REFERENCES

ABB 'Oil and Gas Production Handbook', 2nd ed. Oslo 2009

Cagnon F., Schweitzer J.'GASQUAL project: a step closer to gas quality harmonization in Europe'- International Gas Union Research Conference 2011 paper.

ECG -Briefing Report November 2011.

Einang Per Magne 'LNG Fuelling the Future Ships', MARINTEK, Shanghai LNG Seminar, November 2011.

Graugaard Claus Winter 'LNG as Fuel for Ship Propulsion' Det Norske Veritas, November 2010.

Hughes E. Air Pollution and Climate Change, Marine Environment Division, IMO, June 2011.

Kalli J., Karvonen T., Makkonen T. Sulphur Contents In Ship's Bunker in 2015', Ministry of Transport, Finland, Helsinki 2009.

Osberg Torril Grimstad 'Gas Engine Propulsion in Ships', Det Norske Veritas, April 2008.

Staffel Ian 'The Energy and Fuel Data Sheet'-University of Birmingham, UK, March 2011.

www.helcom.fi/shipping/emissions/en_GB/emisions/

Chapter 3

Sea Port and Harbours Development

The Future of Santos Harbour (Brazil) Outer Access Channel

P. Alfredini & E. Arasaki
Polytechnic School of Sao Paulo University, Sao Paulo, Sao Paulo State, Brazil

A.S. Moreira
Companhia Docas do Estado de Sao Paulo, Santos, Sao Paulo State, Brazil

C.P. Fournier
Baird & Associates Coastal Engineers Ltd., Santiago, Chile

P.S.M. Barbosa
Serviços de Praticagem do Estado de S. Paulo, Santos, Sao Paulo, Brazil

W.C. Sousa Jr.
Aeronautic Technologic Institute, Sao José dos Campos, Sao Paulo State, Brazil

ABSTRACT: Santos Harbour Area (SHA) is the most important marine cargo transfer terminal in the Southern Hemisphere. A long term relative tidal level variability assessment shows a consistent response to relative sea level rise. A wave data base from 1980 to 2012 was validated with wave buoy records and propagated in shallow waters offshore Santos Bay. The assessment of the hydrodynamics extreme conditions and of the sea level rise are presented from the Harbour Pilot's point of view. The current bed level of SHA Outer Channel is − 15.00 m (CD) being considered officially as a two way channel for container vessels, maintained by dredging. According to the cargo throughput forecast, in 2025, the Access Channel will have to be deepened to level − 17.00 m (CD). Hence, the feasibility of that choice is discussed from a technical, economical and conceptual navigation point of view, including the impacts related to the maritime climate changes in course.

1 INTRODUCTION

Santos Harbour Area (SHA) (Fig.1) throughputs approximately 15% of Brazilian maritime exports of more than 800 million tons per year and around 35% of Brazilian Gross National Product of US$ 2.088 trillions (2011 data). These figures put SHA as the fourth largest harbour facility in the Americas and the first in the Southern Hemisphere. In the last decade important oil and gas reserves in the deep Offshore Basin of Santos began to be developed. Santos metropolitan urban area and shoreline is one of the most frequently used, and Santos also has significant tourism from cruise lines. For the economical growth scenario it is essential to have an understanding of the main maritime hydrodynamics forcing processes in tidal levels, tidal currents and waves, induced by sea extreme events and associated hazards influencing the risks upon vessel operations. For official purposes, the current bed level of SHA Outer Channel is − 15.00 m (CD) being considered officially as a two way channel of 220 m in width for container vessels of 9,000 TEUs, which is maintained by dredging. According to the logistic forecast, in 2025 the Access Channel will have to be deepened to − 17.00 m (CD).

2 MATERIAL AND METHODS

Data sets for this assessment were obtained from a physical model of Santos Bay, Estuary and nearby beaches showing the impact of maritime climate changes (Alfredini et al. 2008), wave climate and tidal forcing (Arasaki et al. 2011, Alfredini et al. 2012, Fournier 2012 and Alfredini et al. 2013),

The long term tidal level data variability assessment of the Santos Dock Company tidal gauge, which measured water level fluctuations from 1944 using the same Vertical Datum, provided the possibility to have at least 3 lunar declination periods of 18.61 years each one. The forecasting trends of HHW and LLW depend largely upon metheorological forcing, beyond sea level rise. The use of 19 years mobile average fittings, from 1970 to 2007 and from 1989 to 2007, are consistent showing impressive increasing gradients of relative sea level rise, with century gradient rates shown in Table 1.

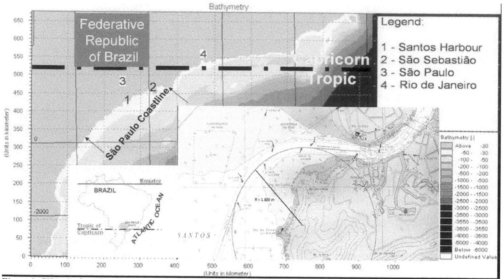

Figure 1. SHA site location

Table 1. 19 years mobile average rates cm/century

Tidal Levels	1970-2007	1989-2007
Highest High Water	50.0	63.2
Mean Sea Level	27.6	57.9
Lowest Low Water	47.4	94.7

Based upon water current surveys it was possible to calibrate and validate the mathematical model of the tidal current charts of Santos Bay using the Mike 21 software (Baptistelli et al. 2008). Figures 2 and 3 show charts for a spring tide in the maximum flood and ebb conditions respectively.

A long term (Jan 1st 1980 – August 6th 2012) wave climate data base (www.OndasdoBrasil.com) was employed to develop an assessment of the characteristics and historical frequency of extreme storm events. The database include definitions of significant wave height (H_{m0}), spectral peak wave period (T_p) and spectral directions as required. The deepwater hindcast was developed with the aid of the WaveWatch III third generate wave model, and was calibrated with Topex Satellite data along Brazilian coastline, and subsequently validated with a directional sea buoy. The deepwater 2D spectra were subsequently transferred to shallow water, spectrally in two dimensions (frequency, direction) to the Santos Bar Channel located at – 10 m (CD). Table 2 classifies Santos wave climate according to PIANC (1997) criteria for Outer Channels nautical projects (L: wave length; L_{pp}: vessel length between perpendiculars; B: vessel beam).

Table 2. SHA additional access Channel width according to long term wave climate (1980 – 2012)

H_{m0} (m) and L:L_{pp}	% per year	Additional Width (xB)
$H_{m0} \leq 1.0$ and $L < L_{pp}$	74.16	0
$1.0 < H_{m0} \leq 3.0$ and $L \sim L_{pp}$	24.62*	1.0
$H_{m0} > 3.0$ and $L > L_{pp}$	1.22**	2.2

Occurrences: * From March to October. ** From April to September.

Seven recent extreme events (2005–2011) occurred in Santos Bay were selected to compare with tidal and wave predictions in terms of expected severity. Figure 4 shows one of the extreme wave climate events from April 26th 2005 (Fournier 2012), which closed Santos Harbour Bar one day (resulting in approximately US$ 1 billion of lost business). Those storms have been more frequent in recent years, perhaps due to more severe wave climate conditions (Arasaki et al. 2011, Alfredini et al. 2012 and Alfredini et al. 2013).

According to Barbosa (2012), in 2011 the Pilot's Association made 13,486 ship manoeuvres through SHA Outer Access Channel, reaching 3 million TEUs and 98 million t of cargo.

Alfredini et al. 2013 presented the frame of SHA tidal and wave changes impacts on maritime structures and possible mitigation works (Fig. 5)

3 PILOT'S ASSOCIATION ASSESSMENT

The SHA Pilot's Association (Barbosa 2012) has made calculations with a Panamax Container Ship (Displacement: 70,055 DWT 5000 TEUs; L_{pp}: 275 m, B: 32.18 m, T: 13.00) for the second class of

Table 2. At 6.5 knots, recommended vessel velocity at the Access Channel, the squat, mean draught increase, in confined shallow waters like those (depth lesser than 1.2 times the vessel draught) is 0.60 m. The draught increase due to a minimum heel of 1°, must also be included, corresponding to 0.37 m, and the pitch of 0,5 m due to a 1 m wave height. Considering an underkeel clearance of 0.30 m with the sand/mud soft bottom, the overall depth necessary for a safe navigation (ODSN) at the tidal level of 0 (CD) is:

ODSN (0) = 13.00+0.60+0.37+0.5+0.3 = 14.77m

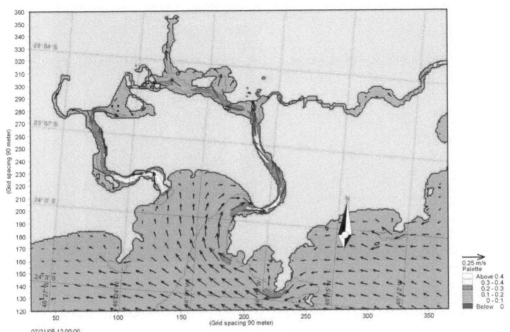

07/21/05 12:00:00

Figure 2. Tidal currents chart in the maximum spring tide flood current in July 21st 2005.

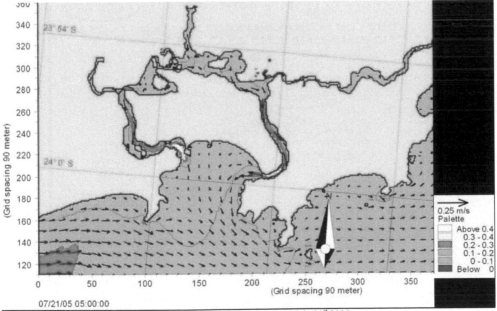

07/21/05 05:00:00

Figure 3. Tidal currents chart in the maximum spring tide ebb current in July 21st 2005.

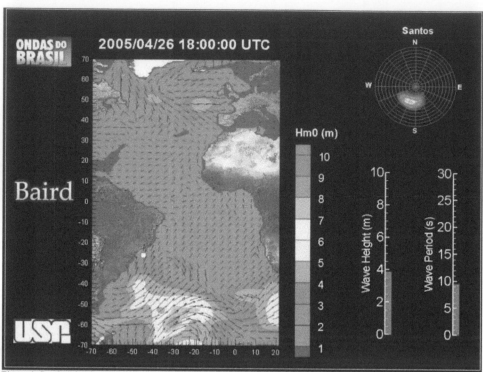

Figure 4. South Atlantic extreme wave climate in the storm peak of April 26th 2005.

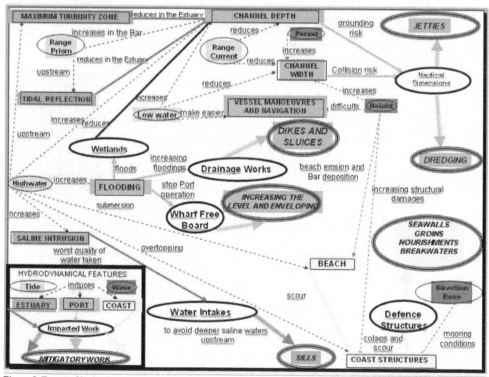

Figure 5. Frame of hydrodynamics for impacts on coastal and estuarines port areas and mitigatory measures and structures.

Considering 2 m wave height and a heel of 2°, ODSN (0) = 15.65 m.

Hence, such a vessel would have time limiting operation due to restrictive depth conditions at tidal level 0 (CD) in the current channel – 15.00 m (CD) dredged depth, mainly in the months from March to October, and in some cases from April to September only would be possible to cross the SHA Outer Channel using high tides.

Considering the aim to receive the Post Panamax Plus (5th generation containers vessels) of 9,000 TEUs; L_{pp}: 335 m, B: 46.00 m, T: 15.00 m, the ODSN would be, in the same previous order: 16.80 m and 17.69 m, considering the same squat. Hence, it is impossible to give warranty to safe navigation except during calm wave conditions. Those vessels will have to wait the next channel deepening scenario to – 17.0 m (CD) and, also in that condition, some operations would have to be cancelled, mainly during April to September period.

The other issue is about the two way channel width of 220 m, proper for Panamax vessels, but would have to be enlarged at least to 315 m (equal 6.8 times the beam) to accommodate the Post Panamax Plus vessels (Fig. 6), and an adequate enlargement in the curve.

Figure 6. Bridge view of a Panamax container vessel that will cross another in the estuarine mouth of SHA.

The last issue to be considered relates to the channel's radius of curvature presently set at 1800 m (Fig. 1), which is small for the Panamax vessel, and too small for the Post Panamax Plus without the aid of tug boats as is currently the practice today, and would have to be redesigned to at least 3300 m, unless the tug use will begin in the Outer Access Channel. That curve is the most accreted area of the Outer Access Channel due to the longshore sand transport along Santos beaches.

4 CONCLUSION ABOUT THE HARBOUR ENGINEERING SOLUTIONS

Considering the necessity to enlarge the geometric dimensions (depth, width and radius) of SHA Outer Channel for the purpose of navigating a two way

fairway with Post Panamax Plus vessels of 9,000 TEUs, two engineering solutions are possible, necessitating an understanding of the coastal engineering issues (waves and sediment transport) and an awareness of the maritime climate change.

The first one, is the dredging maintenance procedure. Using historical data, from 1963, when the Outer Channel SHA dredging was initiated, to 2012 60 million m^3 were dredged to maintain an average bed level – 13.00 m (CD), at an estimated cost of US$ 0.5 billion (present cost). To maintain 30 more years of the existing navigation channel of – 15.00 m (CD), without considering the additional accretion impacts shown in Fig. 5, another US$ 0.5 billion of OPEX maintenance dredging will be required.

Alternatively, a rubble mound structure, such as two jetties with a total length of 9 km (Fig. 7) will provide a solution, significantly reducing the OPEX maintenance dredging requirements, and redesigning the adequate channel dimensions suitable for the larger vessels. The cost estimate of this alternative is approximately US$ 0,2 billion, using quarries accessible by railway and waterway, hence reducing the cost of transport.

With the jetties the downtime due to lack of nautical safety dimensions in the channel would be reduced dramatically by the improvement of their alignment, creating a sheltered navigation channel.

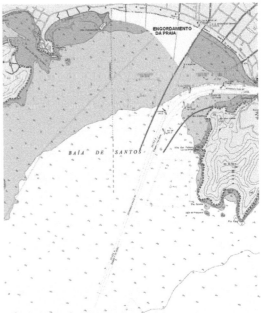

Figure 7. Schematic location of the proposed jetties and probable beach nourishment.

The jetties will provide a sheltered access channel in the critical bar area, retaining the longshore sand transport and naturally promoting the beach nourishment (Fig. 7) in the frequently used beaches that front the city of Santos. That process will enhance the coastal defences against sea level rise flooding due to storm surges.

The cost–benefit of the jetties solution is clearly demonstrated by comparing their cost with the maintenance dredging in a deeper bed level as – 17.00 m (CD) and the losses of US$ 1 billion like occurred in the storm of April 26th 2005.

ACKNOWLEDGEMENTS

This paper has the financial support of CAPES, Human Resources Improvement Agency of Brazilian Government. The authors also want to thank the support of São Paulo University, CODESP, Baird & Associates Coastal Engineers Ltd., Serviços de Praticagem do Estado de S. Paulo and Instituto Mauá de Tecnologia.

REFERENCES

Alfredini, P., Arasaki, E. & Amaral, R. F. 2008. Mean sea level rise impacts on Santos Bay, Southeastern Brazil – Physical modelling study. *Environmental Monitoring and Assessment* vol.144: 377-387. Amsterdam: Springer.

Alfredini, P., Pezzoli, A., Cristofori, E. I., Dovetta, A. & Arasaki, E. 2012. Wave and tidal level analysis, maritime climate changes, navigation's strategy and impact on the coastal defences – Study case of São Paulo State coastline harbour areas. *GeophysicalResearch Abstracts*: vol. 14, EGU2012-10735. Viena.

Alfredini, P., Arasaki, E., Pezzoli, M. & Sousa Jr., W. C. 2013. Impact of maritime climate changes on harbours, land areas and wetlands of São Paulo State coastline (Brazil). *6th International Perspective on Water Resources & the Environment*. Izmir: Environmental & Water Resources Institute of the American Society of Civil Engineers.

Arasaki, E., Alfredini, P., Pezzoli, A. & Rosso, M. 2011. Coastal area prone to extreme flood and erosion events induced by climate changes. In A. Weintrit & T. Neumann (Org.), *Human Researches and CrewManagement* vol. 1(27). London: CRC Press/Balkema.

Baptistelli, S. C., Harari, J. & Alfredini, P. (2008). Using "Mike 21" and "POM" for numerical simulations of the hydrodynamics in Baixada Santista region (São Paulo, Brazil). *Afro-America Global Sea Level Observing System News*: 12 (1).

Barbosa, P. S. M. 2012. Canal do Porto de Santos e segurança da navegação. *Workshop Molhes Guias-Correntes na Barra de Santos – O futuro da acessibilidade marítima ao Porto*. Santos: Associação de Engenheiros e Arquitetos de Santos.

Fournier, C. P. 2012. Ondas Santos. *Workshop Molhes Guias-Correntes na Barra de Santos – O futuro da acessibilidade marítima ao Porto*. Santos: Associação de Engenheiros e Arquitetos de Santos.

PIANC - Permanent International Association of Navigation Congresses, IMPA – International Maritime Pilots Association & IALA - International Association of Lighthouse Authorities 1997. *Approach channels - A guide for design*. Tokyo: Joint PIANC Working Group II-30 in cooperation with IMPA and IALA.

TRB 91st Annual Meeting 2012. The U. S. Marine transportation system – responses to climate changes and variability.

Port Safety; Requirements & Economic Outcomes

M.A. Hassanzadeh
Ports & Maritime Organization (PMO), Tehran, Iran

ABSTRACT: Because of elimination or reduction of tariff barriers, today international trade is more dependent on non-tariff barriers such as transport cost. Taking advantage of economies of scale in shipping, maritime transport handles 8.7 billion tons of cargoes in 2011 and is now considered one of the fundamental pillars of global trade. In such business environment, ports as landside part of maritime transport play a central role in reducing the total transport cost. Providing loading and unloading services to thousands of ships that carry such enormous volume make ports sensitive to any cost rising factors such as accidents. While the accident costs varies from direct, indirect and economic opportunity loss costs, usually such massive economic burdens are not entirely taken into consideration.

Despite all economic burdens of accidents, driving forces for port authorities and service providers to increase workplace safety and prevent accidents were and in some countries are still legal obligations of unions' requirements. Given the total costs of accidents and their impacts on viability of a transport route from one hand and competitiveness of transport services on the other hand, reducing frequency and severity of accidents will save millions of dollars. That is why nowadays safer ports can echo their ports' safety level as a marketing tool to attract shipping lines, global terminal operator[7], investors and transport players. This paper examines the port accidents & their associated costs; identifies the impact of such costs as an externality on trade; explores the market requirements for higher safety level; and finally analyzes the expected change in willingness to pay for safety by port authorities, terminal operators and other service providers at ports.

1 INTRODUCTION

The associated cost of trade is significant and consequently increasing attention has been given to the cost of trade in both the empirical trade literature and in theoretical models of international trade. International organizations such as WTO[8] and WCO[9] are doing their best to harmonize and facilitate trade and governments make every effort to eliminate or trim down artificial barriers to trade. In a world where artificial barriers to trade are becoming less significant, the focus is increasingly being directed towards the role of non-policy barriers such as transport costs and their effect on international trade flows (Jane Korinek, 2009). Since transport cost have become the main natural barrier to trade (Shou, 2003), different stakeholders especially in the maritime industry, as the dominant mode handling international trade, attempt to reduce maritime transport cost. Basically, shipping lines operate gigantic ships to reduce unit transport cost through economies of scale. According to Merk and Dang (2012) ports have to be efficient to handle containers and goods more rapidly, provide more adequate and performing equipment, reduce berth times and delays, enable large storage capacity and ensure multi-modal connections to hinterland. Ports use broad range of strategies to increase efficiency and reduce service costs as a market demand. Several practices that are being implemented in order to achieve such goals include: encouraging public private partnership to manage and operate port terminals, establishing port community systems (PCSs) with integrated and IT based networks, taking advantage of state-of-the-art equipment, and improving customer relationship management (CRM).

As it can be seen in the Figure 1, according to UNCTAD, all the players' efforts have resulted in a reduction of the transport cost as the most significant

---

[7]GTO
[8] World Trade Organization
[9] World Customs Organization

natural barrier to the trade from 8.84 to 7.60 percent of the import price of commodities during the last 25 years (UNCTAD, 2012).

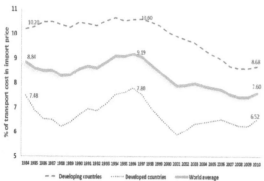

Figure 1. Reduction in percentage of transport cost in import price of commodities for developing, developed countries and world from 1948 to 2010.
Source: (UNCTAD, 2012)

As Merk and Dang (2012) mentioned with growing international sea traffic and changing technology in the maritime transport industry, seaports being forced to improve port efficiency in order to gain comparative advantages that will attract more traffic. Some of the key challenges ports are surmounting to include secure traffic flows and prevent diversion to nearby ports (Merk, 2012). But port accidents challenge the port efficiency and imply various costs on transport chain actors. Lindberg (1999) states that accident costs are one of the most high profile and the highest costs of transport (Lindberg, 1999). As Tongzon (2009) indicated if a port has a reputation that its handling of cargoes is unsafe, this could drive away potential clients and discourage existing clients.

Now the main question is when ports are primarily focusing on infrastructures (berths & channels, transport links), superstructures (equipment & facilities) and infostructures (PCS & CRM) and how investment in port safety for them and port service providers can be justified. The scope of this paper is to point out that port accidents will reduce port efficiency and its position among competitors and show that investing in safety makes good business sense. De Greef, Van den Broek, Van Der Heyden, Kuhl and Schmitz-Felten (2011) mentioned linking occupational safety and health to an economic perspective should therefore be appealing for port management. (European Commission, 2011).

Throughout this paper, in addition to the role of ports in the transport chain, business & economic impacts of port accidents and their costs, as well as new trend emerging in choosing safer ports by shipping lines and global terminal operators will be discussed.

2 PORTS IN THE TRANSPORT CHAIN

The relevance of seaports in the efficient working of an economy cannot be understated since all goods and passengers transported by sea require the use of, at least, two ports. In the majority of countries, most international trade (export/import), and in some cases also large shares of domestic trade, is done through maritime transport (Lourdes Trujillo, 1999). As of 2001, 328 million people passed through EU seaports and the total tonnage of goods handled in the EU was estimated at 3000 million tones. There were 261 maritime ports handling over 1 million tonnes of goods per year. 70% of all trade with third countries was channeled through the ports (Ronza, 2007). To remain competitive in the market and to keep up with trading pace, ports have to be efficient, reliable, secure, and safe in providing services to their customers. Port costs, for example, represent about 8–12 percent of total transport costs from origin to destination. Shippers, who consider port costs as one of the very few, if not the only, controllable costs in the logistics chain, make shipping decisions in part based on those costs (Paul E. Kent, 2004).

3 PORT ACCIDENTS

Rise in idle time, waiting time and service time to vessels and trucks, reduction in ship output per hour and rising damage to cargo are a few out of the widespread consequences of port accidents. Less reliability, lower flexibility and inefficiency are what port users customarily experience from ports' low safety level. High-speed operation in combination with a mixture of equipment, heavy truck and vessel traffic, and dissimilar types of cargo especially dangerous ones are considered safety risk factors causing accidents. The most common origin for a port accident is in the transport of cargo (56.5%), followed by loading/unloading operations (14.9%). The majority (65%) of transport-origin accidents involve ocean-going vessels, followed by pipeline accidents (12%) (Darbra, Ronza, Carol, Vílchez, & Casal, 2005). Such port accidents can be reported in several forms for example per number of port employees and/or tonnage or teu handled in port.

In a survey undertaken by Department for Transport in the UK, it was estimated that 54,000 employees in terms of full time equivalents were working in 2005 on the port estate in directly port related jobs. The accident rate for direct businesses on port was estimated to be 1.2 per 100 employees on average, annually (1.2 per cent) (DfT, 2005). As an example, if accident frequency and its severity reflect safety performance, the number of fatalities per container throughput at the three following container ports in the same period can be taken as an indication of ports' safety level: Shahid Rajae Port (SRP) in Iran, Port Botany in Australia and the Port of Rotterdam in the Netherlands. Accident data for SRP confirm two fatalities whereas the total container throughput has been 2.64 million TEU in 2011. Handling 2.62 million TEU at Port Botany in 2011 resulted in 2 fatalities as well (Jones, 2011). At the same time, during port operations in Rotterdam Port, one person died and another went missing upon handling 11.9 million TEU(The Marinewaves, 2011). For the two former ports, handling an average of 1.3 million TEU implies one fatality while for the latter one it is on 5.9 million TEUs. No matter how to report accidents, they are one of the prominent factors that lead to port inefficiency and a surge in total transport cost. Weaknesses in trade-related transport and logistics, particularly in developing countries, impose costs on producers that erode the intended benefits of trade preferences in major markets, such as the United States and the European Union. In fact, the cost of such weaknesses may be many times greater than the benefit of trade preferences. Paul E. Kent and Alan Fox (2004) described port inefficiencies are among weaknesses that need to be addressed because they may be the most serious and least understood.

4 COSTS OF PORT ACCIDENTS

When it comes to calculating the costs, costs of accidents have to be categorized into three main categories; direct, indirect & opportunity costs or economic loss costs. Direct costs are visible costs incurred immediately following an accident or incident. Damage to ship, damage to or loss of cargo, damage to port equipment and their repair costs, workers' injury and treatment cost are considered part of the direct cost of an accident. Sometimes the direct cost of an accident contributes only 10- 20 % to the total accident cost. But the main part of the accident cost is an indirect one that cannot be determined immediately and need vigilant investigation. Time needed to investigate port or port people after an accident, money needed to train replacement stevedore, crane operator or technician, higher premium that needs to be paid to the insurance company for higher risk at workplace, idle times for port equipment and demurrage to shipping lines for delay in providing services are some of the indirect costs of a port accident.

Estimation of the exact ratio of direct to indirect costs depends on factors such as the nature of business and organizational culture. The results of an investigation by Montreal University (Bernard Brody & Poirier, 1990) indicate that on average, indirect costs of accidents are 83% of the direct costs ($1,156 vs. $1,391). This figure is in line with

other agencies, including the US Occupational Safety and Health Administration, who indicate that indirect costs associated with a workplace injury can be as much as three to five times the direct costs (Michael, 2001). According to Western National Insurance most experts estimate that the indirect costs are 3 to 10 times the direct costs of an accident and an employer's insurance does not pay these costs (Loss Control Consultants, 2009), while The UK HSE found that typically accidents cost companies up to 37 times the directly insured losses (Purdy, 1997).

Alongside this ratio, the opportunity cost and social costs are non-economic or unaccountable. Dorman (1996) describes opportunity costs as those costs which are neglected because either there is no clarification on definition or no formal assessment method to include those hidden costs into direct and indirect cost. The social cost includes the emotional cost to the victim's family and community, and damage to social values like justice and solidarity. Attempts have been made to place monetary values on some of these but at the end no number can capture the losses which money cannot recompense. Costs to the end users that are getting service from a specific port have never been calculated during an investigation. Other opportunity costs may include loss of goodwill and port reputation, loss of a loyal customer, either shipping line or FFs, failure to attract economies of scale advantages, delays in delivery of service and commitment & lower return on investment due to losing ground to competitors.

Direct, indirect and opportunity costs may be difficult to measure, but they are straight results of accidents and are considered external costs. In transport infrastructure, external costs are those imposed by the users of the infrastructure on the others. These costs may take the form of congestion, accidents and environmental costs. This cost can come back in terms of more elevated rates of freight, congestion of the traffic associated with the handling operations, a decrease in the level of security and a loss in terms of competitiveness in the whole region (Abbes, 2007). As an externality, rise in costs of port services lead to various outcomes such as a higher commodity price for the end users that is inevitable. Reduction of market shares for transit and transshipment business that regional ports are competing for and an increase in the entire transport cost that will play as barrier to trade are amid these consequences.

5 PORT INEFFICIENCY & TRADE

Various researchers investigated how transport inefficiencies may affect trade. Among them Henderson, Shalizi, and Venables (2001) explore how transport costs influence trade (María del Pilar

Londoño-Kent, 2003). Hoffmann and Kumar also find a relationship between trade and maritime transport (Hoffmann, 2002). Other studies show the precise impact of high transport costs. Limao and Venables for example, show that increasing transport costs by 10 percent can reduce trade volumes by 20 percent (Limao, 2001). High transport costs, however, can be explained by many other factors, such as lower cargo volumes, trade flow imbalances, inadequate infrastructure, onerous border and cargo processing procedures, and port inefficiencies (Paul E. Kent, 2004). That is why more and more studies of the impact of transportation on trade and welfare are focusing on ports and border crossings. Clark, Dollar, and Micco (2002) declare that port efficiency can affect transportation costs and that an inefficient port can increase the distance to a shipper's export market by 60 percent (Ximena Clark, 2002). In a study of Asia Pacific Economic Cooperation countries, Wilson, Mann, and Otsuki (2003) explored the importance of port efficiency relative to other factors that enhance or constrain trade, such as customs performance, the regulatory environment, and e-business. Calculating indicators for each factor, WMO (2003) finds that improvement in port efficiencies yields the largest increases in trade flows; specifically, an improvement of just 0.55 percent in the port efficiency indicator has the same impact as 5.5 and 3.3 percent improvements in customs performance and e-business indicators, respectively (Paul E. Kent, 2004).

In an attempt to modeling port efficiency and calculating its impact on trade by using gravity trade model, Blonigen and Wilson estimated that improved port efficiency significantly increases trade volumes (Bruce A. Blonigen, 2006). Sánchez, Hoffmann, Miccó, Pizzolitto, Sgut and Wilmsmeier (2002) in their research about Port Efficiency and International Trade found that more efficient seaports are clearly associated to lower freight costs after controlling for distance, type of product, liner services availability, and insurance costs, among others. Port efficiency factors, as explanatory variables, reflect a set of components easily observable in any port terminal. Variables included are: the container hourly loading rate; the annual average of containers loaded per vessel; waiting times, and several others. According to their estimations, a 25% improvement of one efficiency factor implies a reduction of approximately 2% in total maritime transport cost. This result, for a set of low value export commodities, could be the difference between being or not being competitive in the global market (Ricardo J. Sánchez, 2002).

6 PORT USERS' REQUEST

To have other transport actors at port providing a safe workplace is a necessity. As a true example, it is not viable for a shipping line to call at an unsafe port with a mega ship since if an accident happens, not only huge capital will be idled, but also delay in providing transport services to other ports en route will be unavoidable. If shipping lines use middle or even small size vessel to call such ports, then their operational and economic risks will decline. That would be a tragic decision for ports since they will lose advantages resulting from handling mega ship's economies of scale from one side and productivity and efficiency of port's infrastructures, superstructure and infostructures on the other side. In other words, the unit service cost for a port's customers will be higher compared to the past when the port handled mother vessels. In such circumstances, the shipping lines' market share begins to decline and competitiveness of cargo owners in the market will drop. That is why, according to UNCTAD, main shipping lines set a precondition for a minimum level of safety and security for putting a port on their route (UNCTAD, 2007).

The same rationale takes place by global terminal operators (GTOs) and port transnational corporations (TNCs). The globalization of ports and the creation of TNCs have brought with them many opportunities for developing economies. Amid such opportunities are sharing knowledge and expertise in the areas of management and operational techniques, infrastructure planning, methods of international finance, the adoption of tried and tested computer software systems, and the replication of success factors. Interesting to know that during an UNCTAD meeting on globalization of port logistics in 2007, major terminal operators listed inappropriate hinterland connection to be the most important external factor and safety the main internal cause for success or failure of ports in the maritime transport market (UNCTAD, 2008).

7 CONCLUSION

In a port where the safety level is low, incident and accidents are happening often (Hassanzadeh, 2009), unless it handles captive cargo, otherwise as a general rule, its attractive ness for port users i.e., lines, cargo agents, multimodal transport operators and freight forwarders will be low. In spite the fact that ports are one of the important transport nodes in international trade, their inefficiencies resulting from accidents play a countertrade role. To overcome such trade setback, ports have undertaken several measures to cut the frequency and consequences of port accidents throughout the world. Even though being safe and economically viable for the transport operators are considered a prerequisite, it is interesting to know that driving forces to put safety measures into place are primarily legal obligations or mandates from transport unions. Miller (2009) quoted that legal compliance is the most important driver for occupational safety and health. So according to the European Commission, moving beyond legal compliance requires a sound strategy on occupational safety and health tying its outcomes to the overall business outcomes (European Commission, 2011).As stated by Verbeek (2009) calculating the costs and impacts of accidents on efficiency may show their impact on company performance. However, it is much more interesting to know how we can effectively prevent the causes of such accidents and cases of ill-health and how much we can benefit from this prevention in monetary terms (Verbeek, 2009).

In conclusion, an increase of the safety level at ports would be in the interest of all the actors in the supply chain management including ports themselves. So linking port safety with trade either at regional or international level is an important issue and could give a basis for putting forward a strong business case for port safety. A lot of researches on safety subjects at ports have been done, but few of them have tried to find an economic relationship between safety level and trade. Realizing costs incurred by accidents and their negative impact on total transport cost of commodities as externalities could help port authorities and other service providers to invest more willingly on safety. Such level of safety will satisfy workers' unions or lawmakers as well. This new situation would be a win-win situation for port authorities, cargo owners, shipping lines and society on top.

REFERENCES

Abbes, S. (2007). Marginal social cost pricing in European seaports. *European Transport \ Trasporti Europei, n. 36*, pp. 4-26.

Bernard Brody, L. Y., & Poirier, A. (1990). *Estimation of the indirect costs of occupational accidents in Québec.* Université de Montréal.

Bruce A. Blonigen, W. W. (2006). *Port Efficiency and Trade Flow.* Oregon: University of Oregon.

Darbra, R., Ronza, A., Carol, S., Vilchez, J., & Casal, J. (2005). A Survey of Accidents in Ports. *Loss Prevention Bulletin*, pp. 23-28.

DfT. (2005). *PORT EMPLOYMENT AND ACCIDENT RATES.* London: Department for Transport.

Dorman, P. (1996). *Markets and Mortality: Economics, Dangerous Work, and the Value of Human Life.* Cambridge : Cambridge University Press.

European Commission. (2011). *Socio-economic costs of accidents at work and work-related ill health.* Luxembourg: Directorate-General for Employment, Social Affairs and Inclusion.

Hassanzadeh, M. A. (2009). *Safety Management System At Ports* (Vol. 1). Tehran, Iran: Ministry of Road and Transportation.

Hoffmann, J. S. (2002). Globalization,The Maritime Nexus. In *Handbook of Maritime Economics.* London.

Jane Korinek, P. S. (2009). *MARITIME TRANSPORT COSTS AND THEIR IMPACT ON TRADE.* Paris: The Organisation for Economic Co-operation and Development (OECD).

Jones, L. (2011, September 5). *Ninth death at Australian ports in ten years.* Retrieved March 21, 2012, from IFW: http://www.ifw-net.com/freightpubs/ifw/article.htm?artid=20017901201&src=rss

Limao, N. a. (2001). Infrastructure, Geographical Disadvantage, and Transport Costs. In W. Bank, *World Bank Economic Review* (pp. 451-479.).

Lindberg, G. (1999). *CALCULATINGTRANSPORT ACCIDENT COSTS.* Borlänge: HIGH LEVELGROUP ON INFRASTRUCTURE CHARGING.

Loss Control Consultants. (2009, August). *The total cost of accidents and how they affect your Profits.* (Western National Insurance,) Retrieved from www.wnins.com

Lourdes Trujillo, G. N. (1999). *PRIVATIZATION AND REGULATION OF SEAPORT INDUSTRY.* de Las Palmas: Universidad de Las Palmas de Gran Canaria.

María del Pilar Londoño-Kent, P. E. (2003). *A Tale of Two Ports: The Cost of Inefficiency.* The World Bank.

Merk, O. D. (2012). *Efficiency of world ports in container and bulk cargo (oil, coal, ores and grain).* PARIS: Regional Development Working Papers, 2012/09, OECD Publishing.

Michael, R. (2001). *More Liberty Mutual Data on Workplace Safety.* ,.

Miller, P. H. (2009). Why employers spend money on employee health: Interviews with occupational health and safety professionals from British Industry. *Safety Science*, 163-169.

Paul E. Kent, A. F. (2004). *THE BROAD ECONOMIC IMPACT OF PORT INEFFICIENCY.* United States Agency for International Development.

Purdy, G. (1997). Value Added Accident Analysis. *Queensland Mining Industry Health and Safety Conference*, (pp. 60-66). Melbourne.

Ricardo J. Sánchez, J. H. (2002). *PORT EFFICIENCY AND INTERNATIONAL TRADE.* Argentina: Austral University.

Ronza, A. (2007). *Contributions to the risk assessment of major accidents in port areas.* Barcelona: Universitat Politècnica de Catalunya & Escola Tècnica Superior d'Enginyers Industrials de Barcelona.

Shou, M. (2003). *Maritime Transport Economics.* Malmo, Sweden: World Maritime University.

The Marinewaves. (2011, Jan 28). *One dead, one missing in Rotterdam Port accident.* Retrieved 3 4, 2012, from The Marinewaves: www.moneycontrol.com/.../video-one-dead-one-missing-in-rotterdam

Tongzon, J. L. (2009). Port choice and freight forwarders. *Transportation Research*, 186–195.

UNCTAD. (2007). *Globalization of Port Logistics, Opportunities and Challenges for Developing Countries.* Geneva: UNCTAD.

UNCTAD. (2008). *World Investment Report on TNCs in infrastructure.* Geneva: UNCTAD.

UNCTAD. (2012). *Review of Maritime Transport.* Geneva: United Nations Conference on Trade and Development.

Verbeek, J. (2009). Economics of occupational safety and health. *Scandinavian Journal of Work, Environment and Health*, 401-402.

Wilson, John S.,Catherine L. Mann, and Tsunehiro Otsuki. (2003). Trade Facilitation and Economic Development: Measuring the Impact, World Bank Policy Research Working Paper 2988.

Ximena Clark, D. D. (2002). *Maritime Transport Cost and Port Efficiency*. The World Bank.

Method of Assessment of Insurance Expediency of Quay Structures' Damage Risks in Sea Ports

M.Ya. Postan
Odessa National Maritime University, Odessa, Ukraine

M.B. Poizner
R&D Institute of Marine Transport (Chernomorniiproekt), Odessa, Ukraine

ABSTRACT: The article presents a theoretical approach to assessment of insurance risks' expediency concerning the possible damage to ports' quay structures caused by ships' lean. This approach is based on determination of so-called reliability function of the berths via interpretation of port terminal as a queueing system.

1 INTRODUCTION

Quay structures are an important part of the sea ports' infrastructure ensuring the safety of different port operations fulfillment. Particularly, the safety and efficiency of loading/unloading operations in ports depend essentially on current technical state of quay structures' elements. As shows an international shipping practice, most significant damage to berths may occur at moments of ships' lean during their moorage. In spite of fenders and experience of navigators some deformations of quay structure may occur at moment of a ship's moorage [1,2].

Some types of deformations of elements of quay structures may arise also during the ships loading/unloading because of they leans on berth under sea roughness. If a failure of berth construction's element appears under loading/unloading operations it leads up to demurrage of a ship and additional costs for ship-owner or charterer. Therefore, forecasting of possible berth failure and its reliability is an important practical task for the ports authorities.

At present, the risks of quay structures damage caused by ships lean and/or cargo transshipment are usually not insured by ports authorities. It is explained by the following reason: according to international shipping practice the repair-expenditures of the damaged berths are covered by ship-owners or charterers, as a rule. In one's turn, they insure these risks in P&I clubs [3]. However, because duration of reimbursement procedure is often very prolonged, port authorities face the problem of economical substantiation of above risks insurance. If port authorities will insure the risks of quay structures damage, they can receive a corresponding reimbursement from insurers relatively prompt.

In our paper, the method of insurance expediency of quay structures by port authorities is proposed. It is based on the determination of main index of safe operation of a berth, namely, its reliability function.

2 METHOD FOR RELIABILITY FUNCTION OF BERTHS DETERMINATION

At present, the theoretical approach of insurance risk' expediency assessment concerning the possible damage to ports' quay structures caused by ships' lean, which is based on determination of reliability function of the berths is widely well known. The concept has been worked out by the UNCTAD secretariat in the mid of 70th of the XX century. But this concept doesn't allow to determine the reliability function taking into account the specific character of the loads acting on berth at moment of ships berthing and during their loading/unloading.

Our method is based on representation of port's terminal as a queueing system and determination of reliability function of berth in the terms of queueing theory [4]. In the framework of such representation the ships arriving at terminal are interpreted as customers and berths are interpreted as servers [5]. By definition, the reliability function of a berth $R(t)$ is the probability of its operation in good repair (without failure) in time interval $(0,t)$. Note, that from the point of view of actuarial mathematics [6], the failure of berth is associated with accident and corresponding random damage.

For the sake of simplicity, below we'll restrict ourselves only by consideration of impulse loads from ships on a berth arising at moment of their moorage. Term 'impulse' means that we ignore the duration of ships' lean on the berth (see Figure).

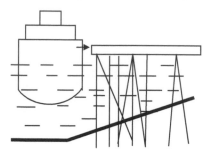

Figure. Scheme of ship's lean on the pier built on metallic piles

Firstly, let us consider the case when port's terminal consists of one berth only and introduce the following designations:

Δ_r is random value of deformation of given constructive element of berth (pile, section of berth, concrete slab, etc.) arising in result of the rth arrived ship's lean at moment of its moorage;

d is the value of limit permissible deformation of given element of berth;

$v(t)$ is the number of ships served at terminal in time interval $(0,t)$;

$\theta(t)$ is the number of ships being at moment t under loading/unloading (it equals to 1 if there is a ship at berth and equals to 0 otherwise).

We assume also that

$$\Delta_r = k_r \varphi(r) Q_r, r \geq 1, \qquad (1)$$

where Q_r is a random value of lean of the rth arrived ship's at moment of its berthing;

$\varphi(r)$ is so-called coefficient of accumulated deformations recurrence (it allows take into account the change of concrete solidity in dependence on number of load's cycle);

k_1 is a coefficient of given berth element's pliancy.

If the given element of berth construction is subjected to most load than other elements, then it is most "important" and in more extent determines reliability of berth. According to the linear law of deformations accumulation the reliability criterion for berth construction is given by the following condition

$$\sum_{r=1}^{\theta(t)+v(t)} \Delta_r \leq d \qquad (2)$$

or, taking into account (1),

$$k_1 \sum_{r=1}^{\theta(t)+v(t)} Q_r \varphi(r) \leq d. \qquad (3)$$

Therefore, for this case from (1) and theorem of total probability, the reliability function is determined as follows

$$R(t) = P\{ \sum_{r=1}^{v(t)+\theta(t)} \Delta_r \pounds\, d\} =$$
$$= P\{\theta(t) = 0, v(t) = 0\} +$$
$$+ \sum_{r=1}^{\yen} P\{v(t) = r, \theta(t) = 0\} P\{\sum_{j=1}^{r} \Delta_j \pounds\, d\,|v(t) = r, \theta(t) = 0\} + \quad (4)$$
$$+ \sum_{r=1}^{\yen} P\{v(t) = r-1, \theta(t) = 1\} P\{\sum_{j=1}^{r} \Delta_j \pounds\, d\,|v(t) = r-1, \theta(t) = 1\}.$$

If for modeling the terminal we take the *Poisson-Erlang* queueing system of M/M/1 type with single server and infinite waiting room, then the following result is valid in equilibrium [3]

$$P\{v(t) = r, \theta(t) = 0\} = \frac{(\lambda t)^r}{r!} e^{-\lambda t}(1-\rho),$$
$$P\{v(t) = r-1, \theta(t) = 1\} = \frac{(\lambda t)^{r-1}}{(r-1)!} e^{-\lambda t}\rho, \qquad (5)$$

where $\rho = \lambda / \mu < 1$; λ is input rate of the *Poisson* stream of ships; $1/\mu$ is mean time of ship's standing at berth under loading/unloading.

As long as $Q_1, Q_2, \dots$ are mutually independent and identically distributed random variables (this is natural supposition), the same will be the random variables $\Delta_1, \Delta_2, \dots$, as well. Let us denote

$$L(y) = P\{Q_1 \leq y\}.$$

Hence we find

$$P\{\sum_{j=1}^{r} \Delta_j \leq d\,|v(t) = r, \theta(t) = i\} = F_1 * F_2 * \dots * F_r(d), i = 0,1, \quad (6)$$

where (see (2))

$$F_r(x) = L(\frac{x}{k_1 \varphi(r)})$$

is distribution function (d.f.) of random deformation of berth's element in result of the r-th served ship's lean; * is symbol of the *Stieltjes* convolution of d.f.

From (4)-(6), it follows that for steady-state regime of terminal functioning

$$R(t) = \qquad (7)$$
$$= e^{-\lambda t}\{1 - \rho + \sum_{r=1}^{\infty} \frac{(\lambda t)^{r-1}}{(r-1)!}[(1-\rho)\frac{\lambda t}{r} + \rho]F_1 * F_2 \dots * F_r(d)\}.$$

If

$$L(y) = 1 - e^{-y/\overline{q}}, y \geq 0,$$

then

$$F_1 * F_2 * \ldots * F_r(d) \equiv E_r(d) = \sum_{j=1}^{r} \frac{\prod_{i=1}^{r} \gamma(i)}{\prod_{\substack{i=1 \\ i \neq j}}^{r}(\gamma(i) - \gamma(j))}(1 - e^{-\gamma(j)d}), \quad (8)$$

where $\gamma(r) = (k_1 \bar{q} \varphi(r))^{-1}$. It is obvious that $E_r(x)$ is the general *Erlang's* distribution of the r-th order.

Put $\varphi(r) = 1/r$. Then from (8), it follows

$$E_r(d) = (1 - e^{-\gamma d})^r, \quad (9)$$

where $\gamma = (k_1 \bar{q})^{-1}$.

After substitution (9) in the right-hand side of (7) and summation the infinite series we obtain

$$R(t) = (1 - \rho e^{-\gamma d}) \exp(-\lambda t e^{-\gamma d}). \quad (10)$$

From the point of view of total stability and strength of berth's construction most dangerous is the case when process of deformations accumulation almost doesn't damp. To take into account this case we can put $\gamma_r = \gamma, r = 1, 2, \ldots$ Then from (8), we get

$$E_r(d) = 1 - e^{-\gamma d} \sum_{k=0}^{r-1} \frac{(\gamma d)^k}{k!} = \frac{(\gamma d)^r}{r!} e^{-\gamma d} + \frac{\gamma^{r+1}}{r!} \frac{d}{\int_0^d} u^r e^{-\gamma u} du.$$

Substituting the last expression in the right-hand side of (7) we find

$$R(t) = (1 - \rho)e^{-\lambda t}[e^{-\gamma d} I_0(2\sqrt{\lambda \gamma d t}) + \gamma \int_0^d e^{-\gamma u} I_0(2\sqrt{\lambda t \gamma u}) du] + \quad (11)$$
$$+ \rho e^{-\lambda t}[e^{-\gamma d}\sqrt{\frac{\gamma d}{\lambda t}} I_1(\lambda t \gamma d) + \gamma \sqrt{\frac{\gamma}{\lambda t}} \int_0^d e^{-\gamma u} \sqrt{u} I_1(2\sqrt{\lambda u \gamma d}) du],$$

where $I_m(z)$ is the modified *Bessel* function of the m-th order. Applying to the formula (11) the *Laplace-Stieltjes* transform after some calculations we get

$$r(s) - \int_{-0}^{\infty} e^{-st} d(1 - R(t)) = \quad (12)$$
$$= \frac{\lambda}{\lambda + s}(1 + \frac{s}{\mu})e^{-\gamma s d/(\lambda+s)}, \text{Re } s \geq 0.$$

From (12), it may easily be found the moments of berth's life-time (τ) d.f., i.e. $1 - R(t)$. For example,

$$E\tau = -r'(0) = \frac{1}{\lambda}(1 - \rho + \lambda d),$$

$$E\tau^2 = r''(0) = \frac{1}{\lambda^2}[2(1-\rho)(1+\gamma d) + \gamma d(2 + \gamma d)].$$

When t takes the large values, for calculation of $R(t)$ may be used the asymptotic formula which follows from central limit theorem for renewal processes [7]

$$R(t) \approx N(\frac{d - \lambda t k_1 \bar{q}}{k_1 \sqrt{\lambda t \bar{q}^{(2)}}}), t \to \infty, \quad (13)$$

where

$$\bar{q} = EQ_1, \bar{q}^{(2)} = EQ_1^2; N(x) = \frac{1}{\sqrt{2\pi}} \int_{-\infty}^{x} e^{-y^2/2} dy.$$

Let us refer to the case of several berths. Generally speaking, this means that to model a port terminal the multi-server queueing systems should be considered. Consider, for example, the n berths located in sea-front. In this case, instead of reliability criterion (2) we can take the following one

$$\sum_{r=1}^{v_1(t)+\theta_1(t)} \Delta_{1r} \leq d_1, \ldots, \sum_{r=1}^{v_n(t)+\theta_n(t)} \Delta_{nr} \leq d_n, \quad (14)$$

where d_i is limit permissible level of deformations accumulation for the ith berth; lower indices in (14) denote the numbers of berths. Particularly, if each berth belongs to a separate terminal and services its own stream of ships, then stochastic processes

$$\sum_{r=1}^{v_1(t)+\theta_1(t)} \Delta_{1r}, \ldots, \sum_{r=1}^{v_n(t)+\theta_n(t)} \Delta_{nr}$$

are mutually independent. Therefore, the reliability function of system of berths is

$$R(t) = \prod_{k=1}^{n} R_k(t), \quad (15)$$

where $R_k(t)$ is the reliability function of the kth berth calculated by formulas like (4).

For many different cases the reliability functions of berths are derived in the book [4].

3 ASSESSMENT OF INSURANCE EXPEDIENCY OF QUAY STRUCTURE'S DAMAGE

From the point of view of insurer and insurants the failure of a berth is a random event which is called insured risk. Mathematically this event may be expressed, for example, as the following condition (for a single berth)

$$\sum_{r=1}^{\theta(t)+v(t)} \Delta_r > d.$$

The simplest criterion for the risk of a berth's failure insurance expediency may be found with the help of following reasoning. Suppose that total value of damage to a berth in result of its failure is d. This damage may be measured by any physical units (e.g., m or m²). It is naturally to assume that sum insured for the risk of a berth failure is proportional to value d, i.e. equals to wd, where w is a value of damage's unit (estimated in money). The port authorities must decide: to insure or not to insure a risk of a berth's failure for the period of insurance contract action T with the gross risk premium rate c?

To answer this question, consider two random variables ξ_1 and ξ_0, where $\xi_1(\xi_0)$ is the value of port's gain in the case of risk of berth's failure insurance (giving insurance up). By definition, the following probabilistic relations are valid

$$\xi_1 = -cT\mathbf{I}(\sum_{r=1}^{\theta(T)+\nu(T)}\Delta_r \le d) + wd\mathbf{I}(\sum_{r=1}^{\theta(T)+\nu(T)}\Delta_r > d),$$

$$\xi_0 = cT\mathbf{I}(\sum_{r=1}^{\theta(T)+\nu(T)}\Delta_r \le d) - wd\mathbf{I}(\sum_{r=1}^{\theta(T)+\nu(T)}\Delta_r > d),$$

(16)

where $\mathbf{I}(A)$ is the indicator of an event A.

From the relations (16), it follows that

$$E\xi_1 = wd - (cT + wd)R(T),$$
$$E\xi_0 = -wd + (cT + wd)R(T),$$ (17)
$$Var\xi_1 = Var\xi_0 = (cT + wd)^2 R(T)(1 - R(T)),$$

where $R(T)$ is the reliability function of berth related to the period T. Therefore, the insurance is expedient if $E\xi_1 > E\xi_0$, or taking into account the relations (17) if

$$R(T) < \frac{wd}{wd + cT}.$$ (18)

Particularly, for the case (10) from (18) we have inequality

$$(1 - \rho e^{-\gamma d})\exp(-\lambda T e^{-\gamma d}) < \frac{wd}{wd + cT}.$$ (19)

Put $cT = 0{,}03wd$, then the right-hand side of inequality (19) equals to 0.97. The values of reliability function (10) calculated for $T = 10$ years, $\mu = 0{,}5\,day^{-1}$ and different values of parameters $\rho = \lambda / \mu$ and γd are given in the Table. One can see that under values $\gamma d < 11$ the insurance of berth's failure risk is expedient for all given values of parameter ρ.

Table

Parameter ρ	Parameter γd				
	7	8	9	10	11
0.5	0.4349	0.7363	0.8934	0.9594	0.9849
0.6	0.3682		0.6925	0.8735	0.9515	0.9819
0.7	0.3117	0.6513	0.8540	0.9437	0.9789
0.8	0.2639	0.6126	0.8350	0.9437	0.9759
0.9	0.2234	0.5762	0.8164	0.9281	0.9729

The similar reasoning may be conducted for several berths, as well, using the criterion (14). In this case the condition of insurance expediency must be modified by corresponding way.

Let us demonstrate this for two independently functioning berths (i.e. belonging to different port's terminals with their own input streams of ships). Let

ξ_{ij} be the port's gain for different policies of the berths insurance determining by the values of indices i and j, where $i,j = 0$ or 1. Values $i,j = 0$ correspond to giving insurance up for the berths, and values $i,j = 1$ correspond to their insurance.

Similarly to relations (16) we find

$$\xi_{ij} = T[(-1)^i c_1 + (-1)^j c_2]\mathbf{I}(\sum_{r=1}^{\nu_1(T)+\theta_1(T)}\Delta_{1r} \le d_1)\mathbf{I}(\sum_{r=1}^{\nu_2(T)+\theta_2(T)}\Delta_{2r} \le d_2) +$$

$$+[(-1)^i c_1 T + (-1)^{j+1}w_2 d_2]\mathbf{I}(\sum_{r=1}^{\nu_1(T)+\theta_1(T)}\Delta_{1r} \le d_1)\mathbf{I}(\sum_{r=1}^{\nu_2(T)+\theta_2(T)}\Delta_{2r} > d_2) +$$

$$+[(-1)^{i+1}w_1 d_1 + (-1)^j c_2 T]\mathbf{I}(\sum_{r=1}^{\nu_1(T)+\theta_1(T)}\Delta_{1r} > d_1)\mathbf{I}(\sum_{r=1}^{\nu_2(T)+\theta_2(T)}\Delta_{2r} \le d_2) +$$

$$+[(-1)^{i+1}w_1 d_1 + (-1)^{j+1}w_2 d_2]\mathbf{I}(\sum_{r=1}^{\nu_1(T)+\theta_1(T)}\Delta_{1r} > d_1)\mathbf{I}(\sum_{r=1}^{\nu_2(T)+\theta_2(T)}\Delta_{2r} > d_2).$$

(20)

From relations (20), taking in view the condition (15), we obtain

$$E\xi_{ij} = 2[(-1)^i(c_1 T + w_1 d_1) + (-1)^j(c_2 T + w_2 d_2)] \times$$

$$\times R_1(T)R_2(T) +$$

$$+ (-1)^i(c_1 T + w_1 d_1)R_1(T) + (-1)^j(c_2 T + w_2 d_2)R_2(T) -$$

$$- [(-1)^i w_1 d_1 + (-1)^j w_2 d_2], \quad i, j = 0{,}1.$$

Hence, the optimal insurance policy of port relative to system of two berths may be determined by calculation of expression $\max_{i,j} E\xi_{ij}$.

4 CONCLUSIONS

Method of assessment of possible failures of berths' insurance expediency proposed above may be realized in practice only in context of general system of evaluation and forecasting of quay structures operational reliability. Such a system supposes the organization of periodic observations for current technical states of main elements of quay structures and is aimed on:
– determination of existing deformations of elements;
– evaluation of current carrying ability of a berth;
– control for bringing in correspondence throughput of port terminal and carrying ability of its berths;
– assessment of damage degree caused to berth in result of the ships leans;
– account of number and characteristics of the ships served by terminal in the given time interval, durability of their standing at berths, etc.

This information must be collected in corresponding data base at technical department of port to be used for necessary analysis, calculations of reliability functions with the standard software, and evaluation of insurance expediency.

REFERENCES

1. Koziol W. and Galor W. 2007. Some Problems of Berthing of Ships with Non-conventional Propulsions. International Journal of Marine Navigation and Safety of Sea Transportation. 1(3):319-323.
2. Artyszuk J. 2010. Simulation of Load Distribution along a Quay during Unparallel Berthing Manoeuvres. International Journal of Marine Navigation and Safety of Sea Transportation. 4(3):273-278.
3. Brown, R.H. 1985-1993. Marine Insurance: Vol. I. Principles and Basic Practice; Vol. II. Cargo Practice; Vol.III. Hull Practice. London: Witherby Publishers.
4. Poizner, M.B., Postan M.Y.1999. Operational Reliability of Quay Structures: A Probabilistic Approach. Odessa: Astroprint (in Russian).
5. Cooper, R.B. 1981. Introduction to Queueing Theory. 2nd Ed. New York: North Holland.
6. Grandell, J. 1992. Aspects of Risk Theory. Berlin Heidelberg New York: Springer.
7. Prabhu, N.U. 1997. Stochastic Storage Processes: Queues, Insurance Risk, Dams, and Data Communications. 2nd Ed. Berlin Heidelberg New York: Springer.

Solid Waste Management: Compliance, Practices, Destination and Impact among Merchant Vessels Docking in Iloilo Ports

B.G.S. Sarinas, L.D. Gellada, M.M. Magramo & D.O. Docto
John B. Lacson Foundation Maritime University-Arevalo, Iloilo City, Philippines

ABSTRACT: There are no or few existing studies exist on the solid waste (plastics, papers and others) management of ships in Fort San Pedro port (FSPP) and Dumangas port (DP), Iloilo, Philippines. Thus, this survey determines the compliance and practices on ship's solid waste management, its impact to crew members on board the vessel and the fate of these solid wastes during docking. Nine ships served as samples, six of which are Ro-ro from DP and three passenger-cargo vessels from FSPP. There were 141 crews were interviewd which comprise of six masters (four masters from DP and two from FSPP), three chief officers (two from DP and one from FSPP), 50 crew from Ro-ro vessels at DP and 82 crew from passenger vessels in FSPP were interviewed on the impacts of solid waste management. The results showed a 100% compliance to solid waste management among vessels in both ports while in wastes practices' on board, specific garbage bins where used by these vessels. Consequently, these vessels upon reaching the receiving port, relinquish their solid wastes to the "Golden Dragon" that collects solid wastes. The present study showed that vessels in both ports observe the Annex V of the MARPOL 73/78 and reveals an eco-friendly shipping.

1 INTRODUCTION

Garbage are present everywhere and their accumulation is unstoppable. They can pose hazards to human and wildlife and aesthetically displeasing (Horsman, 1982). Urban and rural areas are experiencing tons of garbage due to misinformed waste practices and effects to the environment. Most people dumped their wastes or garbage in the bodies of water such as freshwater or to the marine environment. Gomez, Velazquez and Baniela (2004), enumerated various contaminants of the sea. One of the sea contaminants is the accumulation of garbage or solid wastes such as papers, tins, bottles, plastics and tires. Garbage from ships is harmful to the environment (IMO, 2010). Thus, the International Maritime Organization created the MARPOL 73/78, a convention by the International Convention for the Prevention of Pollution from Ships in 1973 which was modified by the Protocol of 1978 (MARPOL, 2006). MARPOL 73/78 regulates and prevents pollution by garbage from ships most especially plastics as amended in Annex V (IMO, 2010). MARPOL Annex V defines garbage as "all kinds of victual, domestic and operational waste excluding fresh fish and parts thereof, generated during the normal operation of the ship and liable to be disposed of continuously or periodically except those substances which are defined or listed in other Annexes to the present Convention" (MARPOL, 2006). In addition, the MARPOL Annex V prohibits all ships to dispose garbage at sea and "specific areas" such as the Mediterranean Sea area, Baltic Sea area, Black Sea area, Red Sea area, Gulfs area, North Sea, Wider Caribbean Region and Antarctic Area (MARPOL, 2006; IMO, 2010).

2 STATEMENT OF THE PROBLEM

This study aimed to determine the solid waste management compliance, impacts, practices and destination among merchant vessels docking in Fort San Pedro port (FSPP) and Dumangas port (DP).

Specifically, it sought answers to the following questions:

1 Do merchant vessels comply with the solid waste management as mandated by MARPOL 73/78?
2 What are the practices observed by the merchant vessels docking in Iloilo ports toward their solid wastes?
3 What are the impacts of solid waste management to the crew of the vessel?

4 To where do merchant vessels relinquish their solid wastes upon disembarkation in Iloilo ports?

3 METHOD

3.1 *Research Design*

This study utilized a survey research design because it involves asking the same set of questions on compliance, practices, perceived impacts and fate of solid wastes on board the vessel to various crew of vessels in FSPP and DP.

3.2 *Procedure*

A total of 141 respondents were used in this study. In FSPP, two master mariner, one chief officer and 82 various crew were interviewed while in DP, four master mariner, two chief officers and 50 various crew were interviewed.

Convenience sampling was employed to select the respondents that are available during the sampling. These merchant vessels were classified as passenger-cargo vessels. The two identified ports in Iloilo were Fort San Pedro port (FSPP) and Dumangas port (DP). Sampling was done in the month of December 2011.

A researcher-made questionnaire-checklist was made to answer the problems of the study. The checklist contained consists of questions if the vessel complies with the solid waste management as mandated by MARPOL 73/78, varied management practices of solid wastes as specified by MARPOL 73/78 (MARPOL, 2006) and impacts of solid waste management to the crew of the vessels. In addition, the master mariner was asked as to where do they relinquish their solid wastes upon disembarkation or upon arrival in the receiving port.

The researchers asked permission from the Philippine Ports Authority (Region 6) to have an access in the said merchant vessels docking in FSPP and DP. The answers of the respondent was collected to determine if the vessels comply with the solid waste management as mandated by MARPOL 73/78, as to what practices was made in the list of garbage specified by the MARPOL 73/78 and the impacts of solid waste management to the crew of the vessel. In addition, the fate of the solid wastes generated from each merchant vessel was determined.

3.3 *Data Analysis*

The descriptive analysis was used to analyze the data gathered, specifically frequency count to answer the research problems.

4 RESULTS AND DISCUSSION

4.1 *Compliance, Practices, Impact and Destination among Merchant Vessels Docking in Fort San Pedro Port (FSPP)*

Three officers from three passenger-cargo vessels were interviewed in Fort San Pedro Port (FSPP) namely: two Masters from NN M/S St. Peter the Apostle and M/V St. Michael the Archangel and one Chief Officer of M/V Filipinas Cebu. The list of garbage is from the list of MARPOL: plastics, paper/cardboard, metals, glass, food wastes, wood, pain containers, fertilizers and pesticide containers, batteries and old medicines.

The result showed that 100% of the vessels comply with the solid waste management. It was found out that each vessel has specified garbage bins for every waste collected. This result is also strengthened by the fact that the practice of these passenger-cargo vessels does not dumped their garbage in the ocean, but instead, they hand over these packed garbage to the port's facility called the "Golden Dragon." This is true for NN M/V St. Peter the Apostle and M/V St. Michael the Archangel except for M/V Filipinas Cebu because they have their own waste services in the port as supervised by the Cokaliong shipping company. Table 1 below shows the solid waste management practices of the three passenger-cargo vessels docking in FSPP.

Table 1. Solid waste practices of M/V St. Michael the Archangel, M/S Saint Peter the Apostle and M/V Filipinas Cebu docking in FSPP.

| Materials | Dumped in the | | |
	Incinerator	Ocean	Ports Facility*
1. Glass			
2. Paper/Cardboard			
3. Metals			
4. Plastics			
5. Food wastes			
6. Wood			
7. Paint containers			
8. Fertilizer and pesticide containers			
9. Batteries			
10. Old medicines			

*The shade indicates the responses of the officers from three passenger-cargo vessels in FSPP.

With regards to the impacts of solid waste management among the crew of three vessels in FSPP, generally, almost all of the crew have the positive impacts of solid waste management and these impacts are: health and sanitation (HS) was observed (100%), showed neatness and cleanliness (NC) which is 98%, pests are prevented (PP) which is 98% and finally, 98% for aesthetic value to the passengers (AEVP). Figure 1 shows the percentage distribution of impacts of solid waste management among various crews in FSPP.

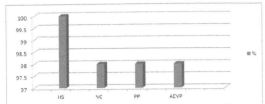

Figure 1. Percentage distribution of impacts on solid waste management among six vessels in FSPP.

4.2 Compliance, practices, impact and destination among merchant vessels (Ro-ro) docking in Dumangas port (DP)

Six officers were interviewed from the six Ro-Ro passenger vessels in Dumangas port (DP) and these comprise mainly of four Masters of M/V Marie Teresa, LCT Holy Family Navistar, Lakbayan Uno and LCT Arc Angel Navistar and two Chief Officers of LCT Sr. Sto. Nino and M/V Maria Beatriz. It was revealed that 100% or 6 out of 6 Ro-Ro vessels comply with the solid waste management as imposed by the MARPOL. This is however, supported by the fact that these Ro-Ro vessels have the corresponding garbage bins for specific garbage type as with the passenger-cargo vessels docking in FSPP. This is to note that these Ro-Ro vessels do not dump their garbage in the ocean except in one vessel. Their paper wastes are incinerated and dumped in the ocean, that is 12 nautical miles away from the land where it is allowed by the MARPOL Annex V of the IMO rules (MARPOL, 2006). Moreover, all of these vessels hand over their on board solid wastes when they reach the receiving port, that is, in DP. It was also found out that this kind of practice was mandated by the Philippine Ports Authority (PPA), Iloilo. This is to fully ensure that plastics should not be dumped in the ocean as strictly imposed by the MARPOL Annex V.

Table 2. Solid waste practices of six Ro-ro vessels docking in DP.

Materials	Dumped in the		Ports Facility
	Incinerator	Ocean	
1. Glass			
2. Paper/Cardboard			
3. Metals			
4. Plastics			
5. Food wastes			
6. Wood			
7. Paint containers			
8. Fertilizer and pesticide containers			
9. Batteries			
10. Old medicines			

*The shade indicates the responses of the 5 officers while the red shade indicates the response of one officer which is different from the rest of the vessels.

During the arrival of the Ro-Ro vessels in DP, the PPA has a waste services facility that mainly collects the solid wastes of these vessels, and likewise with the FSPP, the name is called "Golden Dragon." Table 2 shows the practices of Ro-Ro vessels in DP on solid waste management.

On the other hand, most of the impacts of solid waste management among crew showed positive impacts such as health and sanitation (100%), neat and clean (100%), prevention of pests (100%), adds aesthetic value to the passengers (96%) and deliver good services to the passengers (2%) as an additional impacts of solid waste management. Figure 1 shows the percentage distribution of impacts of solid waste management among various crews in DP.

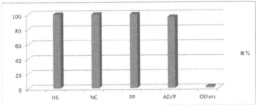

Figure 2. Percentage distribution of impacts on solid waste management among six vessels in DP.

5 CONCLUSIONS

The result of the study showed that the Philippines Ports Authority, Iloilo is strict when it comes to solid waste management. This is to ensure that the MARPOL Annex V of IMO is fully implemented most especially in the dumping of garbage in the ocean. The PPA provides each port with the service facility called Golden Dragon that caters the collection of the solid wastes from the docking vessels. In addition, all sampled vessels have garbage specific bins on board. This simply shows that vessels in Iloilo port (FSPP and DP) operates in accordance with the law and is eco-friendly. For further research, more merchant ships should be included in the study.

ACKNOWLEDGMENT

We would like to thank Dr. Ronald Raymond Lacson Sebastian, Chief Executive Officer, John B. Lacson Foundation Maritime University for the full support extended in this study and to Ms. Minerva R. Alfonso of INTERTANKO for the words of encouragement.

REFERENCES

Fraenkel, J. R. and Wallen, N. E. 2010. How to Design and Evaluate Research in Education. 7th ed. International ed. New York: McGraw-Hill, Inc.

Gomez, J. I. G., Velazquez, O. H. and Baniela, S. I. 2004. The sea and its contaminants. *J. of Mar. Res.* 1(2):85-93.

Horsman, P. V. 1982 The amount of garbage pollution from merchant ships. *Mar. Poll. Bull.* 13(5):167-169.

International Maritime Organization (IMO). 2010. Prevention of Pollution by Garbage from Ships. Retrieved November 2011 from: http://www.imo.org/OurWork/Environment/PollutionPrevention/Garbage/Pages/Default.aspx.

Marine Pollution (MARPOL). 2006. Consolidated Edition. International Maritime Organization, London. MPG-Books Ltd, Bodmin, United Kingdom.

Keeping a Vigilant Eye: ISPS Compliance of Major Ports in the Philippines

R.R. Somosa, D.O. Docto, M.R. Terunez, J.R.P. Flores, V. Lamasan & M.M. Magramo
John B. Lacson Foundation Maritime University-Arevalo, Inc.

ABSTRACT: This study aimed to find out the compliance of the 7 Major International Ports in the Philippines to the requirements of the International Ships and Port Facility (ISPS) Code and the level of knowledge and awareness of the concerned port personnel in the proper implementation, compliance and evaluation of the said code. It also attempted to determine the control and compliance measures initiated by port authorities and the impacts of proper implementation to all concerned parties. Furthermore, it made use of a checklist type questionnaire devised by the researchers. The checklist contains items that pertain to the level of awareness of port personnel and the level of compliance of the Iloilo International Port as a whole. The study is a descriptive type of research utilizing both the quantitative and qualitative method of research.

1 INTRODUCTION

The shipping industry, as it sails towards a more technologically-sophisticated, yet more vulnerable generation, needs to answer to a higher set of security and safety standards and promote awareness and implementation among ship and port personnel. With these complexities that the industry experiences, there is a need for a stricter obedience of rules and regulations, as implemented by the International Maritime Organization (IMO). The International Ship and Port Facility Security Code (ISPS Code) is a comprehensive set of rules and guidelines formulated primarily to enhance the safety and security of complying ships and ports. The Code deals with risk management, its awareness and assessment that could eliminate or lessen threats among ships and port facilities.

This study covers the ports of Iloilo, Subic, Manila, Davao, Cebu, and Cagayan de Oro. These ports are major suppliers of tons of goods needed for the island as well as it is the major exporter of the resources of the country. Thus, these ports play a very important role in the improvement of its economy and livelihood of the citizens. The number of international and local vessels that make use of these ports has become a major concern when it comes to general security and safety while these vessels are berthed at the port. As observed with the international trade, the coming in and out of the vessels follows certain procedures that ensure the security and safety of the said ports. An inspection is conducted upon arrival; proper checking of documents and sanitation within the vessel are few of the major concerns of the port authorities. Same procedures are being observed prior to departure. This greatly depends on the management of the port authorities with the conjunction of the ISPS Code.

Issues dealing with drug transportation, smuggled goods, stowaways, discrepancies in ship organization, falsification of documents, cleanliness within the vessel, control of visitors for the ship and crew's health are some of the problems in which international and local trades are experiencing and the port authorities need to strictly validate. As with other ports all over the world, the vulnerability of these ports against these kinds of problems is not very far to be experienced.

The study dealt with the proper implementation of control and compliance measures as required by the ISPS Code for all concerned vessel within the port's area of responsibility. Following the tragic events on September 11, 2001 when international terrorists has threatened the security of one of the most developed nations in the world, the International Maritime Organization (IMO) agreed to develop security measures which are applicable to ships and port facilities. These measures have been included as amendments to the Safety of Life at Sea (SOLAS) Convention. On December 9 to 13, 2002, contracting governments to the SOLAS Convention finalized of the text on preventative maritime security regime. This was held at the IMO's headquarters in London. This conference adopted the acceptance procedures as established in SOLAS and also ensured that the maritime security measures would be accepted internationally by January 1, 2004, and in force after six months, by July 1, 2004. (ref:http://www.infrastructure.gov.au /transport/security/maritime/isps/index.aspx).

In the Philippines, the Philippine Ports Authority (PPA) has undertaken steps in relation to IMO Resolution No. A. 924(22), particularly the review of measures and procedures in the prevention of acts of terrorism which threaten the security of passengers and crew as well as the safety of ships. Accordingly, the IMO Security-General, within the Integrated Technical Cooperation Program was requested to take appropriate action and to assist government in strengthening maritime and port security.

A Maritime Security Seminar-Workshop and Conference was then initially conducted in October 10 to 11, 2002 at the Port District Office (PDO) in Southern Mindanao, participated by all Port Managers, Security Staff Officers and Station Commanders. The conference aimed to familiarize the participants with the concepts and principles of maritime and port security, and thus ensure understanding of the methods of conducting a port facility security survey.

PPA Operations Memorandum Circular No. 02-2002 dated November 5, 2002 was immediately issued thereafter to Port District Managers and Port Managers regarding the Conduct of Comprehensive Security Survey/Plan, particularly directing all concerned to conduct security survey of the base ports and terminals under their respective jurisdiction. This memorandum called for a series of presentations submitted by the different Port District Officers (PDOs)/ Port District Managers (PMOs) for approval and validation by the Office of the Assistant General Manager for Operations (AGMO) during the period from April 10, 2003 to May 6, 2003.

The approved and validated PMO Port Security Surveys/Plans were then implemented following the issuance of PPA Memorandum Order No 971-2002, dated November 28, 2002. This memorandum also designated PMO-PPD Managers/Station Commanders as Port Security Officers of their respective PMO.

Following that, PPA Memorandum Circular No. 10-2003, dated March 29, 2004, regarding the Conduct and Submission of Comprehensive Security Survey/Plan, and Designation of Port Facility Security Officer, was issued to all private port operators and owners.

Last December 3, 2003, during the Seminar on Maritime Security: Countdown to ISPS Code Deadline at the Mandarin Oriental Hotel in Makati City, all the PMOs were required to submit their revised comprehensive Port Facility Security Plans in accordance with the ISPS Code before the end of February 2004. Approval by the General Manager was scheduled early March 2004.

With these preparations, it is expected that ports under the Philippine Ports Authority will be ISPS Code Compliant by March 2004. This mandate from the Philippine Ports Authority (PPA), therefore will serve as this study's conceptual framework, assuming that because it was so mandated, all ports will be compliant to the ISPS Code.

2 STATEMENT OF THE PROBLEM

The International Ships and Port Facility Security (ISPS) Code was introduced and mandated after the 9/11 terrorists attacks to the World Trade Center Twin Towers in New York City, USA. On the 11th of September 2001, thousands of lives and millions of properties were lost and it affected not only land-based businesses but also the maritime industry. In order to alleviate the vulnerability of the industry to terrorism, the IMO mandated ISPS as a major security protocol among the ships in the maritime business. As the years passed it has proven to lessen both major and minor terrorist attacks. This study aims to determine the extent of implementation in the different ports in the Philippines. This study aimed to find out the level and over-all compliance of Port Authorities in the major ports in the Philippines as to the requirements of the International Ships and Port Facility Security (ISPS) Code.

2.1 Significance of the Study

This study aims to determine the level of compliance of the Iloilo International Port with the requirements of the International Ships and Port Facility Security (ISPS) Code in general. Specifically, it intends to measure how knowledgeable and aware the port authorities are in implementing the ISPS Code among all concerned party/parties making use of the port. It will also reflect the procedures that they are doing and how the ship takes action with this implementation. Through this, port authorities and crew personnel will be more aware and more vigilant in keeping an eye for port and ship's security and safety. This will also be a refreshing call for our port authorities to enhance their system of management in dealing with the vessel and its crew. This study will be a great benefit to our local government specifically the port authorities, to ship's personnel, to maritime students and to future researches in dealing with the International Ship and Port Facility Security Code in promoting safety and security as well as enhancing knowledge and awareness.

3 METHOD

3.1 *Research Design*

The study is a descriptive-non experimental research which used a one- shot or cross sectional survey design. The descriptive research attempted to describe the existing phenomena by gathering information through the library method and data through the use of a researcher-made questionnaire.

3.2 *Respondents of the Study*

The respondents of the study were the Personnel and Staff of the major ports in the Philippines. The total numbers of respondents who were given the chance to answer the researcher made questionnaire were 21 port personnel, all engaged in security and safety operations of the ports of Subic, Batangas, Manila, Cebu, Cagayan de Oro, Davao and General Santos City. The number of personnel was determined from the matrix of organization and as directed by the Port Manager of each port that were covered in the study.

4 RESULTS

4.1 *Descriptive Data Analysis*

Descriptive Data Analysis shows the responses of the respondents from the Personnel and Staff of the Iloilo International Port, particularly those engaged with the security and operations of the port. In this study, frequency count was used to determine the level of awareness, compliance, security measures and its impact. Percentage was also taken in each questions or category. Mode was also used to determine the over-all assessment of awareness. Rank was also used to determine the most frequent reduction of maritime threats in accordance with the implementation of ISPS Code.

Table 1 shows the frequency count and the percentage in each level of awareness or compliance. Table 2 shows the over-all assessment of the data collected.

When the respondents were asked regarding the strict implementation of ISPS Code and as well as the plans and designations of the Iloilo International Port, the following data were gathered: there is a strict implementation of the ISPS Code (19 or 90% said yes) and the plans or designations the Iloilo International Port has/have (Port Facility Security Officer/s, Port Facility Security Plan/s and Port Facility Security Contingency Plan). Table 3 presents the data.

Table 1. Level of awareness among port personnel and Level of ISPS Code Compliance of the port in each area/system

RESPONDENTS LEVEL OF AWARENESS OF THE ISPS CODE, ITS IMPLEMENTATION, COMPLIANCE AND EVALUATION

	Highly Aware		Moderately Aware		Slightly Aware		Not at all	
	F	%	F	%	F	%	F	%
Awareness Of ISPS Code	15	71.4	6	28.6	0	0	0	0

PORT AREAS/ SYSTEMS LEVEL OF COMPLIANCE TO THE REQUIREMENTS OF THE ISPS CODE

	Highly Compliant		Moderately Compliant		Slightly Compliant		Not at all	
	F	%	F	%	F	%	F	%
Organizational	3	14.3	15	71.4	3	14.3	0	0
Administrative	18	85.7	2	9.5	1	4.8	0	0
Physical	6	28.6	12	57.1	3	14.3	0	0
Technical Systems	8	38.1	11	52.4	2	9.5	0	0

Table 2. Over-all Awareness and Compliance based to ISPS Code on the most number of frequency count

RESPONDENTS LEVEL OF KNOWLEDGE & AWARENESS OF THE ISPS CODE

	Frequency-Percentage	Level of Awareness	Numerical Value	Assessment
Awareness Of ISPS Code	15-71.4%	Highly Aware	1	Highly Aware

PORT LEVEL OF COMPLIANCE WITH THE REQUIREMENTS OF THE ISPS CODE

Organizational	15-71.4%	Moderately Compliant	2	
Administrative	18-85.7%	Highly Compliant	1	
Physical	12-57.1%	Moderately Compliant	2	
Technical Systems	11-52.4%	Moderately Compliant	2	
		Mode	2	Moderately Compliant

Table 3. Implementation of ISPS Code, plans and designations of the Different International Ports

	Yes	
	F	%
Implementation	19	90

Plans or designations Iloilo International Port has:

	Yes	
	F	%
Port facility Security Officer/s	21	100
Port Facility Security Plan/s	21	100
Port Facility Security Contingency Plan	21	100

4.2 *Control and Compliance Measures*

The data collected were presented in a tabular from and analyzed using frequency count and percentage. In this category, it shows how the routines were being done and the mostly information asked from the vessel. The frequency count and its corresponding percentage in every category show the over-all assessment. The number of yes is dependent on the officer or personnel implementing or conducting the ISPS Code. Table 4 presents the data.

Table 4. Routines done and security information ships are required to provide

Routines	Yes	
	F	%
Ship in Control	21	100
Validation of ISSC	17	81
Required Information	21	100
Information that applies:		
Valid ISSC	21	100
Current Security Level	20	95
Security Level from previous port	15	71
Security measures taken	12	57
Security procedures conducted	14	67
Last 10 port of calls	12	57

4.3 Accordance and Impact of Compliance

In this category, the data were presented in a tabular form and analyzed using frequency, percentage and rank. Rank is used to determine the great impact or threat reduction in corresponding to the implementation of the ISPS Code. The top three threats that had a number of reductions were: Cargo Theft, Contraband Smuggling, Stowaways and refugees and Delays. Table 6 presents the data.

Table 5. Impacts of the compliance to the ISPS Code of the Iloilo International Port

Threats Reduction	Yes		
	F	%	Rank
Cargo Theft	21	100	1
Contraband Smuggling	20	95	2
Delays	18	86	3.5
Stowaway and refugees	18	86	3.5
Piracy and Armed Attacks	17	81	5
Terrorism	9	43	6
Collateral Damage	5	24	7
Hi-jacking	0	0	8
Hostage Taking	0	0	8
Mutiny	0	0	8

Findings of the study
1 The study shows that the Port Authorities of the major ports in the Philipines are highly aware of the ISPS Code and generally, the ports are highly compliant with the requirements of the Code.
2 As per result, high percentage of answers from the respondents confirmed that there is a strict implementation of the ISPS Code.
3 There are plans and designations required by the ISPS which are being utilized in the Major ports in the Philippines particularly on: Port facility security officer/s, Port facility security plan and Port facility security contingency plan.

4 The major ports in the Philippines implements the ISPS Code by initiating necessary control and compliance measures such as ensuring ships when in port are in control and checking if they have valid International Ship Security Certificate. Moreover, authorities require ships to provide security information like the current security level the ship is operating with, security level from previous ports, security procedures conducted and the information of the last 10 ports of call.
5 The proper compliance yielded positive impacts to the port itself, the ships, the shipping companies, charterers and other concerned parties primarily because of the reduced possibility of the occurrence of various maritime threats.

5 CONCLUSION

Being aware of the consequences of the non-compliance of the ISPS Code, the port authorities are strictly implementing the provisions and they are also taking all necessary precautions in order not to repeat the experiences of September 11, 2001.

6 RECOMMENDATIONS

A continuous and sustainable effort in implementing strictly the ISPS Code and the high awareness among port personnel nowadays and in the coming years will assure a safe and secured fort facility. It is attainable through enhancement of skills and knowledge of employees and by supervising trainings, seminar-workshops, drills and research amongst them. If there are some amendments to the code in the future, the port personnel should adjust and implement them on customary or strict basis if necessary just like what they are doing in the present. The Port shall also strive hard to have a highly compliant level in conjunction with the ISPS Code compliance.

The researchers would like to commend the proper compliance of the major ports in the Philippines to the requirements of the ISPS Code, in doing so; there is a strict implementation of the rules and regulations as per ISPS Code manual within its portals. Moreover, port personnel are doing their best in monitoring the stature of ships within their area of responsibility as to the aspect of ISPS Code while giving the researchers recommend to keep up what the port and its personnel are observing to implement the ISPS code properly and provide a world class service among vessels and crew using the port at any given time.

REFERENCES

Philippine Ports Authority. www.ppa.com.ph

HabagatCentral.Com. 2011. Iloilo Ports and Shipping (thread at Skyscrapercity.com). Online Available: http://www.skyscrapercity.com/showthread.php?p=88770815

National Geospatial-Intelligence Agency. 2010. Sailing Directions (Enroute): Philippine Islands. The United States Government. Bethesda, Maryland. Online Available: http://library.csum.edu/navpubs/pub162bk.pdf

http://en.wikipedia.org/wiki/Port_of_Iloilo#cite_note-ILOppa-0

http://en.wikipedia.org/wiki/International_Ship_and_Port_Facility_Security_Code

http://en.wikipedia.org/wiki/Isps_code

http://en.wikipedia.org/wiki/International_Convention_for_the_Safety_of_Life_at_Sea

The Philippine Star. 2004. 22 Philippine ports meet international security guidelines. Online Available: http://www.philstar.com/Article.aspx?articleId=256146

http://en.wikipedia.org/wiki/Philippine_Ports_Authority

http://www.infrastructure.gov.au/transport/security/maritime/isps/index.aspx

PPA Annual Report 2006: Bringing ports closer to the clients. 2006. Philippine Ports Authority. Online Available: www.ppa.com.ph/annualreport/PPA_R_06.pdf

PPA Annual Report 2010: Together we make things happen. 2010. Philippine Ports Authority. Online Available: www.ppa.com.ph/annualreport/PPA_AR10.pdf

The Using of Extruded Fenders in Yachts Ports

W. Galor
Maritime University of Szczecin, Poland

ABSTRACT: The continuous development of water tourism is observed. It causes an increase in the number of newly constructed and modernize yacht ports and marinas. Independent of kind of these ports, it is necessary to provide the proper safety of floating units. The basic condition is berthing and mooring without of damage of hull yacht and quay. The fenders are devices to protect against of yachts impact on quay. The fender construction is based on resiliency property and their shape. Most of them are extruded fenders as a hollowed bars. Actually, the polyurethane elastomers are used to produce of the fenders. The papers present the problems of safety maneuvering of floating unit in yachts ports and application of extruded fenders.

1 INTRODUCTION

There is a clear development of inland and sea tourism, which causes an increase in the number of newly constructed yacht port (called also as marinas). In particular, one can observe an increase the marinas as a ports to ensure the safe mooring of yachts in a specified period of time (from several hours to several weeks).

The yacht port, sea or inland, is a set of port water areas, port hydro structures, lands and building technical equipment, to ensure the safe mooring and maintenance of yachts and other recreational or leisure units and floating equipment. For the fulfillment of their tasks marina must be properly shaped water areas sheltered from the effects of winds, waves, currents and ice movement, as well as the right size port land areas (within the harbor), allowing through appropriate equipment, services the sailors and other users of the marina, that be planned in range of area plan of marina (Mazurkiewicz 2010). Figure 1 shows an example of the marina.

Marina is the yacht port combined with complementary residential buildings. For a complementary residential buildings should be intended as pavilions and other buildings, providing space for hotels, shops, bars, cafes and restaurants, and all the functions required by the temporary or permanent residents. The marina can also include recreation centers.

Another modern concept of the yacht port is so called sailor marina village. It is characterized by the fact that around the marina are integrated with the residential marina. This contributes to a significant enhancement of the marina.

Regardless of the type of the marina, it is necessary to take the appropriate steps to ensure the safety of individuals. The basic requirement is berthing and mooring without damaging the hull and the quays.

Figure 1. The marina in Athens (Greece)

2 THE YACHTS BERTHING TO THE QUAY

The maneuver of berthing a yacht to a port structure (a quay, a pier) is the final stage of the navigational process. An ideal maneuver would be consisted in a total loss of speed at the moment the unit makes contact with the quay. However, in reality, a dynamic of unit interaction takes place, that causes a deformation and stress of the hull and the fender (when applied). Fenders improve the safety of berthing operations by partially absorbing the kinetic energy of the ship. It consists in an elastic deflection (shape elasticity) of the material the yacht is made of, and the energy of berthing turns into work of deflection.

The fender absorbs a part of yacht's kinetic energy. The remaining part of the energy is absorbed by the hull structure and the port structure. The conditions of yacht's safe berthing are as follows (Galor 2002):

$$E \leq E_d \quad dla \quad p \leq p_{dop} \tag{1}$$

where $E=$ absorbed by the berth-fender-yacht's system; E_d = yacht's admissible kinetic energy absorbed by the berth-fender-yacht system; p = maximum pressure of individual fender on yacht's hull ; p_{dop} = admissible pressure of individual fender on yacht's hull.

Admissible pressure of an individual fender on the floating unit depends on its size and design.).

When determining yacht's kinetic energy to be absorbed by the fenders the following factors should be taken into account:
− yacht's canting when approaching the berthing line,
− location of the hull contact point with fender in relation to the ship's centre of gravity,
− elasticity properties of the fenders, the shell and the port structure,
− resistance of water between the yacht and the port structure, the shell plating and the fender.
− surface friction in way of contact point between

The above factors cause the kinetic energy of the ship to decline and as a result the fender is forced to absorb part of the entire energy which is understood as the effective energy. The effective energy while berthing the floating unit to a port structure (knocking into the fenders) will equal:

$$E = E_K \cdot C \quad [J] \tag{2}$$

where E_K = yacht's maximum kinetic energy; $C=$ total coefficient of energy loss.

3 THE QUAY FENDERS DEVICES

Quay fender is a device that protects port structure and sailing craft during its berthing and mooring to the structure. Yacht's berthing maneuver is actually the last stage of navigation process based on safe guidance of the craft from the starting to the point of destination.

There is a dynamic affect of yacht on fenders at berthing, causing deformation and tension of the device, ship and wharfs. The proper selection of fender assures that tensions are resilient in admissible range. The design and choice of the fenders should correlate with the construction of wharf to be protected. This choice depends substantially on the ship size and operational conditions in which the berthing maneuver is performed. These conditions consist of:
− hydro meteorological restrictions (wind, current),
− ship's maneuvering properties (as a power of main propulsion),

When selecting a fender, it is necessary to identify all limitations connected with preliminary choice of fender, from the vantage point of reaction force value transmitted from yacht to the quay, by means of fenders and their number. The protective action of fenders on the hull of yacht and the wharf depends on absorbing parts of kinetic energy (Galor 2007a). These devices take over the struck energy at the moment when the yacht touches the fender. Then the fender undergoes a resilient deformation and ship's energy bounding changes in work of resilience. During this process, the fender absorbs reaction of ship impact power which is transmitted to wharf construction. It means that fender transforms the overtaking energy of ship which transmits its energy to wharf construction. The fender transforms the energy to work on resilient units of the fender. The fenders play a great role in safety assurance during ship berthing maneuver. Reduction of the damage risk of the ship or wharf depends on the proper selection of fenders.

The basic parameters fenders are:
− reaction force in function of deflection,
− energy absorbed in function of deflection,
− admissible deflection.

There are also dimensions of fender (length, width, height) and sorts of fastening as additional parameters.

Basic parameters depend on the size, shape and sort of material used in fenders making. If the material has clear properties, then the fender parameters depend only on size and shape (Galor 2008). An efficient berth fender should be characterized by possibly high absorption of the energy of a mooring yacht and low pressure on the hull and on the berth structure. In this respect the most effective are devices using elements of high elasticity. Materials used previously, comprising wood, solid rubber, used tires and rubber cylinders do not satisfy these requirements. Fenders of the latest generation are manufactured of properly chosen rubber and polyurethane elastomers that have exceptionally elastic properties and feature high parameter repeatability. Nowadays, the most frequently used substances absorbing energy are rubber or plastic (mainly polyurethane elastomers). Fenders made of these materials can be used in full sections and combination of small elements in whole units (buckling units). Buckling fender consists of specifically joint elements that give us a fender with proper characteristics. It depends on obtaining quick growing reaction at not large deflection, and then on maintenance of almost constant reaction at further deflection. With the same amount of material used to build the fender it can get even four times more efficient (Fig. 2).

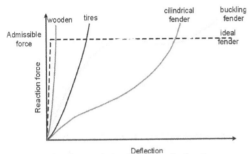

Figure 2. The characteristic of different fenders type

Designing and choosing the right type of device fender should be taken into account not only the energy absorption capacity of the yacht berthing, but also the structural strength of the yacht hull. This statement follows from the analysis of the damage during berthing. However, it can ascertain that in the case of small craft type barges, tugs, port fleet units, tourist passenger ships, recreation craft (yachts), etc., the basic condition safety is not exceeding the limit of deformation of the hull by absorption of kinetic energy during berthing maneuver. Thus, the selection of fenders for this type of unit should pay special attention to this element

4 KINEMATIC ENRGY OF SHIPS BERTHING

The amount of kinetic energy of a yacht's approach to a port structure can be generally expressed as dependence:

$$E_k = 0.5 \cdot m_v \cdot V_s^2 \quad [m/s] \qquad (3)$$

where m_v = yacht's virtual (total) mass;

$$m_v = m_s + m_w \quad [kg] \qquad (4)$$

where m_s = yacht's mass depended on its dimensions; m_w = mass of added water;

$$m_w = C_m \cdot m_s \quad [kg] \qquad (5)$$

where C_m = added water mass coefficient.

Mass of added water can be described as:

$$m_w = 0,25 \cdot \Pi \cdot T^2 \cdot L_{pp} \cdot g_w \quad [kg] \qquad (6)$$

where T = yacht's draught [m]; L_{pp} = length between perpendiculars [m]; g_w = density of water [kg/m³].

Added water mass coefficient Cm can be determined according to many dependencies. The most frequent are :

$$C_m = 1 + 2 \cdot T / B \qquad (7)$$

where B = breadth of the yacht [m].

The weight of the yachts largely determine its dimensions (4). According to the recommendations (Mazurkiewicz 2010), it can be assumed division into three units: sailing yachts (Table 1), motor yachts (Table 2) and trimarans and catamarans (Table 3).

Table 1. Sailing yachts

Group	L_C	B_{rml}	TJ_{ml}
AS	$L_C < 6.0$	< 1.8	< 1.3
BS	$6.0 < L_C < 8.0$	< 2.8	< 1.5
CS	$8.0 < L_C < 10.0$	< 3.2	< 1.6
DS	$10.0 < L_C < 12.0$	< 3.6	< 2.0
ES	$12.0 < L_C < 14.0$	< 4.2	< 2.4
FS	$14.0 < L_C < 18.0$	< 4.6	< 2.6
GS	$18.0 < L_C < 24.0$	< 5.6	< 3.2
HS	$L_C > 24.0$	> 5.6	> 3.2

L_c = overall length [m]; B_{rml} = extreme beam [m];
TJ_{ml} = maximal draft [m];
AS = Small centerboard yachts;
BS = Small sailing yachts , sword- ballast and removable ballast (centerboard race class yacht);
CS = Small sailing yachts
DS = Average sailing yachts;
ES = Larger sailing yachts;
FS = Large sailing yachts;
GS = Very large sailing yachts;
HS = Sailing ships and mega yacht.

Table 2. Motor yachts

Group	L_C	B_{rml}	TJ_{ml}
AM	$4.5 < L_C < 8.0$	< 3.3	< 1.0
BM	$8.0 < L_C < 10.0$	< 3.8	< 1.1
CM	$10.0 < L_C < 12.0$	< 4.2	< 1.2
DM	$12.0 < L_C < 14.0$	< 4.8	< 1.4
EM	$14.0 < L_C < 18.0$	< 5.2	< 1.6
FM	$18.0 < L_C < 24.0$	< 5.2	< 2.0
GM	$L_C > 24.0$	> 5.6	> 2.0

AM = Very small motor yachts;
BM = Small motor yachts;
CM = Average motor yachts ;
DM = Larger motor yachts;
EM = Large motor yachts;
FM = Very large motor yachts;
GM = Ships and motor mega yachts.

Table 3. Catamarans and trimarans

Types of yachts	L_C	B_{rml}	TJ_{ml}
small	$6.0 < L_C < 8.0$	< 4.8	< 0.6
average	$8.0 < L_C < 10.0$	< 6.0	< 0.8
large	$10.0 < L_C < 12.0$	< 9.0	< 1.2
very large	$L_C > 12.0$	> 9.0	> 1.2

5 EXTRUDED FENDERS

Extruded fenders are destined for mounting on quays and watercraft hulls. They are should be protect the hulls of small craft type barges, tugs, port fleet units, tourist passenger vessels, yachts, inland recreational crafts, etc., and a different ports structures. They are made as a bars in the different shapes: square, conical, circular and D-shaped (Galor & Galor 2007). For each shape also can be done inside the holes of varying diameter and shape (Fig. 3). Bar without the internal bore is used as a protective fenders on the quays and floating crafts, which require low energy absorption, but there is constant contact the hull with port structure. They are resistant to high impact. Square or trapezoid bars with a hole have applying on quays and dolphins to absorb energy during berthing of ships. They are also used for bows, sides and sterns of crafts. Particularly applicable of bars of shaped double trapeze shape are on bows and sterns of tugs. Fenders with circular hole (cylindrical) are most commonly used to protect the quays. They are also used on small ships, boats, yachts and tugs. Additional holes in the side wall fenders make them more "soft", which is a desirable feature for units with little reliable sides (glass-fiber, aluminum alloy).

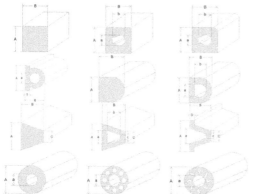

Figure 3. The most frequent of using the extruded fenders

The D-shaped fenders are commonly used on port quays, marinas and small vessels, dredgers, fishing boats al. This profile has a flat base and a rounded front surface providing gentle initial contact between the crafts side and the quay. The extruded fenders can be fixed mounted vertically, horizontally and diagonally. They can be regularly arranged to form a grid on the wall of quay, which facilitate berthing and mooring of floating crafts units of different sizes. Figure 4 presents an example of yacht's quay.

Figure 4. The yacht's quay with extruded fenders

6 CONCLUSION

Berthing maneuver of the yacht to the port structure (quay, pier, dolphin) is the last step in the process of navigation. During this maneuver, the dynamic affect caused strain and stress of the craft hull, quay and fender (if applicable). Quays fenders are special devices to protect port structures and floating craft during its berthing and mooring. Effective fender should have a possibility of high energy absorption and low pressure to the hull and structure. Currently, the newest generation of fender is built based on correctly selected components of polyurethane elastomers which is characterized by unique elastic properties and the ability to achieve high reproducibility parameters. An example of such elements are extruded fenders, which are designed for mounting on port structures and hulls of watercraft. These fender bars can be mounted vertically, horizontally or diagonally on quay walls. They can be regularly arranged to form a grid on the quays, which facilitate berthing and mooring floating crafts of different sizes. This is particularly useful the yacht ports and marinas.

REFERENCES

Galor, W. 2002. Bezpieczeństwo żeglugi na akwenach ograniczonych budowlami hydrotechnicznymi, Szczecin: FRWSM.(in Polish)
Galor, W. & Galor, A. 2007: The fenders in safety system in inland shipping. *Polish Journal of Environmental Studies. Vol.16,No 6.*
Galor, W. 2007a. The modeling of buckling fenders to protect the ship berthing process. *The Journal of KONES, Vol.14, No 3.* Warsaw.
Galor W. 2008. Odbojnice z elastomerów poliuretanowych *Inżynieria Morska i Geotechnika, Nr 6.* (in Polish).
Mazurkiewicz B. 2010. Porty jachtowe i mariny. Projektowanie Gdańsk: FPPOiGM. (in Polish).

The Positive Implications for the Application of the International Ship & Port Facility Security and its Reflects on Saudi's Ports

A. Elentably
Faculty of Maritime Studies, King Abdul-Aziz University, Saudi Arabia

ABSTRACT: The research paper summed up on the requirements of the application code security and safety of ships and ports (ISPS) and the technical aspects necessary for the application by the Saudi marine Ports. The requirements of the international code of safety and security of ships and ports such as:
- Additional tasks to be undertaken by port management - Activities and tasks that will port authorities
- The impact of the elements of the maritime transport - Application and amendments to the deck
- Government requirements - Special requirements for the management of ships
- Application optimized for the requirements of the code ,Also interested in the research paper the mechanism of how to put these requirements into effect and the positive impact associated with the application. And also the requirements of the bridge on the ship ,beside clarification of the interconnections between the parties to the transfer process, such as administration of the commercial maritime fleet operations, control to the owners and how the administrative process for the crew to apply the appropriate code on the deck of ships and mutual relations with the insurance and chartering operations as well as the role of the port facility, to arrived How can the ports of Saudi Arabia to benefit from the positive application of code requirements and to enable these requirements with the parties to the process of maritime transport

1 MARITIME SECURITY IN SAUDI'S PORTS

1 The Scopes which Saudi's efforts should be coverage
Important Tasks for Maritime Administration
2 Ideal Means of Stable Marine Transport in future
On the Assurance of Stable International Marine Transport
3 Efforts to assure the global competitiveness of Saudi's ocean-going shipping service operators and a planned increase of Saudi's -flag ships and Saudi's seafarers [study of tonnage-geared standard tax system and so forth],Consultation in search for the "Ideal Means of Stable Marine Transport in future" was Conducted by the Minister of Land, Infrastructure and Transport to the Council of Transport Policy on February 8, in order to ask the Council to discuss the means of assuring stable marine transport indispensable for Japan to accomplish continued sustainable growth as a maritime and trading nation in a global international economic community. Following the consultation, an "International Marine Transport Task Force" was established, composed of members with wide knowledge,

representing various circles, such as individuals of experience and academic standing in the areas of trade of resources and energy, finance, traffic economy and so forth.
The policy aims were drastically condensed into "Assurance of the global competitiveness of Saudi's ocean-going shipping service operators" and "Securing Saudi's -flag ships and Japanese seafarers", while as for measures to achieve the policy aims concerned, the introduction of the laws for, among others, the introduction of a tonnage-geared standard tax system, securing of Saudi's -flag ships and Saudi's seafarers and so forth were enumerated. Hereafter, the decision was taken to work on constructing an institutional framework to target stable international maritime transport based.
4 Efforts to Secure and Nurture Human Resources for the Sound Development of the Maritime Industry
Efforts to gather, nurture seafarers and target their career development to support them in transforming themselves into land-based ocean engineers Marine transport, which is indispensable for the society and economy of

Japan as a maritime state, is supported by seafarers engaged in ship navigation and ocean engineers who manage and support it on land. In securing the safety and stability of marine transport, the role played by seafarers (ocean engineers) as the human infrastructure is considerable. Since the valuation related to the navigational safety of Saudi's -flag ships and ships served by Saudi's seafarers on board is extremely high in these days, the government should positively promote:

- Efforts to secure and nurture excellent Japanese seafarers (ocean engineers). With this in mind,
- The Human Infrastructure Task Force was established within the Maritime Affairs Subcommittee of the Traffic Policy, which investigated and discussed an ideal maritime policy to secure and nurture human resources in the field of maritime affairs, focusing on securing and nurturing excellent Saudi's seafarers (ocean engineers).Subsequently, an interim to the effect that efforts were required mainly for four measures, namely nurturing seafarers, gathering them, targeting their career development and supporting their transformation into ocean engineers on land, with the necessary institutional revision and so forth scheduled to be carried out in future. Moreover, with a view to nurturing young seafarers, who will play a key role in the Saudi's marine transport of the next generation.

Support program to develop next-generation human resources in the shipbuilding industry
Since nearly half the skilled technical experts for shipbuilding in the Saudis shipping industry are over 30 years old, an unprecedented rapid and large-scale alternation of generation will take place in the coming decade. If effective countermeasures are not taken under such circumstances, the level of technique at manufacturing sites, which has underpinned the international competitiveness of the Japanese shipbuilding industry to date, will be abruptly degraded, which might lead to the loss of such competitiveness. With such conditions in mind, an intensive training project commenced from fiscal 2004 to ensure "expert workman techniques related to shipbuilding, could be smoothly passed on to the younger generation. Beside that there are a huge governmental toward marine educational such as establish separate faculty specialized in ports and maritime transport, navigation, surveying and marine engineering

2 ASSURANCE OF SAFE, SECURE AND ENVIRONMENTALLY-FRIENDLY MARINE TRANSPORT

2.1 *Reinforcement of Safety Assurance Measures*

Reinforcement of the audit of safety management and seafarer's labor / guidance system In recent years, there have been intense efforts to ensure navigational safety in the form of the appropriate navigational control of ships and improved working environment of seafarers, Since accidents involving ships, including coastal freighters or ultrahigh-speed vessels, have been Occurring. The safety assurance of vessel navigation is the responsibility of the Inspector for Safety Management and Seafarers Labor, who is appointed in each regional transport bureau and so forth, after the unification of the Inspector of Navigation in charge of inspection on safety Management of passenger boats as well as the freighters and the Inspector of Seafarer's Labor in charge of the working conditions of seafarers. Therefore, an efficient and agile audit can be performed by the executive officer, who has a wide supervisory authority related to the business laws (Maritime Transportation Law, Coastal Shipping Business Law) and seafarers-related laws (Seafarers Law, Seafarers Employment Security Law, Law for Ships' Officers and boats 'operators).

Moreover, the training system has been reinforced, and a new audit system has been Constructed, capable of checking the past audit status, record of contraventions and so forth any time on the spot during the audit, in order to enhance accuracy when the Inspector for Safety Management and Seafarers Labor is executing duties over a wide area, in order to conduct unified planning / gestation and guidance for the services to be provided by the Inspector for Safety Management and Seafarer's Labor.

Measures to prevent recurrence when a serious accident occurs When a serious ship accident occurs, measures are taken, with the cooperation of the Saudi's Coast Guard and so forth, such as prompt inspection, an examination to find out the cause, reprimand or guidance of the party concerned, in accordance with the laws for reconstructing the safety management system, and the implementation of thorough safety management in order to prevent the recurrence of similar accidents on a nationwide basis and so forth oceangoing vessel grounding accident in the offing of the Kashima port and so forth.

Measures to ensure the safety of ultrahigh-speed vessels In recent years, accidents caused by the collision of hydrofoil type ultrahigh-speed vessels, navigating at a high velocity of about 40 knots, and whales and the like have occurred one after the other in the seas around Japan. In consideration of such circumstances, the Ministry of Land, Infrastructure and Transport established the "Safety Measures Advisory Committee for Ultrahigh-Speed Vessels" in April, 2006 to study how to ensure the safety of hydrofoil type ultrahigh-speed vessels and finalized an interim summary report in August the same year.

Moreover, it was decided that unified guidelines for the content of training and the training period for the navigation personnel of hydrofoil type ultrahigh-speed ships should be provided, and it has also been decided that the "Guidelines for the Training of Navigation Personnel of Hydrofoil Type Ultrahigh-Speed Vessels" are to be formulated to improve the training level by the end of fiscal 2007.

Introduction of transport safety management system "Law Revising a Part of the Railway Business Law etc. for the Improvement of the Safety of Transportation" (Law No. 19 of 2006) was approved in the Diet and put into effect in October, 2006, to deal with circumstances whereby the trust in the safety of public transportation facilities for the nation was seriously eroded and seek to enhance the safety management system. Thereby, the transport safety management system was applied to the marine transportation field in addition to traffic fields, such as rail and air transport.

Implementation of The Voluntary IMO Member State Audit Scheme In the wake of large-scale accidents involving oil spillages from tankers, there has been an increasingly urgent need to eliminate substandard vessels. The background involves the present situation having been illustrated, in which the government of the flag state has failed to satisfactorily meet obligations to supervise and oversee ships of its own flag, to ensure they observe the international standard. which audit scheme by International Maritime Organization (IMO) on the enforcement of the conventions by the flag states, and after considerations under IMO to seek a means to have the government of the flag state meet its obligations under the conventions and subsequently to introduce the audit scheme, the implementation of the audit scheme was adopted at the 24th Session of the IMO Assembly in December, 2005, and has started since September, 2006.in recognition of the fact that the operation is conducted comprehensively and efficiently to meet obligations under the international conventions, from all the viewpoints of flag, port and coastal state, including the construction of the "Maritime Affairs Quality Management System", the nurturing of inspectors, and the establishment of the system of Port State Control (PSC) implementation and so forth.

Drastic reform of the pilotage system As Saudis' seafarers have become increasingly scarce in recent years, a shortage of pilots with sea captain experience is anticipated in the near future, raising apprehension of a potential inability to maintain smooth shipping traffic operations. Furthermore, in view of the increasing demand for improved operational efficiency / accuracy of the piloting service forming part of the port service, and based on the perspective of strengthening the international competitiveness of Saudi's ports, the Pilotage Service System" was established, within which discussions concerning the desirable nature of the pilotage system took place, and a bill partially amending the Marine Pilot Law ("Bill for the Partial Amendment of the Port Law and Others for Strengthening the Basis of Maritime Distribution").

Reinforcement of safety and security measures in the Straits of Malacca and Singapore In order to promote the measures against piracy and armed robbery against ships, the guideline was compiled in March 2006. Based on this guideline, Ministry of Land, Infrastructure and Transport has decided to promote various measures even more strongly in order to reduce the number of the incidents by pirates and armed robbers, through efforts for cooperation with related agencies and shipping industries, and through enhancement of maritime security in international society. As part of efforts for international cooperation in the Straits of Malacca and Singapore, survey of traffic volume was conducted to gage how many ships were actually navigating in the Straits. The results of a survey made clear that beside Saudi, many other countries were the beneficiaries in various ways from the passage through the Straits. for enhancing safety of navigation and environmental protection were proposed by the littoral states, and the "Kuala Lumpur Statement" was adopted, outlining cooperation and so forth, toward establishing mechanism to provide funding for the projects such as replacement and maintenance of Aids to Navigation. In order to establish a new framework for international cooperation including foundation of Aids to Navigation Fund, Japan, as one of the major user states of the Straits, contribute proactively to the future progress of discussions at international conferences and so forth.

3 TACKLING ENVIRONMENTAL PROBLEMS

1 Countermeasures against global warming In order to attain the targets for reduction accordance Protocol through the promotion of a modal shift from transportation by truck to coastal shipping and so forth, the targeted goal in the maritime transportation-related sector is a reduction of around 1.4 million tons in the CO_2 emission volume by fiscal 2010, and the Maritime Bureau is implementing "Comprehensive Measures for the Greening of Maritime Transportation" in order to attain the said reduction target. In addition, in order to prevent any increase in the CO_2 emission volume from the transportation

sector, such as from automobiles and ships, using petroleum and similar fuels, the Energy Saving Law was revised in fiscal 2005 (put in force on April 1, 2006), which obliges shipping service operators with a transport capacity exceeding a certain scale (holding ships with gross tonnage of 20,000 tons or more) in the maritime transportation-related sector.

2 Tackling ship recycling system at an international level, Since the poor conditions of the related labor environment, sea pollution originating from recycling yards and so forth are viewed as problems related to ship recycling (the dismantlement of ships) conducted in developing countries, in recent years, a study is underway in international organizations, such as the United Nations Environmental Programme (UNEP),International Maritime Organization (IMO), International Labour Organization (ILO) and so forth to try and solve such problems. In particular, the IMO has decided to formulate a new convention concerning ship recycling in 2008-2009, and discussion of the convention draft is progressing.

3 Efforts to ratify the ILO Maritime Labour Convention
At the 94th(Maritime) Session of the International Labour Organization (ILO) Conference held in February, 2006, the Maritime Labour Convention 2006 was adopted, which consolidates all the 60 or so conventions and similar bodies that have been adopted to date since the 1919 establishment of ILO, to ensure they reflect the present era, and simultaneously improve their effectiveness. It has been decided that future efforts for the preparation and study required to ratify this convention, such as the arrangement of domestic laws, an enforcement system and so forth for governing inspections of flags state or PSC, etc. will be advanced, and, at the same time, coordination and cooperation with the countries in the Asia region will also be promoted in order to expedite ratification by the same.

4 TACKLING MARITIME POLICY AND REGIONAL REVITALIZATION AND SO FORTH IN THE AREA OF MARITIME AFFAIRS

Dealing with Basic Act on Ocean Policy Basic Act on Ocean Policy, which contains the basic concepts of ocean policy, government responsibility, local public bodies and so forth, as well as basic measures etc., (put in force on July 20, 2007). While the Maritime Bureau has been promoting such various measures to date, including improvement in the environment of international competitiveness,

assurance of stable transportation, promotion of the marine business and support for various kinds of research and development as well as nurturing and securing human resources, it has been recognized, in view of the enforcement of the "Marine Basic Law", that various measures toward the realization of a sea-oriented state shall be promoted concentrically and comprehensively in future, as in the past and the decision has also been taken to diligently strive for the further development of the marine industry as a whole and reinforcement of its international competitiveness.

Efforts for regional revitalization In view of the severely worsening circumstances surrounding public transport in local areas, the "Act for Revitalizing and Reviving Local Public Transport" was enacted in May, 2007 for the purpose of implementing measures for the smooth introduction of a new form of passenger transport service suited to local needs, as well as comprehensive government support for the joint efforts of related local parties led by the municipality, so that they may create attractive regions through the revitalization and revival of local public transport.

5 EFFORTS EXPLOITING THE ADVANTAGE OF MARINE TRANSPORT

Enhancing the appeal of voyages by sea and the promotion of coastal passenger ships, including encouragement of sightseeing tours to and from remote islands In view of the interim proposal compiled at the "Roundtable Conference for Reviving the Attractiveness of Voyage by Sea" in June, 2006, the topics of "Enhancing the appeal of Voyages by Sea" and "Promotion of Sightseeing Tours to Remote Islands" have been positioned as the most important measures of fiscal 2006. With a view to enhancing the appeal of "Voyages by Sea in Casual Wear".

Promotion of a future business model for coastal shipping the coastal shipping industry has faced various problems, such as securing seafarers, building ships for replacement and safety assurance. However, under present circumstances, it is difficult for coastal shipping operators, who are mostly medium, small and micro enterprises, to work on these problems individually. Under such circumstances, a movement for the loose grouping of coastal shipping operators, utilizing ship administration companies, is attracting attention. It is important to promote these grouping movements as a new business model of coastal shipping for the future, in order to ensure stable marine transport and revitalize coastal shipping. For this purpose, the national government has positively started striving for its propagation and promotion.

6 EFFORTS TO PROMOTE THE CONSTRUCTION OF NEW COASTAL VESSELS TO REPLACE OLD ONES

Coastal shipping is one of the trunk distribution industries in Japan which supports its economy and national life, accounting for about 40% of domestic distribution, and in particular, about 80% of transport of fundamental goods for industry (steel, petroleum, cement and so forth).In recent years, the tendency toward an "aging population combined with diminishing birthrate" has advanced rapidly in the coastal shipping sector, which supports the above-mentioned activities. Given the importance of revitalizing coastal shipping in order to realize the construction of new coastal vessels replacing old ones on a stable and adequate scale, an "Action Plan for Promoting the Construction of New Coastal Vessels to Substitute Old Ones" was formulated in March, 2006 to solve those problems.

7 MARITIME SECURITY IN PORTS

7.1 Examples of maritime terrorist

There are some examples of maritime terrorist attacks that we all remember such as e.g.:
- The USS Cole bombing was a suicide attack against the US Navy guided missile destroyer USS Cole (DDG 67) on October 12, 2000 while it was harboured in the Yemeni port of Aden. A small craft approached the port side of the destroyer and an explosion occurred, putting a 35-by-36-foot gash in the ship's port side. The blast hit the ship's galley, where crews were lining up for lunch. Seventeen sailors were killed and 39 others were injured in the blast.

Figure 1. USS COLE

Figure 2. USS COLE - impact of small suicide launch

- On October 6, 2002, the m/t Limburg was carrying 397,000 barrels of crude oil from Iran to Malaysia, and was in the Gulf of Aden off Yemen to pick up another load of oil. It was registered under a French-flag and had been chartered by the Malaysian petrol firm On October 6, 2002, the Limburg was carrying 397,000 barrels of crude oil from Iran to Malaysia and was in the Gulf of Aden off Yemen to pick up another load of oil. It was registered under a French flag and had been chartered by the Malaysian petrol firm Petronas. While it was some miles offshore, an explosives-laden dinghy rammed the starboard side of the tanker and detonated. The vessel caught fire and approximately 90,000 barrels of oil leaked into the Gulf of Aden .A 38 year-old Bulgarian crew members was killed and 12 other crew members were injured.

Figure 3. Offshore location of m/t Limburg attack

Figure 4. Impact of dinghy on m/t Limburg

By far the most lethal maritime terrorist incident this millennium was the attack on the m/v Super ferry 14 in Manila by the Abu Sayyaf Group on 27 February 2004. Just after midnight local time, a bomb exploded on board the passenger ferry, which had left Manila Bay two hours earlier. The resulting fire caused the ship to capsize and more than 116 people were killed in the attack. On that day, the 10,192 ton ferry was sailing out of Manila with about 900 passengers and crew. A television set filled with 8 lb (4 kg) of TNT had been placed on board. 90 minutes out of port, the bomb exploded. 63 people were killed immediately and 53 were missing and presumed dead.

Figure 5. m/v Superferry 14 after bombing

Figure 6. m/v Super ferry after capsizing

– The November 5, 2005, pirate attack on the Seaborne Spirit cruise ship 100 miles off the Somali coast was the 25th such incident in the last six months. Six vessels are currently being held by pirates, one of them captured at a distance of 120 miles from the coast. The Seaborne Spirit managed to evade being boarded by two boatloads of pirates on inflatable speedboats armed with grenade launchers and machine guns. The ship, with more than 300 people on board, was on its way to the Kenyan port of Mombasa where it was due to pick up more passengers, including Australians. The ship came under attack at 5.30 a.m. as the pirates approached in at least two speedboats shooting at the ship with grenade launchers and machine guns. They were repelled by the ship's crew who set off electronic countermeasures, described as "a loud bang" by one of the passengers. One crew member was slightly injured in the early-morning incident. There was at least one RPG that hit the ship, one in a stateroom. There were calls for a naval task force to try to stop attacks in Somali waters - among the most dangerous in the world.

Figure 6. Workers are seen painting the flank of the US-owned Seaborne Spirit ship docked in Port Victoria in the Seychelles archipelago on November 7, 2005 after experts removed an unexploded grenade embedded in a passengers cabin. (AFP photo)

– In Southeast Asia in particular, since the September 11 attacks a number of worst case scenarios have been postulated by the media and academics alike. The formation of a terrorism-piracy nexus was, and still is, seen as a potential alarming development. It was believed that given the high rates of piracy seen in the region's waterways, coupled with the valuable knowledge and skills of the pirates, it was only a matter of time before terrorists teamed up with pirates. The possibility of terrorists blocking strategic waterways like the Malacca and Singapore Straits was also seen as a real threat. Predictions were made that militants could sink a large vessel at a narrow chokepoint in one of the region's waterways, block the passage of shipping and cause widespread economic chaos. Despite these isolated incidents of maritime terrorism and the predictions of worst case scenarios, maritime terrorist attacks are, and have remained, quite rare. They constitute only two percent of all international terrorist incidents over the last three decades. While there is no doubt that a number of terrorist organizations have the desire or motivation to carry out attacks of this kind, in general there is still currently a lack of capability in this area of operation and it is likely to remain the case in the immediate future. Attacks against maritime targets require specialized equipment and skills; they also might require some knowledge of local shipping patterns, boat operation and maintenance, and boarding techniques. Even the attack involving the USS Cole, conceivably one of the simplest methods of attacking a maritime target, failed in its first attempt. The original intended mark was in fact the USS The Sullivan. However, in their first try at launching the suicide boat, the al-Qaeda operatives underestimated the weight of the explosives they were carrying on board and the boat sank as it entered the water. Although, at present, the probability of a large-scale maritime attack is low, the threat of maritime terrorism must not be ignored altogether. There is evidence that preliminary steps have been made by the al-Qaeda network in particular to develop some competency in this area. Recently, a basic diving manual was recovered in Kandahar in Afghanistan and it is believed that this is evidence of a larger plan to set up and run a diving school. J.I. (Jemaah Islamiyah's) has also been conducting training in the southern Philippines in order to develop underwater destruction capability. In addition, J.I. and a number of other jihadist groups based in Indonesia already fully exploit the maritime domain for the purposes of transporting people and arms to and from the Philippines. The threat of terrorist acts against the shipping and port industry is real and not imaginary. It is for these reasons the Assembly of IMO, in November 2001, decided that the organization should review measures and procedures to prevent acts of terrorism that threaten the security of passengers and crew and the safety of ships. It is also obvious that the Contracting Governments to the 1974 SOLAS Convention, when they adopted the special measures to enhance maritime security in December 2002, were well aware of potential threats.

8   MARITIME TRANSPORTATION SECURITY ACT OF 2002 (MTSA)

After the terrorist attack of 9/11 in 2001 on the WTC twin towers the fear of imports of mass destruction weapons or terrorists transported by ships in containers was imminent and the US took measures by installing the Maritime Transportation Security Act of 2002 (MTSA) in January 2002. The goal of MTSA is to prevent a Maritime Transportation Security Incident (MTSI) with:
– loss of life
– environmental damage
– transportation system disruption
– Economic disruption to a particular area.

MTSA calls for a series of plans on the national, port and individual vessel/facility level - this "family of plans" concept worked well for oil spill response and was used to increase MTSA awareness throughout the maritime community to coordinate information and to deal with potential threats. Vessels and facilities that load/carry certain dangerous cargoes (flammable, potentially explosive, caustic or environmentally hazardous) must have individual security plans that address fundamental security measures such as access controls, communications, restricted areas, cargo handling and monitoring, training and incident reporting.

The "port plan" called the Area Maritime Security Plan covers facilities and waterway venues such as parks or public piers that are not required to have individual security plans. The AMS plan is developed and implemented by an Area Maritime Security Committee with representatives from federal, state, and local governments as well as industry and the public sector. These Committees and the AMS plans are the backbone of communicating and coordinating surveillance and preparatory measures as threats to our maritime infrastructure warrant.

9   CSI/CTPAT (SUPPLY CHAIN SECURITY)

One of the Customs and Border Protection (CBP) programmes is the Container Security Initiative (CSI) programme for CBP-inspectors at large overseas ports. The duty of such inspectors is to pre-screen cargo containers being shipped to the United States, i.e. identify and inspect high risk containers before they are loaded on ships at their port of origin. The programme focuses on four core elements:
1 Using automated information to identify and target high-risk containers
2 Pre-screening containers as high risk before they arrive at a US port

3 Using detection technology to quickly pre-screen high-risk containers

4 Using smart, tamper-proof containers

Companies and organizations become participants in the programme by defining and implementing a formal internal supply chain security programme based on a self assessment against guidelines provided by the CBP that address various items such as procedural security, physical security, education and training, access controls, manifest procedures and conveyance security. Customs and Border Protection instituted the 24-Hour Rule, which requires information on cargo destined for the United States to be submitted through the CBP Automated Manifest System (AMS) by the carrier or by a "non-vessel operating common carrier" if they are AMS certified. The rule requires detailed descriptive information for all cargo. It requires cargo vessels entering ports to provide a cargo manifest 24 hours before leaving their last foreign port. A "Do Not Load" order may be issued for the carriers at the foreign port for cargo that does not meet the 24-Hour Rule. Some new programmes focus on point-to-point verification of the global supply chain. Operation Safe Commerce (OSC) and Safe and Secure Trade Lanes (SST) both aim at finding reliable and cost effective procedures and technologies to track containers from their point of origin to their final destination. Operation Safe Commerce (OSC) is a public/private partnership implemented by the Transportation Security Administration. OSC is dedicated to finding methods and technologies to protect commercial maritime shipments from the threat of terrorist attack, illegal immigration and other contraband while minimising the economic impact on this critical transportation system. It is a federally funded programme providing a test-bed for new techniques to enhance the security of containerized shipping, from the overseas point of origin throughout the supply chain to the US point of distribution. Those security techniques that prove most successful under the programme will then be recommended to create international standards for secure and efficient containerized shipping.

10 CSI: CONTAINER SECURITY INITIATIVE (12-04-2002)

Containerized shipping is a critical component of international trade. According to the CBP: about 90% of the world's trade is transported in cargo containers almost half of incoming US trade (by value) arrives by containers onboard ships nearly seven million cargo containers arrive on ships and are unloaded at US seaports each year. As terrorist organizations have increasingly turned to destroying economic infrastructure to make an impact on nations, the vulnerability of international shipping has come under scrutiny. Under the CSI programme, the screening of containers that pose a risk for terrorism is accomplished by teams of CBP officials deployed to work in concert with their host nation counterparts.

CSI consists of four core elements:

1 identify high-risk containers. CBP uses automated targeting tools to identify containers that pose a potential risk for terrorism, based on advance information and strategic intelligence.

2 pre-screen and evaluate containers before they are shipped. Containers are screened as early in the supply chain as possible, generally at the port of departure.

3 use technology to pre-screen high-risk containers to ensure that screening can be done rapidly without slowing down the movement of trade. This technology includes large-scale X-ray and gamma ray machines and radiation detection devices.

4 use smarter, more secure containers that will allow CBP officers at United States ports of arrival to identify containers that have been tampered with during transit.

The initial CSI programme has focused on implementation at the top 20 ports shipping approximately two-thirds of the container volume to the United States. Smaller ports, however, have been added to the programme at their instigation and participation is open to any port meeting certain volume, equipment, procedural and information-sharing requirements. Future plans include expansion to additional ports based on volume, location and strategic concerns. The CSI programme offers its participant countries the reciprocal opportunity to enhance their own incoming shipment security. CSI partners can send their customs officers to major US ports to target ocean-going, containerized cargo to be exported from the US to their countries. Likewise, CBP shares information on a bilateral basis with its CSI partners. Japan and Canada are currently taking advantage of this reciprocity. CSI has also inspired and informed global measures to improve shipping security. In June 2002, the World Customs Organization unanimously passed a resolution that will enable ports in all 161 member nations to begin to develop programmes according to the CSI model. On 22 April 2004, the European Union and the US Department of Homeland Security signed an agreement that calls for the prompt expansion of CSI throughout the European Community.

11 C-TPAT - CUSTOMS-TRADE PARTNERSHIP AGAINST TERRORISM

C-TPAT is a joint government-business initiative to build cooperative relationships that strengthen overall supply chain and border security.

C-TPAT recognizes that Customs can provide the highest level of security only through close cooperation with the ultimate owners of the supply chain: importers, carriers, brokers, warehouse operators and manufacturers. Through this initiative, Customs is asking businesses to ensure the integrity of their security practices and communicate their security guidelines to their business partners within the supply chain.

C-TPAT offers trade-related businesses an opportunity to play an active role in the war against terrorism. By participating in this first worldwide supply chain security initiative, companies will ensure a more secure and expeditious supply chain for their employees, suppliers and customers. Beyond these essential security benefits, CBP will offer benefits to certain certified C-TPAT member categories, including:

– a reduced number of CBP inspections (reduced border delay times)
– priority processing for CBP inspections (front-of-the-line processing for inspections when possible)
– assignment of a C-TPAT Supply Chain Security Specialist (SCSS) who will work with the company to validate and enhance security throughout the company's international supply chain
– potential eligibility for CBP Importer Self-Assessment programme (ISA) with an emphasis on self-policing, not CBP audits
– Eligibility to attend C-TPAT supply chain security training seminars.
– International Ship and Port Facilities Security Code (ISPS code)

The ISPS code is limited to ships over 500 gt. the main objectives of the ISPS code are as follows:

– To detect security threats and implement security measures
– To establish roles and responsibilities concerning maritime security for governments, local administrations, ship and port industries at national and international level
– To collate and promulgate security-related information
– To provide a methodology for security assessments so as to have in place plans and procedures to react to changing security levels.

In Belgium they installed a central Federal Committee for the Security of Port Facilities (FCSPF) and a Local Committee for the Security of Port Facilities (LCSPF) for each seaport. The members of these committees are shown below:

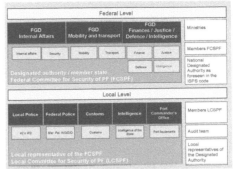

Figure 7. Federal and Local Committees for the Security of Port Facilities

The process and flow chart used by the port authority of Ghent is illustrated below:

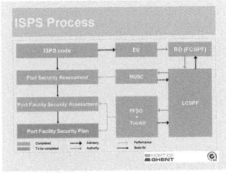

Figure 8. ISPS process

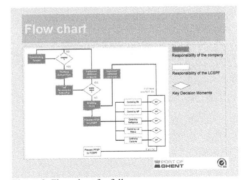

Figure 9. Flow chart for follow-up

A ship has to give his security level (SL 1,2 or 3) 24 hours before arrival in port. It is the designed authority of the government of the flag state that decides about the SL of its ships. The master of the ship can take extra security measures if he wants, but he cannot put his SL on a higher level. Only emergency safety measures can overrule the security measures provided in the SSP (Ship Security Plan). If the SL of the port facility is higher than that of the ship, then the SSO (Ship Security Officer) has the obligation to equalize the security level of the ship. When the ship has a higher security level than that of the port facility, the SSO together with the PFSO (Port Facility Security Officer) have to make up a DOS (Declaration Of Security).

It is the designed authority of the government of the port that decides about the SL of the port facilities in the port area. The security measures for each SL are written in a PFSP (Port Facility Security Plan).

The security measures are more severe when the SL is higher and also depend on the risks that can be encountered at the port facility. For instance a dangerous goods terminal, a passenger terminal or a container terminal will be fenced and guarded. An open bulk terminal, with no dangerous commodities, will only be fenced and guarded on the ship/shore interface when we have an SL 2 or 3. At security level 1 we have business as usual, without any economic constraints. The security measures can be visualized in the following templates:

Figure 10. Non sensitive terminals – OBC – SL1

Figure 11. Non sensitive terminals – OBC – SL2

Figure 12. Non sensitive terminals – OBC – SL3

Figure 13. Sensitive terminals – containers – SL1

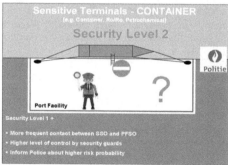

Figure 14. Sensitive terminals – container – SL2

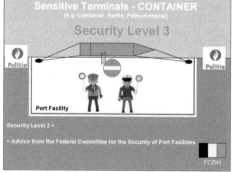

Figure 15. Sensitive terminals – container – SL3

Figure 16. Passenger terminal – SL1

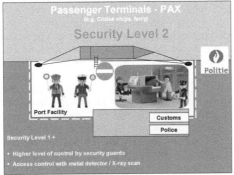

Figure 17. Passenger terminal – SL2

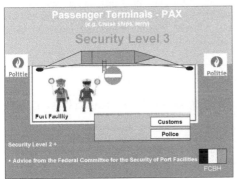

Figure 18. Passenger terminal – SL3

There is always access control to the ship on each security level. This can be done by means of authorised passes delivered by the port authority or the federal government (e-id card) or by the PFSO (e.g. alfapass), be it an authorised visitor badge as described in the PFSP.

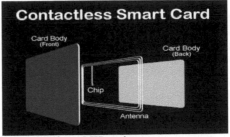

Figure 19. Alfapass RFID-card port

12 CONCLUSION

The Saudi government should have taken implementation various security measures, the government's Maritime Transportation Security Act of 2002 (MTSA) provides additional security to Saudi ports. The International Convention for the Safety of Life At Sea (SOLAS) and the International Ship and Port Facility Security Code (ISPS) provide further security. and should concern with the following aspects:

A ship security plan must outline measures to prevent weapons and other items that could be used to harm passengers and crew from being brought aboard the ship, unless carried by authorized personnel.

A ship security plan has to list restricted areas on a ship and how access to those areas will be deterred. How unauthorized ship access will be prevented also must be detailed.

The plan should include what response measures will be taken when there is a security breach or threat, including maintaining ship operations. How responses to security commands from government agencies will be handled also must be included, as well as how the ship will be evacuated if necessary.

The plan will list security responsibilities of the crew, including auditing security, training for implementation of security measures, reporting security problems and reviewing and updating the plan periodically.

The plan must indicate when security equipment is tested, where security system activation locations are and procedures and training regarding the security system.

The security plan must designate who the security officer is on the ship. The security officer is responsible for ensuring the ship's security and that the plan is carried out. The security officer also oversees security maintenance and training. A second security officer from the company owning the ship also is mandated, with that person working to develop the plan and having it approved by the company before implementation. The company security officer acts as a liaison with the ship security officer.

ISPS or International Ship and Port Facility Security, code implemented a ship security plan and provided preparation for action in the event of a terrorist attack on a ship. The Act requires commercial yachts heavier than 500 GT to be certified. ISPS security training also enacted mandatory training against piracy attacks and includes both ship and port personnel.

Several categories of job types are required to receive training in order to comply with safety requirements. This include the company security officer, the vessel security officer, the head company/vessel/ship security officer, the facility security officer, facility or vessel personnel with specific security duties, port facility security officer and other ship or port facility personnel with specific security duties.

A wide variety of specific training subjects prepare personnel to address safety issues. Classes include general maritime security, port awareness, tactical operations, tactical boat operations, safe boarding techniques and procedures, incident response, tactical underwater operations, emergency medical procedures, basic and advanced fire fighting, crowd control, crisis management, advanced sea survival, basic first aid, first aid care at sea, personal survival techniques, personal safety and liability and medical refresher.

The MTSA requires SOLAS-certified vessels that are over 300 gross tonnage (GT), to carry an automated identification system. This rule also applies to small passenger vehicles that are certified to carry more than 150 passengers. The original version of the rule also applied to specified commercial vessels on international voyages, as well as certain other commercial vessels, but the government rethought this revision after the vessel operators complained about the costs of installing such systems.

Owners of foreign SOLAS vessels do not have to submit security plans to the Saudi Roles for approval. However, under MTSA regulations, non-SOLAS foreign vessels still have to submit security plans accordance Saudi Roles for approval. They may also comply with an alternative security plan, or with measures suggested in another bilateral or multilateral agreement. The Saudi Authority should have to examine and enforces a vessel's compliance with international security regulations and may deny non-compliant vessels entry to Saudi ports.

The Saudi Authority maritime security requirements allow some flexibility for non-SOLAS vessels and port facilities to participate in alternative security programs if they wish to. This allows them to tailor security measures to the requirements of their industries. However, all vessels must follow security plans.

This paper is intended to serve as a conceptual piece that draws from the interplay between engineering and supply chain approaches to risk in the context of recent maritime security regulations. It is hoped that cross-disciplinary analysis of the perception and impact of the security-risk will stimulate thinking on appropriate tools and analytical frameworks for enhancing port and maritime security. In so doing, it may be possible to develop new approaches to security assessment and management, including such aspects as supply chain security. The framework and methods reviewed in this paper could serve as a roadmap for academics, practitioners and other maritime interests to formulate risk assessment and management standards and procedures in line with the new security threats. Of particular importance, new relevant approaches can be developed to assess the reliability of the maritime in the context of the complex network theory (Bichou, 2005; Angeloudis et al.,2006; Bell et. al, 2008). Equally, further research can build on this to investigate the mechanisms and implications of security measures on port and shipping operations, Companies, ports and other parties active in today's international supply chains face a large number of regulations and private initiatives prescribing measures to be taken in order to raise the level of supply chain security. These measures range from putting up a fence around the terminal facilities at a seaport, to establishing a certified security programme at the production facility ('point of stuffing') in order to be admitted to the "green lane".

REFERENCES

[1] 1MARITIME SAFETY AND SECURITY.APRIL 16-20, 2007
[2] American Port Security Cooperation; as adopted by CECIP/RES.12 (VI-04) in Managua, in December 3, 2004.
[3] A Hemisphere port security conference holds in Venezuela in 2006.
[4] Accorsi, R, Apostolakis, G, Zio, E, 1999, Prioritising stakeholder concerns in environmental risk management, Journal of Risk Research, 2 (1), 11-29
[5] Angeloudis, P, Bichou, K, M.G.H Bell and Fisk, D, Security and reliability of the liner container-shipping network: analysis of robustness using a complex network framework, In: Bichou, K, Bell, M.G.H. and Evans, A (2007), Risk Management in Port Operations, Logistics and Supply Chain Security, Informa: London
[6] Babione, R, Kim, C.K, Rhone, E and Sanjaya, E, 2003, Post 9/11 Security Cost Impact on Port of Seattle Import/Export Container Traffic, University of Washington: GTTL 502 Spring Session 2003.
[6] Bedford, T and Cooke, R, 2001, Probabilistic Risk Analysis: Foundations and Methods, Cambridge University Press
[7] Bell, MG, Kanturska, U, Schmocker, JD, 2008, Attacker-defender models and road network vulnerability, Philos Transact A Math Phys Eng Sci, 366, 1893-1906

[8] Bichou, K, Bell, M.G.H. and Evans, A, 2007a, Risk Management in Port Operations, Logistics and Supply Chain Security, Informa: London

[9] Bichou K, Lai K.H., Lun Y.H. Venus and Cheng T.C. Edwin, 2007b, A quality management framework for liner shipping companies to implement the 24-hour advance vessel manifest rule, Transportation Journal, 46(1), 5-21

[10] Bichou, K and Evans, A, 2007c, Maritime Security and Regulatory Risk-Based Models: Review and Critical Analysis. In: Bichou, K, Bell, M.G.H. and Evans, A (2007), Risk

[11] Management in Port Operations, Logistics and Supply Chain Security, Informa: London Bichou, 2008b, „Security Of Ships And Shipping Operations', In Talley, 2008 (eds.), Ship Piracy and Security, Informa, 73-88

[12] Bichou, K, 2004, The ISPS code and the cost of port compliance: an initial logistics and supply chain framework for port security assessment and management, Maritime Economics and Logistics, 6 (4), 322-348

[13] Bichou, K, 2005, Maritime Security: Framework, Methods and Applications. Report to UNCTAD, Geneva: UNCTAD, June 2005.

[14] Bier, V.M, 1993, Statistical methods for the use of accident precursor data in estimating

[15] SECURITY AND RISK-BASED MODELS IN SHIPPING AND PORTS:REVIEW AND CRITICAL ANALYSIS,Khalid BICHOU,Centre for Transport Studies.Imperial College London.United Kingdom.December 2008

[16] International Code for the Security of Ships and of Port Facilities (ISPS) Code Overview

Chapter 4

Dynamic Positioning and Offshore Technology

Dynamic Positioning and Offshore Technology
Maritime Transport & Shipping – Marine Navigation and Safety of Sea Transportation – Weintrit & Neumann (Eds)

Verifications of Thrusters Number and Orientation in Ship's Dynamic Positioning Systems

J. Herdzik
Marine Power Plant Department, Gdynia Maritime University, Gdynia, Poland

ABSTRACT: The paper presents a probe of correctness selection of the number and orientation of thrusters in ship's dynamic positioning systems. The design of DP system takes into consideration required the DP system class, the requirements of adequate classification society under supervision the ship is built. It is necessary to fulfill the system redundancy. The design calculations of position keeping possibilities in considered propulsion system concern the specified position of ship centre of gravity. The change of the centre of gravity position caused of its exploitation states has the essential effect on ship positioning possibilities. It requires the system verification of already built ship in various exploitation conditions. The sea trial is only a part of DP system verification process.

1 INITIAL REMARKS

For multipurpose vessels the Dynamic Positioning System is a standard now. The DP system is designed to keep the vessel within specified position and heading limits.

Dynamic positioning may either be absolute in that the position is locked to a fixed point over the bottom, or relative to a moving object like another ship or an underwater vehicle. They may also position the ship at a favourable angle towards wind, waves and current, called weathervaning.

The main challenge for DP system is the optimum propulsion arrangement, number and types of thrusters, location of thrusters in the hull, their maximum power and thrust, especially the direction and resultant thrust of working thrusters [1]. The boundary conditions for DP vessel are generally designed to its operating in and surviving at extreme sea environmental conditions, although statistically these conditions occur very rarely. The DP vessel runs at median or lower power levels during the majority of operation [5].

For class 2 and 3, IMO requires an online consequence analysis during DP operation. This function must continually perform an analysis if the vessel is able to maintain its position and heading (course) after a predefined single worst case failure during operation. Possible consequences are based on the actual weather condition, enabled thrusters and power plant mode. The DP system needs to fulfill the classification society requirements and next sea trials to receive the DP certificate. For DP class 2 and 3 it is a necessary to have more than 2 independent thrusters in the propulsion system (Fig.1), more often 4 up to 8. The redundancy in propulsion system is obligatory in accordance with classification society rules [3,4,9,11]. For DP class 3 the power plant is located in 2 or more (now up to 4) engine rooms divided by watertight bulkheads [2,3].

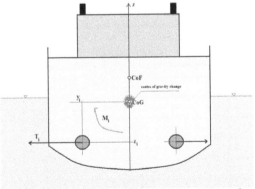

Figure 1. An example of vessel forces and momentum from propulsion system and its orientation.

The propulsion system and its integration with other DP equipment is vital for correct operating of multimode vessels. The system must be able to handle transient conditions such as changes in external forces, failure of a signal from sensors and position measurement equipment and system hardware failures. The next important function is to control the vessel so as to minimize fuel consumption and to keep the thruster wear to a minimum.

2 PROPULSION ARRANGEMENT AND LOCATION

2.1 *Propulsion arrangement*

Propulsion system solution, propulsion arrangement and its location are important problems during design and later in vessel operation. Improving safety and reliability of ship propulsion system during DP operations of multipurpose vessels is an essential advantage.

The limits of continuous bollard pull for different types of thrusters are presented in table 1. Knowing the power delivered to the thruster it is possible to estimate the thrust and vice versa.

Table 1. Bollard pull for different thruster types [5].

Type of propulsion	Continuous bollard pull [N/kW]
Open propeller Azimuth thruster	130
Contra rotating Azimuth thruster	140
Azimuth thruster with nozzle	170
Tunnel thruster	123

There are two types of power transfer to thrusters: mechanical and electrical. Sometimes it may be met partly mechanical partly electrical propulsion. For transit it may be used more often mechanical for station keeping electrical.

According to the DPS class, the vessel work area, the type of vessel etc. it is possible to calculate the needed thrust and its orientation to counteract the vessel position or heading change for all weather conditions. DPS must suppress the low-frequency motions, keeping the mean position of the vessel as close as possible to the desired point. The minimum thruster configuration for DP system is presented on figure 2.

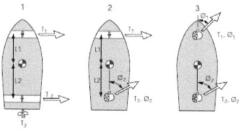

Figure 2. Minimum thruster configuration for DP system [10]:
1. Two tunnel thrusters and controllable pitch propeller;
2. A tunnel thruster and azimuthing thruster;
3. Two azimuthing thrusters.

Other more complicated possibilities of thrusters arrangement are presented in Fig.3.

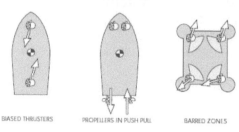

Figure 3. Other possibilities of thrusters arrangement [10].

It is important that the thrusters are symmetrically orientated to the centre of vessel gravity.

2.2 *Propulsion location*

The station keeping needs an active propulsion. It often uses azimuthing thrusters or azimuthing podded thrusters. Sometimes the tunnel thrusters or CP propellers or retractable thrusters are used together with azimuthing ones [6,7]. The number of thrusters for DP depends on DP class. From DP class 2 it is minimum four, the maximum number is not specified but due to increasing investment costs it is about eight. The minimum recommended DP equipment class according to vessel type is shown in table 2.

Table 2. Minimum recommended DP equipment class according to vessel type [2,3,4,9].

Vessel type	Recommended DP equipment class
Drilling	3 earlier 2
Diving	2
Pipelay	2
Umbilical Lay	2
Lifting	2
Accomodation	2
Shuttle Offtake	1
ROV support	2 open water 1
Floating production	2
Well stimulating	2
Logistic operations	2

The power plant may be located in one engine room up to DP class 2. For DP class 3 it is minimum two. For drilling vessels four engine rooms divided by watertight bulkheads are met. In every engine room is complete power system from engines, switch boards, control systems to propulsion units. The connections (tie breakers) among the switchboards are needed to ensure the electrical power for the most number of thrusters in emergency situations.

2.3 Thrusters orientation

The azimuthing thrusters have the possibility of 3600 rotation round own vertical axis. During DP operation they may interact with each other. The problem was presented in table 3.

Table 3. Force correction factors for thruster-hull interaction TDF2, current inflow velocities Va and thruster housing TDF1 for drill vessel "Glomar C.R. Luigs".

Va [knots]	Thrust Tn [kN]	Tn/T0 [-]	TDF1	TDF2	EF=Tn*TDF1* TDF2
0	877.4	1	0.99	0.95	825.2
1.94	779.0	0.888	0.97	0.95	717.8
3.92	687.0	0.783	0.96	0.95	626.5
6.0	600.0	0.684	0.94	0.95	535.8

The thrusters often work in pairs (symmetrical to the longitudinal vessel axis or to the centre of vessel gravity) where the first one is in opposite in thrust direction to the second one. It allows in some cases to restrict the work angle of each thruster to about 90^0-270^0. It decreases the time of reaction to achieve the needed thruster orientation.

For CP propellers the thrust region is limited to possible cooperation between the propeller and the rudder. The tunnel thruster give the thrust only in its main axis.

An example of DP capability plot for modeled vessel is presented in Fig.4. The propulsion system consists of bow tunnel thruster, three azimuthing thrusters and two controlled pitch propellers.

a)   b)

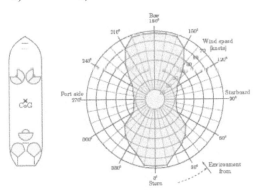

Figure 4. An example of DP capability plot [4]:
a) the modeled thrust region from each thruster;
b) the corresponding DP capability plot.

Figure 4a shows the limitation of thrust direction (region) from each thruster. The modeled DP system allows station keeping up to 68 knots of wind speed from bow and stern direction but only to about 22 knots from starboard or portside direction.

3 THE CHANGES IN DP SYSTEM DURING VESSEL EXPLOITATION

3.1 The change of position the centre of gravity

The centre of gravity of designed vessel may change the position due to shipbuilding faults like asymmetrical hull. The centre of gravity on all vessel changes due to:
- the change of cargo mass onboard;
- the change of quantity of fuel, water etc. during voyage;
- the change of ballast water;
- the faults in cargo loading etc.

The vessel ready for sail ought to be without heel (the maximum acceptable heel is 0.5^0) and the best on an even keel, acceptable is more draft on stern.

The thrust vectors plane never goes through the centre of gravity. It results in moment formation from thrusts, mainly heeling moments.

The situation of asymmetrical vessel orientation to water plane requires the correction in thrust distribution from propulsion system.

Let's take into consideration the drill vessel "Glomar C. R. Luigs" with DP system class 3.

The principal particulars of drill vessel "Glomar C.R. Luigs" are:
- Length overall   231.3 m;
- Length between perpendiculars   210.0 m;
- Breadth (molded)   36.0 m;
- Depth to main deck at side   17.4 m;
- Block coefficient at 11.0 m   0.81;
- Load line draft (molded)   11.0 m;
- Load line displacement   68,043 tonnes;
- Height of drill floor above base line   34.1 m.

The position of propulsion units in hull was presented in table 4.

Table 4. The position of propulsion units in hull of drill vessel "Glomar C.R. Luigs" with DP system class 3.

Item to	Thruster No.	Distance to A.P. (AP=frame 0) [m]	Distance to rotary table [m]	Distance to hull centerline [m]
1	Bow	201.61	95.71	0.0
2	Starboard bow	165.21	59.31	6.45
3	Port bow	165.21	59.31	-6.45
4	Centerline stern	30.51	-75.39	0.0
5	Starboard stern	0.34	-105.56	7.76
6	Port stern	0.34	-105.56	-7.76

The schematic thruster location of propulsion system of mentioned vessel was shown in figure 5.

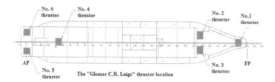

Figure 5. The thruster location of propulsion system of drill vessel "Glomar C.R. Luigs".

The distance to the hull centerline is rather small in comparison to the vessel length overall. It means that the force moments from thrusters are too small to counteract the environmental forces from waves and wind from close to portside or starboard direction (see figure 4). The influence of environment condition is easier to counteract by change the vessel orientation but in many cases it is impossible to do it due to work requirements.

3.2 *The change of centre of flotation vessel position*

The change of: centre of gravity, the vessel orientation to water plane, the vessel draft caused the change of vessel centre of flotation. The vessel transverse and longitudinal metacentre is very important parameter for vessel safety. The crew may change the position of centre of gravity and centre of flotation by adequate ballasting to receive the needed metacentre height.

3.3 *The changes influence on DP vessel system*

For some vessels the distribution of mass changes during work (see Fig.6), especially for heavy-lift, pipe-layer, ROV vessels etc.

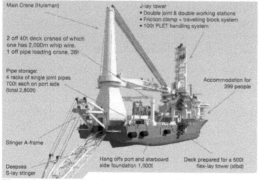

Figure 6. Masses changing the vessel position in relation to water plane (water surface) of pipe-layer VB Seven Borealis [12].

All mentioned changes have an influence on DP system. The designed DP system and the calculations of it don't include the effects of exploitation changes.

The DP system recognizes that changes as a type of disturbances from environment and tries to prevent them. The problem is solved like the other DP system faults, disturbances, inefficiencies etc.

The forming forces and moments from asymmetrical vessel orientation and position may be compensated by thrust change of specified thruster or the change of orientation of specified thruster. It can be done quicker and better by changing thruster orientation and this mode is more simple for DP system control. The angle of change is often only about 5^0-10^0, the needed change of thrust may be about 10-20% of nominal thrust.

3.4 *The possibility of minimizing the consequences of changes*

All actions to fulfill the DP system requirements would be achieved by minimum summary thrust with the constraint that any thruster wouldn't be overloaded [7]. It leads in the end to minimize the fuel consumption.

Reliable and efficient thrusters are essential to effective DP vessel control [9]. The selection of appropriate thrusters for a vessel will be based on;
– the size of the vessel;
– the type and role of the vessel;
– the operating condition in which it will be required to work.

If the thruster demand cannot be fulfilled in all axes, the priority given to the axis is usually:
– first priority to the heading;
– second priority to the sway.

The thruster set required for DP control of a vessel must satisfy the following conditions [9,11,13]:
– the thrusters must provide independent control in all axes;
– the thrusters must be sufficiently large to counteract the waves, wind and current;
– the thrusters must respond promptly and accurately to the DP system commands;
– the predicted output thrust under the stated conditions must be actually achieved.

The weather definition (environmental condition) is expressed in terms of:
– wind speed and direction;
– wave height, period and direction;
– current speed and direction.

To simplify the calculation several assumptions can be made:
– the waves are a fixed proportion of the "fully risen set" for each wind speed and with Jonswop spectrum;
– the current speed is fixed more often to 1 knot and its direction is fixed on the vessel's beam.

The calculation is made only for low-frequency spectrum of wave because the disturbances from low-frequency are dominated and ought to be eliminated first [8,10].

The challenge for DP systems are squalls because the direction and force of squalls are impossible to predict. The DP system ought to be ready for quick and adequate reaction for that disturbance in the limit of change a vessel position and orientation.

A DP controlled vessel can only function if the thrusters respond promptly and accurately. Inaccurate setting or slow response can cause the vessel to oscillate. The table 5 shows the usual acceptable performance levels.

Table 5. Thruster response and accuracy [3,4,5,9].

Speed or pitch response	Zero to full thrust within 8 seconds for small thrusters to 15 seconds for large main propeller
Azimuth rate	180^0 of rotation within 15 seconds and 2 RPM
Speed or pitch	$\pm 2\%$ of maximum
Azimuth accuracy	$\pm 1.5^0$

3.5 Thrusters modes

There are three basic types of thruster mode or pattern:
- bias;
- fixed;
- push/pull.

Where a thruster can operate in its full range is usually referred as free.

There are many possible bias modes, some of them are presented in figure 7. In that mode portside, starboard or no motion (station keeping) are provided.

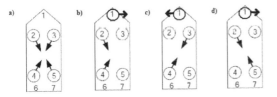

Figure 7. Different bias thruster's modes.

Fixed mode is a situation with fixed orientation of a thruster. A part of azimuth thrusters may work in fixed mode the other part in bias mode. This simplifies the thruster control system and time of reaction.

In figure 7 the thruster no. 1 is a tunnel thruster, no. 6 and 7 are CP propellers worked in push/pull mode.

The types of modes available on a vessel are determined by the available types of thrusters.

4 FINAL REMARKS

The choose of thrusters number, their position and orientation is vital for DP systems. During the design process the propulsion system is fixed. So this is a moment for final decision having an influence on later vessel exploitation.

In a practice, vessels equipped with DP system and DP supporting system, give a crew an enhanced comfort of work during operations because of their reliability and redundancy with required accuracy. Diesel-electric propulsion gives higher overall reliability and high flexibility due to arrangement possibilities. This way is improving the safety of DP operation.

One of the most important issue is to minimize fuel consumption for the actual operation profile and service speeds. Station keeping mode necessary for DP operation wastes a lot of energy, especially during bad weather condition. Using suitable solutions in the generating and delivering power it allows better utilization of propulsion system during DP operation.

REFERENCES

[1] Dietmer D. 1997. Principal Aspects of Thruster Selection, Dynamic Positioning Conference, Houston, USA.
[2] Dynamic Positioning Systems – Operation Guidance, DNV-RP-E307, 2011.
[3] Germanischer Lloyd 2010, Rules for Classification and Construction Ship Technology, chapter 15, Dynamic Positioning Systems.
[4] Guide to Dynamic Positioning of Vessels, ALSTOM 2000.
[5] Halstensen S.O. & Nordtun T. 2009, Improving Total Efficiency and Safety during DP-operations, First International Symposium on Marine Propulsors, Trondheim, Norway.
[6] Herdzik J. 2011. Metoda szacowania wymagań stawianym układom napędowym statków, Logistyka 3/2011, 891-898.
[7] Herdzik J. 2011. Propozycje wykorzystania parametrów aktywnych pędników w podniesieniu bezpieczeństwa żeglugi, Logistyka 3/2011, 899-907.
[8] Holvik J. 1998. Basics of Dynamic Positioning, Dynamic Positioning Conference, Houston, USA.
[9] International Guidelines for The Safe Operation of Dynamically Positioned Offshore Supply Vessels, IMCA M 182, 2009.
[10] Nienhuis U. 1992. Analysis of Thruster Effectivity for Dynamic Positioning and Low Speed Manoeuvring, Doctor Thesis, Delft.
[11] Operator Manual Kongsberg K-Pos DP Dynamic Positioning System, Release 7.0. Kongsberg Maritime AS, Norway.
[12] VB_Seven_Borealis 2012, www.subsea7.com
[13] Zalewski P. 2010. Models of DP system in full mission ship simulator, Zeszyty Naukowe Akademii Morskiej w Szczecinie, 20(92), 146-152.

Underwater Vehicles' Applications in Offshore Industry

K.A. Wróbel

Gdansk University of Technology - Faculty of Ocean Engineering and Ship Technology, Gdansk, Poland

ABSTRACT: As the global economy – even despite financial crisis – grows, the demand on fossil fuels gets bigger and bigger. To satisfy this demand, oil companies look forward for new offshore opportunities. However, oil and gas fields at shallow waters are getting depleted and that forces the mankind to seek hydrocarbons in the very new area of deep seas. That requires new technologies to be developed, with remotely operated underwater vehicles (ROVs) and autonomous underwater vehicles (AUVs) amongst them. In this paper, some of ROVs and AUVs applications in offshore industry are described.

1 INTRODUCTION

People were dreaming about submerging into deep sea since 16th century, when Leonardo da Vinci designed a submarine prototype. However, there are rumours that Alexander the Great of Macedon tried to use diving bell to explore Mediterranean.

Nowadays, it is believed that everywhere except the navy, a human being underwater causes more technological problems than it actually solves. For instance, life support system must be installed. That is the reason to design, construct and operate unmanned vehicles such as ROVs or AUVs. The first type consists of a vehicle connected to the surface mother ship by umbilical cable, through which an electric power and commands are being transmitted. ROV is controlled by a pilot, who sits in a comfortable, air-conditioned room on mother ship's board and – in opposite to submarines' crews – does not need to be afraid of some 100 bar or so hydrostatic pressure.

The next stage of underwater vehicles development is an AUV, where a real time operator is completely eliminated. AUVs are programmed *a priori* to complete their mission while still on surface, put in the water and forgotten until the mission is successfully completed, then retrieved. They are autonomous, which means that no one actually controls them: once in water – they are on their own. However, currently some of them may be operated remotely by hydroacoustic link instead of umbilical cable and then they are called 'untethered'.

There are of course advantages of each of those technologies and disadvantages as well. They are briefly described in Table 1.

Table 1. Advantages and disadvantages of ROVs and AUVs

ROVs' advantages	ROVs' disadvantages
Directly controlled by human, External power supply reduces weight and allows greater powers, Real time data reliability control.	Presence of umbilical cable reduces manoeuvrability, The more ROVs required, the more pilots we need.
AUV's advantages	AUVs' disadvantages
'Deploy and forget', Lots of AUVs can be used at once without pilots' interference, Lack of umbilical cable.	Internal power source required, Poor adaptability to environment, Low reliability.

There are many applications of both ROVs and AUVs: marine research, submarine search and rescue, illegal drug trafficking, seabed mapping, counter-mine warfare, diamonds mining and oil industry operations. Let's focus on the last one.

2 HIDE & SEEK

The very first step of oil or gas field development is to actually locate the deposit of hydrocarbons. The preliminary analysis is based on geological survey, which gives the geophysicists data regarding structure of seabed. Thence, they are capable of nominating the areas, where presence of fossil fuels is most probable. The next step should be drilling a test hole in seabed to check these predictions. The problem is that hiring a drillship and drilling a hole is an extremely expensive idea, which means that additional check is to be made. Here comes a seismic vessel.

Seismic vessel is a ship equipped with devices to produce low frequency seismo-acoustic wave, sending it into the sea bottom, receiving reflected signal and processing it in order to gather data regarding structure of the sea bottom. These devices consist of an acoustic wave source (e.g. sparkers, pingers etc.) and receivers (hydrophones), placed in few nautical miles-long umbilicals called 'streamers' together with gyrocompasses and GPS devices, delivering data of its orientation in sea water. This system proved to be working well, but – as Polish proverb says – 'better is good's enemy' and everything can be improved.

Some problems are experienced when seismic vessels yaw at high seas. They are often equipped with dynamic positioning systems and maintain their course even in severe weather, but the streamers do not – even if special 'winglets' are used. Another thing is the noise, produced by ship's propellers and jamming the acoustic wave. The possible solution to this problems are AUVs (Cydejko, J. & Puchalski, J. & Rutkowski, G. 2011.).

The concept is based on idea to deploy as many AUVs in the water as it is practically reasonable after pre-programming them to spread all over survey area and place them deep inside the ocean. After they have finished this part of task, a single seismic vessel without streamers could proceed with her mission to cover the whole area, using acoustic waves. Those waves, reflected by sea bottom layers, will then be received by AUVs and stored in databanks. The next step for AUVs is to resurface and be collected by assisting vessel from sea surface, data processed. Using this method, one can get all the information required to produce 3D model of sea bottom:

- seismic vessel's position, course and speed in selected time period;
- acoustic signals sent by vessel's sparkers etc.;
- AUVs' position, obtained from inertial dead reckoning or acoustic navigation systems;
- signals, received by AUVs in specific time.

Using this solution, the non-linearity of ship's movements is eliminated, as long as background noises from the propeller. Of course, some noise will still exist, but as the distance between the propeller and AUVs' hydrophones rises, they will be less disturbing. AUVs can also get closer to the sea bottom than streamers, which means that reflected acoustic wave, received by them is less dispersed.

The biggest problems with this solution is deploying big quantities of AUVs in specified area

and picking them up from water after survey is completed. There is always a concern that some part of vehicles can lose the capability of precise navigation and get lost forever. Economies of scale could influence the cost of single vehicle and the operation as a whole, so that some losses will become acceptable. Another problem is successful navigation of large quantities of unmanned vehicles in a way no collisions occur. This may be achieved by highly sophisticated team navigation modules and use of acoustic communication between each member of AUV group like for example the one called GREX. This system is aiming in integrating navigational sensors of a single vehicle with network of other AUVs, so that every single one of them will possess information about others' movements and collisions will be unlikely to happen (Engel, R. & Kalwa, J. 2007.).

3 SEABED MAPPING

After the hydrocarbons deposit is located and estimated to be possibly profitable, oil companies usually wish to make profit as quickly as possible. However, before wells are drilled and oil rigs placed, a seabed must be checked for presence of various hazards like wrecks, mines, ammunition or uncharted rocks. A proper way to do this is deploying AUV or ROV equipped with multi-beam echosounder or sidescan sonar. Those sensors are used to create 3D and 2D image of seabed respectively as shown on Figure 1.

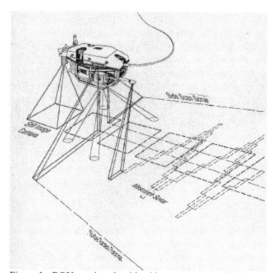

Figure 1. ROV equipped with sidescan sonars, still image camera, scanning sonars and Doppler sonar beams to be used for seabed mapping (Whitcomb, L.L. & Yoerger, D.R. & Singh, H. 1999.).

AUVs can operate in specified area in order to cover the entire surface of seabed on their own, while ROVs must be operated by hand, which allows pilot to take closer look at particularly interesting objects in real time. That saves time required for processing data, acquired by AUVs and deploying them again to check on those objects. So far, a technology to make robots understand what is interesting from engineering point of view and what is not is yet to be developed. That means that still, at some point, a human being must get into action with interpreting data provided by underwater vehicles' sensors.

Information gathered by underwater vehicles' sensors is used to create detailed map of seabed with emphasis on bathymetry and obstructions. This map can also be displayed as an information layer of electronic chart to provide personnel with a full view of oil field's environment.

Another application of sidescan sonars is the inspection of pipeline route. In the design phase of pipeline construction project, designers must have as accurate data, regarding seabed on route, as it is possible, to make sure that pipeline can be constructed and operated safely. They also want to cover a few routes with survey to decide which of them fits safety conditions in the best way. The following aspects of seabed condition are being monitored in particular:
- seabed's ability to support the pipe and additional weight;
- presence of man-made hazards, such as wrecks, dangerous materials, uncharted cables and pipelines etc.;
- topography of seabed – presence of faults or rocks.

Pipeline route is an object to classification societies' regulations. For example, Det Norske Veritas states that surveys shall be carried out along the total length of the planned route with variable accuracy, depending on many factors, such as variability of seabed topography and presence of hazards. Using ROVs, this accuracy can be achieved and guaranteed by pilot's control over the system.

4 DRILLING THE WELL AND STARTING PRODUCTION

Commencement of oil production is preceded by drilling wells, placing blowout preventers, manifolds and other infrastructure. Unmanned vehicles, particularly ROVs, can assist at those operations, as long as divers on shallow waters.

Unmanned vehicles can enable supervision and manipulator operations at one time. ROV is then operated by at least two pilots – one of them takes care of proper navigation and orientation of the vehicle and the other one is operating the manipulators. Modern finger-like manipulators can achieve all 6 degrees of freedom and can be very sensitive, so that many different actions can be performed. Valves, gauges, quick couplings and other equipment can be operated. Nowadays, many of those are being located at special ROV panels – gathered together to facilitate operations.

For example, when the blow-out preventer is placed over the well, some of the valves must be opened or closed manually, as long as some pipes must be connected one to another. Same things happen when some of remote safety systems fail and intervention is required in order to prevent leaks. It is obviously impossible to send divers in waters 2000 meters deep, so that ROV is to be deployed. Well-trained pilot can do his job very fast – sometimes even faster than divers would do this as he does not need to consider his own safety: equipment, although sometimes extremely expensive, is replaceable; people are not.

ROV manipulators are pretty similar to those used in outer space to transport brand new satellites and International Space Station modules brought into orbit. The Canadarm, which was being used during space shuttle program, was equipped with special iris handle in opposite to the finger-like one, which was caused by the fact that once one lets go any object in zero-g conditions, it's never going to come back. Iris manipulator – inspired by irises used in cameras - allows better reliability of hold and – after some modifications - could be also used in ROVs to increase reliability of handling objects in areas distant from seabed. So that iris manipulators might be used to avoid dropping heavy objects which could damage the infrastructure below.

Manipulators can also be equipped with special tools like drills, pneumatic or electric screwdrivers, cleaners, chopsaws, hydraulic guillotine shears and others to perform even more specific tasks underwater. Modern manipulators installed on vehicles allow operators to perform almost-miracles. One day, possibly, sensitivity of operations performed by ROVs can be similar to those by divers.

Even if some operations are performed in shallow water by divers, ROVs and AUVs can still be useful providing them with light or power connections and TV imaging to their supervisors.

5 INSPECTING SUBSEA STRUCTURES

After blowout preventers, oil rig structures, pipelines and other oil field appliances are successfully completed, they should be inspected once in a while and sometimes repaired. Some of them are placed in waters up to 2000 meters deep, well below diving limits. An only way to perform structural inspections is to use underwater vehicles. Most of modern ROVs can submerge up to 3000 meters below the surface, some of them even up to 7000 meters. They can be operated directly from oil rigs, FPSOs and other installations permanently placed in oil field area or ships like AHTS (Anchor Handling Tug Supply Vessels), which are often equipped with proper cranes and can be easily adapted to handle umbilical cable.

So far, using ROVs to inspect underwater structures is far more common than AUVs. It is caused by a poorer reliability of the last ones and the fact that pilot can easily focus on inspecting particular parts of structure as long as it is required to get sufficient information without need to waste time for emerging the vehicle. The presence of human operator also makes ROVs use the cameras rather than sonars. Although visibility in deep waters is worse than ability to create 3D images acoustically – human brain seems to handle visuals better than acoustic image. It is probably connected to colours distinguishing – ability that almost every human being possesses since childhood. Even perfectly visualised data from sonars with its characteristic reds and blues causes our brains need more time to process the image. So that, it is useful in post-processing, but not in a real-time operations.

Advantage of using AUVs to inspect subsea structures is that many of them can be deployed, covering greater area in shorter period of time. Then, after collected data is reviewed, ROV can be sent to check on particularly interesting parts of structure.

Following deficiencies may be detected during an underwater inspection by ROV:
– leaks;
– structural breakdowns;
– mechanical damage;
– corrosion;
– ropes and nets screwed in equipment;
– cover of marine life;
– fish stuck in equipment,
and many others.

With special equipment installed, ROVs can also perform some of non-destructing testing (NDT) tasks. NDT methods, which can be used in wet environment underwater, are: magnetic particles test, ultrasonic thickness, ultrasonic shear wave (LeHardy, P., & Elliott, K. 2010.).

6 LEAKS FIGHTING

A 2010 *Deepwater Horizon* disaster proved leak-fighting techniques used in that time to be insufficient. Offshore industry is still experiencing criticism and environmental concerns, which were always present, but became particularly strong since 2010. Images of oil spills and oil-covered birds do not really work perfectly as an advertisement or money-saver for any oil company, which makes

every single one of them need redundancy in preventing oil spills.

Underwater vehicles' role in leak-fighting is based on reducing the flow by operating ROV panels, but can also be extended to stopping the leaks in other places of pipelines or risers. A solution consists of – for example - an elastic but strong titanium band, which can be wrapped around the damaged pipeline by ROV. The vessel, however, must be equipped with powerful propellers in order to resist flows of hydrocarbons or other liquids with speed greater than usual current. A band, mentioned above, is a temporary solution which can be only used to control environmental conditions in vicinity of leaking pipes and should be replaced by more reliable solution as quickly as possible. Another advantage is that it reduces leaking, so less hydrocarbons pollute waters around.

7 DECOMMISSIONING

Legal regulations in some countries require all offshore structures to be removed and scrapped or refurbished after production is finished instead of sinking them in deep waters. While floating constructions are simply being moved to another place of operation, fixed platforms create bigger problems. The most commonly used way to remove them from the oilfield is cutting its supports, sometimes using explosives. While some of those operations can be performed by divers in shallow waters on continental shelf, in deeper waters only ROVs can operate. AUVs cannot be used as there is no certainty that proper structure will be cut.

Usually, offshore structure is being cut as close to the seabed as it is possible, using chopsaws or reciprocating guillotine saws, which can cut structure up to 32 inch in diameter. This shall not be done by divers as there is always a danger that a whole structure will collapse after being cut.

8 PROBLEMS, REGARDING UNDERWATER NAVIGATION

Despite the very fast development of underwater vehicles technology, there are still some problems to be solved before that technology can be considered to be fully reliable. This concerns AUVs in particular, as ROVs' reliability in most cases can be assured by human operator.

Firstly, as there is no direct control over autonomous vehicles, there is always a possibility that some kind of fatal error will occur and AUV is never to be recovered or found. The same fact makes using AUVs in particularly sensitive tasks, such as valves operating, unpractical – no one can be sure that AUV will actually operate proper valve until the vehicle is recovered and data checked, but then it could be too late for fixing the problem.

Secondly, lack of adaptability of machines to environment is a problem – only living creatures possess that ability, thank God. Even highly sophisticated artificial intelligence cannot prepare technical system for unpredicted conditions with parameters off the scale. In predictable future, ROVs will be used in many ways while AUVs' performance will be limited by artificial neural networks abilities.

Accuracy of underwater navigation is another problem. GPS system does not work in water as its signal is dispersed, as long as other satellite navigation systems'. So that different positioning systems had to be developed. One of them is 3D inertial dead reckoning that can work without any external devices anywhere in the ocean. In this method, accelerometers and gyroscopes, gathered together in inertial measurement units (IMUs – Figure 2.), are used to calculate the position and orientation of a vehicle in relation to an initial position. Its limitations are obvious – accuracy of dead reckoning decreases as time since the last position fix increases. As a research, conducted on board of Polish Navy submarine, shows – error in determining of position can be as high as 8 cables after 12 hours of navigation (Felski, A. & Nowak, A. 2009.). It means that this system has to be improved before implementing in long-time missions. Of course, other navigational devices, like for example magnetic compasses, can be used to correct some errors.

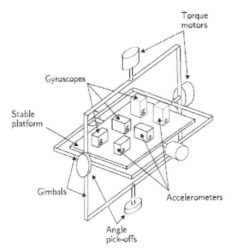

Figure 2. Inertial measurement unit (Woodman, O.J. 2007.)

Another solution of underwater navigation problem is using hydroacoustic systems like Kongsberg Maritime's High Precision Acoustic Positioning - HiPAP® (Uriasz, J. 2008.). The operational principle of such systems is similar to this of GPS, but with use of acoustic signal instead of radio. A network of acoustic devices is placed on a seabed or under the bottom of surface vessel and on board the submersible. Position of underwater vehicle is obtained as a distance to reference station, calculated by measuring differences in phase or time in which the acoustic signal is received. There are various concepts of those systems, in which the processing unit can be located on board the surface vessel or submersible and many different types of acoustic devices can be applied, such as beacons, transponders etc. (Rowiński, L. 2008.). Their performance is limited by need of installing transducers on seabed or bottom of surface vessels, but accuracy of position fix is better than in most of other systems and does not depend on time of operation.

Speaking of underwater navigation, collision avoidance must be considered. As it was already mentioned, visual means of observation are practically useless underwater. Thus, acoustic methods must be again applied in order to ensure safe navigation. In most cases, submersibles are equipped with forward looking 3D sonars on their bows. The range of such sonar can be divided into sub-ranges and whenever an object is detected, its location can be identified in relation to the vehicle (Bowles, I. & Markowski, Z. 2010.). Then, vehicle must alter its course or change operational depth in order to avoid collision, constantly controlling the area ahead. The difficulty of that operation increases as it is to be performed by autonomous vehicle in presence of multiple moving objects. In this case, special algorithms must be implemented. In the worst of scenarios, when a large obstacle completely filled up sonar's coverage, AUV may be forced to abort its mission, withdraw to the initial position and search for another path to reach the goal (Kim, S. & Kim, Y. 2008.).

Finally, energy supply is a serious problem to AUVs. It does not concern ROVs as they can be powered by surface generators. There are few types of power sources for autonomous vehicles – chemical or isotope batteries, solar panels, hydraulic accumulators etc. (Rowinski, L. 2008.). All of them have some advantages and disadvantages, but it is obvious that the longer or more demanding the mission is, the more power is required and it is more difficult to provide vehicle with that power. There should be also special algorithms to estimate the remaining power and to initiate and complete resurfacing procedure before black-out – otherwise AUV can be lost. One of the new trends in underwater engineering is constructing special garages – 'tophats' – for housing vehicles. Those structures are connected to the surface installations by umbilical cable, transmitting power and data. ROVs or AUVs can be stored and powered there, which can save time for pulling them out of the water and then launching back into the sea (Bingham, D. & others. 2002.). At any moment when they are required to do any job, ROV pilot or AUV programmer can order them to leave the shelter and proceed with action.

9 FIGHT THE FUTURE

Subsea technologies are developing fast as there is a bigger and bigger demand on offshore fossil fuels and other materials, such as manganese nodules. Many problems have been already solved, many other to go. The most important of them are: autonomous underwater navigation precision, team navigation, power supply, highly-effective hydroacoustic communication systems and reliability of AUV operations. Most of them can only be solved by cooperation of ocean engineers, control theory engineers, programmers and many others. But, as Leonardo da Vinci would not even imagine modern underwater vehicles' abilities, we may be very surprised by its development in a few years to come.

10 SUMMARY

Unmanned underwater vehicles are widely used in offshore industry in all phases of oilfields development: from seismic survey through production and decommissioning. They can perform various operations in areas where – for any reason – divers cannot get. However, despite constant development of subsea technology, there are still many aspects of underwater operations to be improved. The greatest barrier is, as usually, money – ROVs without manipulators prices start from USD 30,000. Costs of improvements must be high enough to successfully prevent scientists and engineers from inventing and implementing new solutions. But again, it is only matter of time to discover new possibilities and applications and make them real.

REFERENCES

Bingham, D. & Drake, T. & Hill, A. & Lott, R. 2002. The Application of Autonomous Underwater Vehicle Technology in the Oil Industry – Vision and Experiences. *FIG 2002 International Congress Proceedings*. Washington, DC: International Federation of Surveyors.

Bowles, I. & Markowski, Z. 2010. 3d Sonar For Navigation and Obstacle Avoidance. *TransNav International Journal on Marine Navigation and Safety of Sea Transportation* vol. 4 no. 1: 61-64. Gdynia: Gdynia Maritime University.

Cydejko, J. & Puchalski, J. & Rutkowski, G. 2011. *Statki i technologie offshore w zarysie*. Gdynia: Trademar.

Engel, R. & Kalwa, J. 2007. Coordinated Navigation of Multiple Underwater Vehicles. *Proceedings of the Seventeenth International Offshore and Polar Engineering Conference*. Lisbon: International Society of Offshore and Polar Engineers.

Felski, A. & Nowak, A. 2009. Metoda Weryfikacji Systemu Nawigacji Inercjalnej na Okręcie Podwodnym. *Zeszyty Naukowe Akademii Marynarki Wojennej* 178: 7-16. Gdynia: Akademia Marynarki Wojennej im. Bohaterów Westerplatte.

Kim, S. & Kim, Y. 2008. An Autonomous Navigation System for Unmanned Underwater Vehicle. *Underwater Vehicles* 279-294. Vienna: InTech.

LeHardy, P. & Elliott, K. 2010. *Underwater Welding and Non-Destructive Testing in Support of US Navy Ship Repair Operations*. Phoenix: Phoenix International Holdings Inc.

Rowinski, L. 2008. *Pojazdy glebinowe – budowa i wyposazenie*. Gdansk: Wydawnictwo WiB.

Uriasz, J. 2008. Metody pozycjonowania w nawigacji precyzyjnej. *Proceedings of the Satellite Methods on Determining Position in Contemporary Geodesy of Navigation*. Wojskowa Akademia Techniczna www.wig.wat.edu.pl. retrieved: 12-01-2013.

Whitcomb, L.L. & Yoerger, D.R. & Singh, H. 1999. Combined Doppler/LBL Based Navigation of Underwater Vehicles. *Proceedings of the 11th International Symposium on Unmanned Untethered Submersible Technology*. Durham, NH: Autonomous Undersea Systems Institute.

Woodman, O.J. 2007. *An introduction to inertial navigation*. Cambridge: University of Cambridge.

Coordinated Team Training for Heavy Lift and Offshore Crane Loading Teams

A. Oesterle & C. Bornhorst
Rheinmetall Defence Electronics GmbH, Bremen, Germany

ABSTRACT: Rheinmetalls Heavy Lift and Offshore Crane Simulator HLS7 supports to train cargo handling teams, who are involved in loading or unloading processes, including installation of offshore wind parks. The bandwidth of training starts with individual crane operator training up to sophisticated cargo response management, which includes pre-planning, team building, communication, decision making and reaction to emergency situations. The scenarios can be configured from simple beginner training up to complex loading scenarios e. g. tandem crane operation. The simulator may comprise workplaces for cargo loading officers, crane drivers, bridge workplace, ballast station, exercise planning, briefing and de-briefing.

1 INTRODUCTION

During the past years the market for heavy lift and offshore has grown rapidly. The constant growth of the economic caused also an increase of dense cargo transportation at sea. In addition progressive development of oil and gas reserves exploration in deep water regions and the boom in offshore wind energy requires more and more crane operation and has opened up excellent job opportunities for crane operators, but also led to a shortage of experienced and qualified personnel.

Every loading operation with heavy or bulky load independent where it takes place is a big challenge for people and technology involved and it is always some degree of risk related to it. In maritime business, environmental effects like wind, waves and heaving platforms makes it more difficult and it needs a lot of experience to manage these challenges in order to operate large crane systems safely without damages for live, cargo and ship.

How to get experienced operators? Practical training is time consuming and therefore cost intensive. It involves always a real crane and the possibilities for such kind of training are limited. Training onboard of ships is nearly unrealistic, short harbor times and applicable safety regulations leave no room for extensive extra training and especially training of abnormal situations like malfunctions or emergency situations cannot be shown always sufficient in practice.

Altogether, there is a lack of qualified personal and a lack of close to reality training facilities for heavy lift crane operators and/or complete cargo handling teams.

2 GENERAL REQUIREMENTS FOR TRAINING OF HEAVY LIFTS

2.1 International regulations

During the last years, the heavy lift and offshore market grew remarkably. The on-going exploitation of gas- and oil resources in deep-water regions as well as the boom in the offshore wind energy opened good business opportunities. This increase of the market results in a lack of qualified crews. In combination with all difficulties to provide professional training, e. g. in lack of training facilities, missing international standards for crew education come on top. A mismatch between recruitment demands and available qualified crews causes higher costs and less productivity.

The base for education of seafarers is STCW 95 [1] and its amendments. In this standards competence, knowledge, understanding and proficiency as well as the methods for demonstrating are described.

Table A- II/2 provides one of the main subjects in "Cargo handling and stowage at the management level" with the following general competence listed:

"Plan and ensure safe loading, stowage, securing, care during voyage and unloading of cargoes"

The requirements are quite clear, but how can a shipping company, a training institute or a university fulfil this requirement?

Nowadays handling of heavy cargo increases nearly every year and goes up to two thousand tons in tandem crane operation. But it is not only is the heavy weight which creates problems, also extreme dimensions of the cargo makes the process difficult. Especially in the offshore market new challenges have to be solved and cargo worth of multi-millions have to be loaded and unloaded safely.

In addition to STCW, national or individual industry standards have to be fulfilled, for example introduced by OPITO [2] or LOLER [3] which provides guidelines for special demands of the heavy lift and offshore industry.

2.2 *Special demands on heavy lift training*

Analogue to Bridge Team Management, Cargo Handling Management is a team task too.

Typically, the following persons are involved in a heavy lift operation:
- master
- cargo loading officer
- 1 or 2 crane drivers
- chief engineer
- carrier`s supercargo
- cargo owner surveyor
- planning staff
- decks crew

They all have to work together in a competent and efficient way in order to handle the cargo in a safe and effective manner. For that they need:
- a proper understanding of the task
- practical experience
- proper communication skills
- team player attitudes

In general the team has to handle individual cargo, typically no cargo is equal and therefore each loading operation is unique and has to be planned separately and accurately, independent of earlier operations. Some examples for heavy loads are:
- all kind of industrial parts like chemical reactors, harbour cranes, locomotives, ships and barges or transformers etc.
- material for the oil industry like pipes, drilling piles etc.
- parts for wind parks like groundings, turbines or airfoils etc.

During planning and execution of heavy lift operations a lot of the challenges have to be solved and overcome, some are:
- operation of heavy lift and offshore cranes within operational limits

- careful interaction during tandem crane operations
- special demands of cargo like drift of centre of gravity
- permanent observation of ships stability e.g. trim, heel, bending and sheer forces
- keep up always a clear and strict communication between all involving parties

On the other hand, a lot of factors can lead to fail a loading or unloading operation. There are many circumstances and disturbances which can influence the success. A lot of them can be limited by sufficient training, but some come unexpected and are not plannable. For these problems the team behaviour has to be trained again and again in order to avoid damages to involved persons, cargo or equipment. Typical problems may occur by:
- crew with insufficient experiences in handling heavy loads
- missing competence in decision making in the moment a critical situations happens
- unclear or wrong communication between team members
- difficult weather conditions and changes of weather conditions during the lift
- insufficient team spirit
- unexpected malfunction of ship equipment

Apart from the fact, that a crew, involved in loading/unloading operations has to have the skills, attitudes and competence to execute the tasks, intensive and frequent training, provided by professional training institutes are essential. Among others, following items may improve by simulator training:
- demonstration of competence in analysis and pre-planning of loading/unloading processes
- proper communication in maritime English
- professional cooperation among all loading team members
- training to overcome unexpected and critical situations

3 SIMULATOR TRAINING – THE BEST SOLUTION TO IMPROVE THE CARGO TEAM

As mentioned, lack of well trained crews results in longer loading/unloading processes, higher risk of damages and therefore in higher costs. A Heavy Lift and Offshore Simulator is a reliable tool to reduce costs and risks by using it for flexible, safe, reproducible and well-structured training at any time, under any conditions whenever training is required. Using a simulator provides the possibility to simulate difficult and even extraordinary cargo operations for individual training as well as for team training.

In principle heavy lift training can be structured in the following modules:
- case analysis
- briefing
- simulation exercise
- debriefing and evaluation

Like in reality the simulator exercise starts with a case analysis which includes a check of the cargo documentation and the preparation of the loading process. Cargo details, lifting arrangement and storage position in the cargo hold has to be checked. When the documentation is ready, the next step is a briefing with the cargo loading team. The team can consist of the master, cargo loading officer, ballast operator and crane driver(s).

After that the simulator exercise can begin. Due to a complete recording of the exercise, a detailed debriefing can be conducted afterwards, each individual step can be discussed and accessed.

3.1 Simulator structure

Today STCW relevant training is supported by a huge variety of simulators for example ship handling, ship engine or oil and gas cargo handling simulators.

For heavy lift and offshore training, for example for the wind mill transportation and installation training no adequate team training simulator was available on the market in the past.

Now, with the new Rheinmetall Heavy Lift and Offshore Simulator HLS₇ the first loading/unloading simulator was introduced to the market, based on virtual reality scenarios and made for sophisticated team training of ultra heavy or bulky cargo. With this new simulator from Rheinmetall it is possible that a crew can be prepared and trained before the real loading/unloading process starts in reality.

The HLS₇ consists and is structured as follows:

Figure 1. Exemplarily Overview about a Full Mission Heavy Lift and Offshore Simulator

A simulator like shown above is installed and in operation by a German training centre. The configuration includes workplaces for cargo handling operation for crane drivers, cargo loading officer, bridge team and ballast operator.

Sustained training can only be achieved by state of the art simulators with utmost accuracy in mathematical models combined with high-end projection systems which guarantees "Look and feel" close to reality. A short time of familiarization for the trainees is essential, different scenarios have to be available and can be matched to the individual skills of the trainees and the possibility to access the exercise afterwards is an absolute must for modern simulators. Therefore a full replay of each training session supports a convenient usage of the simulator for debriefing.

Figure 2. View from the crane cabin during a heavy lift operation

Figure 3. View from the back at a crane cabin installation in front of the projection dome

3.2 Simulator highlights

The Heavy Lift and Offshore Simulator HLS₇ offers a wide range of training opportunities with practical knowledge and soft skills like:
- early identification of critical situations and problems with the loading and unloading of project cargo already in a planning phase
- consideration of environment influences such as wind and sea conditions during the lift
- consideration of the ship characteristics such as residual stability, shear forces and bending moments to determine the limits for the stability

173

- consideration of crane operating parameters like load restrictions, constant tension, various load levels, overload functions or delays caused by the insufficient hydraulic pressure
- optimization of loading and unloading process through the use of the ballast system
- training in a team mode with all workplaces combined or in single mode for special training for example for crane drivers
- a comfortable malfunction simulation prepares the crew for critical situations
- simulation and evaluation of case studies from reality
- familiarization with different cargo types
- familiarization with different ships and crane types for single or tandem operation

Using a simulator for training provides crews, capable in handling cargo loading/unloading processes before they start with their job on board.

Figure 4. Training of Offshore Windmill installation with a 1500 t crane

The training can be structured to the needs of the trainees. Beginners have the opportunity to familiarize with the operation of cranes and different cargo types. Advanced training provides team training with complex cargo and loading/unloading scenarios. In these complex exercises all team members have to work together properly in order to solve sophisticated tasks. As an add-on, the instructor can run the exercises under different weather conditions or with advanced stress levels due to malfunctions or safety reasons.

The software includes several high accurate simulation models, each tailored and matched to ship and crane type. Depending on the configuration, the two cranes can work in single or in tandem operation.

All crew members are represented by computer generated persons (avatar) which can be moved all over the vessel or the platform.

Rheinmetall is a part of the **Rheinmetall AG** group and is located in Bremen in the northwest of Germany. The company develops and produces nearly all kind of simulators from flight to power plant simulators. Maritime simulation was started about 40 years ago and has experiences in navy simulation as well as in all kind of ship handling and ship technical simulation. With the experience of more than 1000 simulators supplied all over the world, Rheinmetall is the second largest producer for simulators in Europe.

4 CONCLUSION

Rheinmetall **HLS₇** is the most advanced team trainer to train loading/unloading operations on the market today. Without the risk of lifes, ship and cargo damages or for the environment the trainee and its team can learn loading/unloading processes under different conditions before starting it in reality. With a simulation close to the reality the Heavy Lift and Offshore Simulator **HLS₇** from **Rheinmetall** crews can be trained under reliable, reproducible conditions in a safe and cost effective way.

REFERENCES

[1] STCW - Standards of Training, Certification and Watchkeeping for Seafarers 95, including Amendments
[2] OPTITO – Oil and Gas Academy, Aberdeen / UK / www.opito.com
[3] LOLER - International Project Researching Lifting Operations and Lifting Equipment Regulations / www.loler.co
[4] Rheinmetall HLS₇ Product Description PD

A Proposal of International Regulations for Preventing Collision between an Offshore Platform and a Ship

P. Zhang
Shanghai Maritime University, China

ABSTRACT: Casualties and incidents of one kind or another are bound to occur from time to time in the navigation and operation of ships and offshore drilling platforms. There is no doubt that International Regulations for Preventing Collisions at Sea (1972) has played a significant role in ensuring a safe navigation and operation of ships. However, offshore drilling platforms are excluded out of the system. With more and more offshore drilling platforms rising at sea, it has been a most urgent task to give a serious consideration on how to avoid collisions between offshore drilling platforms and ships.

When one offshore platform is under drilling operation, she cannot move from one place to another. If the risk of collision occurs between one ship and her, normally actions to avoid collision are relied heavily on the ship itself. Then the situation would be very dangerous if the ship is out of control or lacking proper watchkeeping, which is very common at sea. It can be immediately seen that offshore platform's actions to avoid collision are of vital importance in such a circumstance. As we all know, the devastation to the environment and marine life resulted by offshore drilling platforms would have such a far-reaching effect that deserves any serious attention.

The authors start from introducing special construction and capability features of offshore drilling platforms which are distinctive to navigational ships, then discuss a number of issues about collision between them, and finally provide some legislative proposals to set up a system in respect of international regulations for preventing collisions between offshore drilling platforms and ships.

1 INTRODUCTION

The Convention on the International Regulations for Preventing Collisions at Sea (1972) (hereinafter called "COLREG") has played an important role in ensuring a safe navigation and operation of ships. However, the danger of collisions between offshore drilling platforms and ships has been never become less because of the COLREG. As another kind of manufactured unit at sea, which is never less dangerous than ships, offshore drilling platforms have caused casualties and incidents of one kind and another until now. When a collision occurs between an offshore drilling platform and a ship, technically, it is not always the ship only to blame. In some cases, the measures taken by the offshore drilling platform are valuable and significantly necessary to avoid an imminent collision.

2 THE CONCEPT AND CHARACTERISTIC OF OFFSHORE DRILLING PLATFORM.

The concept of "offshore drilling platform" refers to a mechanical facility on which offshore drilling operations take place. These include combined drilling and production facilities, such as floating or bottom founded platforms, bottom founded drilling rigs (swamp barges and jackup barges), and deepwater mobile offshore drilling units (MODU) including drill ships and semi-submersibles. These are capable of operating in deep or shallower waters. When they are operating in shallower waters, the mobile units are anchored to the seabed. However, in deeper water (>5,000 ft) the semisubmersibles or drillships are maintained at the required drilling location using dynamic positioning.

First of all, a marine drilling platform, as a tool for oil and gas extraction, is complex in structure with high technology content and very expensive in cost[10]; secondly, although it is a movable property, it must be treated as an immovable property like a vessel. The obtaining, setting, transfer and extinction of the ownership, the mortgage and even the charter of the offshore drilling platform should be registered just like an immovable property. To a large extent, it is legally personified; thirdly, it is very dangerous to work on offshore drilling platforms and special risks and liabilities may exist. When it undertakes drilling operation at sea, besides some usual marine risks, the offshore drilling platform may also encounter such special risks as blowout and blast distinctive to the drilling work. It is often very hard to handle these risks by the ordinary civil law and even by the existing marine legal system; furthermore, once an accident happens to a marine drilling platform, the consequence is often extremely serious. Apart from the huge casualties and property losses, there would be disastrous environmental pollutions.

When offshore drilling platforms shift from one place to another, they are very similar to navigational ships. The laws applied to offshore drilling platforms are normally the same as that are applied to navigational ships. For example, the definition of "ships" in the Maritime Code of the People's Republic of China (hereinafter called "Maritime Code") means sea-going ships and other mobile units. Obviously, the oil drilling platforms being tugged, as one kind of mobile units, are included in the scope of ships in the Maritime Code.

When offshore platform is under drilling operation, she cannot move from one place to another. Therefore, she would be excluded out of the scope of ships under maritime law. Offshore drilling platforms are motionless when they are under operation, they cannot take action to avoid collisions by changing their position, like navigational ships. COLREG, as a law to adjust action between navigational ships, will be not applicable to offshore drilling platforms.

3 THE NECESSITY TO SET UP A SYSTEM TO COORDINATE ACTIONS BETWEEN OFFSHORE DRILLING PLATFORMS AND SHIPS WHEN RISK OF COLLISION OCCURS.

When offshore drilling platforms are transferred from one place to another, they are very similar to navigational ships. However, according to the COLREG, "the word VESSEL includes every description of water craft, including non-displacement craft and seaplanes, used or capable of being used as a means of transportation on water".

Marine drilling platforms are excluded out of the definition of VESSEL, despite when they are moving at sea. Therefore, a problem will be that what kind of actions should be taken when an offshore drilling platform and a vessel are approaching one another at sea. Obviously, COLREG is not comprehensive enough to satisfy that requirement, a new system need to be set up to solve the issue.

Most of time, however, offshore drilling platforms are in another situation. When offshore drilling platforms are under operation, they need to stay still and motionless. In this situation, offshore drilling platforms are to some extent similar to anchored vessels. When there is a collision risk between an offshore drilling platform and a vessel under way, normally, the vessel under way will take action to avoid collision. However, is it the vessel's sole obligation to avoid collision under any circumstance? The answer to the question is negative.

3.1 *Collision risk between offshore drilling platform and a vessel out of control*

There are many reasons causing a vessel out of control, such as that propeller cannot work normally due to main engine failure; the vessel cannot change course due to steering engine failure or rudder blade missing, etc. No matter what kind of reasons that make the vessel out of control, they will be reduced or deprived the ability to avoid collisions with another object. If a collision risk occurs between an offshore drilling platform and a vessel out of control, it is clear that some measures should be taken by the offshore drilling platform to prevent a collision that will happen immediately, instead of waiting for the result of collision without doing anything.

3.2 *Collision risk between offshore drilling platform and vessel underway*

In this section, vessels underway refer to those who are sailing in normal condition, excluding those who are out of control, dragging her anchor or restraint in her ability to maneuvering.

Comparing with offshore drilling platforms, the vessels underway are at the advantageous position to avoid a collision. In the case that offshore drilling platforms are under normal operation situation, the underway vessel should take measures to keep clear of her. However, under the circumstance that it is impossible to avoid an immediate collision by the actions of the underway vessel alone[11], then the

[10] The cost of a drilling platform at the same level as Deepwater Horizon rig in the Gulf of Mexico is about 600 million to 700 million US dollars. China's first independently developed 3000m deep-water drilling platform "Haiyang Shiyou 981" costs about 6 billion RMB.

[11] Please refer to COLREG Rule 17 (b): When, from any cause, the vessel required to keep her course and speed finds herself so close that collision cannot be avoided by the action of the give-way vessel alone, she shall take such action as will best aid to avoid collision.

offshore drilling platform has obligation to take actions as will best aid to avoid the collision or reduce the consequence of collision[12]. If she would not take any positive actions or behave negative to let the collision happen during the immediate risk of collision, it is regarded as "negligence" and the offshore drilling platform needs to burden the subsequent collision liabilities[13].

3.3 *What kind of action could be taken by offshore drilling platform to avoid an immediate collision*

Collision always happens when two objects occupy a certain place at the same time. So the best way to avoid a collision is to change its position as soon as possible. When the offshore drilling platforms are under operation, they cannot transfer their position from one place to another. However, the above conclusion does not mean that there is nothing to do with the offshore drilling platform to avoid or reduce the consequence of collision. First of all, each offshore drilling platform should maintain proper watchkeeping to identify the risk of collision as early as possible, which is the primary obligation for the offshore drilling platform as strict as to a navigational vessel. Aside from that, each offshore drilling platform should be equipped with a reasonable number of tug boats around her to keep guard. If another vessel approaching her without an intention of altering course to avoid collision, the tug boats nearby should be instructed immediately to take any actions necessary to prevent a collision, such as alerting the vessel by any visible or audible signal, so far as to push the approaching vessel to eliminate the risk of collision.

4 IN VIEW OF THE ABOVE REASONS, IT IS NECESSARY AND PLAUSIBLE TO SET UP A SYSTEM TO PREVENT COLLISIONS BETWEEN OFFSHORE DRILLING PLATFORMS AND SHIPS. THEREFORE, INTERNATIONAL RULES FOR OFFSHORE DRILLING PLATFORMS TO PREVENT COLLISIONS AT SEA SHOULD BE ENACTED AS EARLY AS POSSIBLE.

The rules should clarify their application scope and supervisory authority. Watchkeeping should be maintained by watchstanders who are properly trained and certified. It is suggested that training and certification of watchstanders be carried out by Maritime Safety Authority, which has experiences in the supervision of seafarers. In addition, it is the Maritime Safety Authority's responsibility to ensure safety at sea, so they would have efficient ways to supervise and manage offshore drilling platforms.

The most important part of the rules should be the principles and rules in respect of what actions should be taken by the watchstanders to keep proper lookout and to avoid collisions. There should be clauses to stipulate arrangement of watchstanders and watchkeeping plans. In order to keep proper lookout, there must be ample qualified watchstanders equipped to an offshore drilling platform. Like requirements for navigational ships, 24 hours continuous watchkeeping should be maintained and fatigue duty should be avoided. When making watchkeeping plans, a comprehensible consideration should be put on emergency situation that may encountered at sea, requirements of watchkeeping should be well enough to identify the risk of collision as early as possible and emergency action could be taken to avoid an imminent collision.

Offshore drilling platform should be equipped with appropriate aids to navigation such as navigating marks and signals. Such marks and signals should be distinct and obvious enough to alert other vessels the position of offshore drilling platforms. In addition, to ensure external and internal communication, each offshore drilling platform should be provided with various communication equipments. The external communication includes communication among offshore drilling platforms or between offshore drilling platform and other vessel, or shore base. The internal communication includes communication in the offshore drilling platform, such as between watchstanders and tug boat.

There always should be a number of tug boats with ample horsepower standing nearby each offshore drilling platform to keep safeguard. These tugboats are very useful if it is necessary for them to take actions to avoid another vessel colliding with offshore drilling platform or reduce the consequence

---

[12] Please refer to *Marsden on collision at sea*. Simon Gault, Steven J. Hazelwood, A.M. Tettenborn Page.125: A ship at anchor which obstinately refuses to move from her anchorage where she necessarily endangers other craft may be held in fault for a collision that follows. An anchored vessel may reasonably be required to do what she can to assist the other to clear her, either by sheering with her helm, using her engines, paying out chain or in any way possible and failure to do so may be held to be negligence. She must, however, beware of acting too soon.
[13] See the sellina (1920)5 LI.L.Rep.216, The offin (1921)6 LI.L.Rep.444.The Defender (1921) 6 LI.L.Rep.392.See Paragraph.6-46.

of a collision. At the time when supply ships or transport ships get alongside or cast off, these tug boats can concert their actions and avoid misoperations.

5 CONCLUSION

It may be of a small probability for an offshore drilling platform to collide with a navigational ship. However, if such an accident happens, the consequence is very terrible. Until now, the provisions for offshore drilling platforms in respect of watchkeeping and collision prevention are very scattered, and it is very difficult to find a complete and integrated rule to set a certain standard for regulating the actions of offshore drilling platforms. The authors hope this article could contribute the research in this area.

REFERENCES

[1] Simon Gault, Steven J. Hazelwood, Andrew Tettenborn and Glen Plant. Marsden Collisions at Sea. London. Athenaeum Press Ltd. 2003.
[2] SI Yu-zhuo. Maritime Law Monograph [M]. Beijing: China Renmin University Press, 2007.
[3] SI Yu-zhuo. Maritime Law [M]. Beijing: China Renmin University Press, 2008.
[4] WANG Feng-chen, GU Wen-xian, ZHENG Jing-lue. Vessel Control and Collision Avoidance [M]. Beijing: China Communications Press, 1987.
[5] ZHENG Zhong-yi, WU Yao-lin. Decision-making for Vessel to Avoid Collision [M]. Dalian. Dalian Maritime University Press, 2000.
[6] ZHAO Jin-song. Vessel Collision Avoidance Theory [M]. Dalian: Dalian Maritime University Press,1999.
[7] JIA Lin-qing. Maritime Law [M]. Beijing: China Renmin University Press, 2001.
[8] FU Ting-zhong. Maritime Law and Practice [M]. Dalian: Dalian Maritime University Press, 2001.
[9] WU Xian-jiang. Concept and Analysis of Vessel Collision [EB.OL]. available at www. chinacourt.org, 2006-10-11.
[10] Rao Zhong-xiang. Discussion on New Concept of Vessel Collision and Practice Value [J]. Annual of China Maritime Law, 1992.

Other than Navigation Technical Uses of the Sea Space

Z. Otremba
Gdynia Maritime University, Gdynia, Poland

ABSTRACT: Marine areas, apart from their traditional use (maritime transport, fishery, navy), becomes suitable for other technical activities and investments (e.g. gas and electricity transmission, wind farms, gas and oil extraction). These activities interfere with marine environment as well can interfere with navigation. In order to recognize the scale of the problem and to assess possible environmental effects, the most important technical developments had been identified.

1 INTRODUCTION

Intensive navigation and technical large scale objects located in the marine areas modify already existing natural physical fields and distributions of other physical values in marine space which may affect aquatic organisms and consequently negatively alter marine biocenosis. At the same time, individual types of technical activities can interfere with each other.

Maritime management, including spatial planning [Hajduk 2009], should take account of modification of the natural features of marine space. Additionally there is not excluded that in future the need of developing of appropriate standards for marine environmental impact assessments (just as on land) will appear.

In this paper examples of marine space disturbances caused by typical, but different than traditional navigation, current technical activities in these areas are pointed out.

2 TECHNICAL ACTIVITIES IN THE SEA AREAS

2.1 *Offshore wind energy*

Marine shallow areas (up to depth of several dozen meters) appear suitable for installation numerous power stations for conversion wind energy into electricity. Present individual offshore wind turbine can generate electrical power up to 6 MW, which means that 100 windmills can produce energy similarly to large power unit of classical electrical power plant. Rapid growing of electrical power production by offshore windmill generators occurred in the last decade in EU countries – from 500 MW to 5000 MW [OffshorePoland 2012a)].

In the Polish Exclusive Economic Zone suitable areas for offshore wind farms are slopes of Słupsk Bank (proper Słupsk Bank is restricted as the Protected Area), Southern Middle Bank as well as Odra Bank. Coastal areas are protected as environmentally important, therefore cannot be consider as areas for technical use for now. There are several dozen applications for permission to build windmill farms in the Polish Marine Areas (Fig. 1).

Figure 1. Polish Exclusive Economic Zone with areas planed for marine wind farms (updated 2012 on the base of OffshorePoland 2012a. Can be compared with the data from 2008 [Cieślak 2008]).

One can point up various manifestations of maritime wind farms from the point of view of human activity in the sea areas (navigation, fishery, aviation, tourism, military, terrorist threat) and from the point of view of natural environment proper functioning (influence of noise, electromagnetic waves in air, magnetic field in water masses). Additionally cable network between wind turbines and land energetic system is substantial. Above mentioned noise (modification of natural underwater acoustic field) is expected during installation phase (mainly pile driving), in exploitation phase (interaction between wind and wings of rotor, work of the generators) and during removal phase (explosive materials use). For now there is no clear in what range wind farm space can be used for other human activities, for example for fishery.

2.2 Underwater electricity transfer

Marine water masses can be used as a medium transporting electrical energy. There are quite a few energy transfer systems installed in marine areas. If the Baltic Sea region is considered, since sixtieth of XX century several High Voltage Direct Current (HVDC) systems have been installed in which coastal electrodes introduce electrical current into water masses (Fig. 2). Apart electrodes, the single core cable operating with electrical potential of several hundreds of kilovolts in relation to surrounding water must be installed in the seabed.

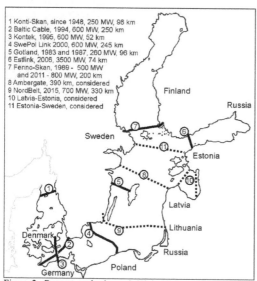

1 Konti-Skan, since 1948, 250 MW, 96 km
2 Baltic Cable, 1994, 600 MW, 250 km
3 Kontek, 1995, 600 MW, 52 km
4 SwePol Link 2000, 600 MW, 245 km
5 Gotland, 1983 and 1987, 260 MW, 96 km
6 Estlink, 2006, 3500 MW, 74 km
7 Fenno-Skan, 1989 - 500 MW
    and 2011 - 800 MW, 200 km
8 Ambergate, 390 km, considered
9 NordBelt, 2015, 700 MW, 330 km
10 Latvia-Estonia, considered
11 Estonia-Sweden, considered

Figure 2. Present and planned High Voltage Direct Current (HVDC) electric power cables in the Baltic Sea (after numerous sources).

Single HVDC system is able to transfer electricity up to 600 MW (Andrulewicz at al. 2003). After installation of planned wind farms (mentioned

in section 2.1), probably numerous of HVDC systems will appear to transporting generated energy to the land electric grids. Furthermore trans-Baltic electric duct will operate using HVDC technique probably, despite the fact that world-electrical-grid operate with the alternating current (AC). DC energy transfer is due to possibilities of present technique economically better than traditional AC method (lower loss of energy along energetic line).

Recently installed HVDC systems (for example the *SwePol Link*), instead of electrodes, uses so called "return cable solution" (simply, electrodes are replaced by cable operating with low electrical potential).

2.3 Oil and gas extraction

"Offshore drilling, deep sea mining, oil platform, jack-up rig, drillship" – there are popular terms appearing in oil and gas production in marine areas. There are a lot of marine technical systems - floatable and fixed – which are in use for hydrocarbons extraction from under the bottom deposits (Fig. 3).

Figure 3. Types of offshore oil and gas structures include: (1, 2) conventional fixed platforms, (3) compliant tower, (4, 5) vertically moored tension leg and mini-tension leg platform, (6) Spar (deepest: Dominion's Devils Tower in 2004, 1,710 m/5,610 ft GOM); (7,8) semi-submersibles, (9) floating production, storage, and offloading facility, (10) sub-sea completion and tie-back to host facility (modified after NOAA 2010).

Extraction devices are accompanied by ships, whose mission is supervision for safety, technical supply, collection and transport of excavated material. Brazilian coast, Gulf of Mexico, Norwegian Sea, North Sea, west coast of southern Africa - are areas of greatest production of crude oil, whereas northwest of Australia, Israel-Egypt coast, Venezuela coast, Brazil Coast and Norwegian Sea – areas of gas production (diffuse sources).

2.4 Gas pipelines

The seabed of above mentioned areas of hydrocarbons extraction is covered by dense pipeline grids.

Not only for hydrocarbons extracted in the sea areas must be transported though pipelines, because marine areas appears suitable for pipeline transport of hydrocarbons from land to land, even it is economically absurd – huge pipelines system between Russia and Germany (*NordStream*) is more expensive like the land solution. Admittedly *NordStream* is invisible for sailors, but enhanced military presence along the route of the pipeline system can be inconvenient for navigation.

Figure 4 shows the route (more than 1000 km) of *NordStream* gas transporting system, together with pipeline connecting the oil/gas rig *Baltic Beta* with the Polish coast (to power station in Władysławowo).

2.5 *Traffic connections*

The world's longest marine traffic bridges are planned (2015) between Qatar and Bahrain (40 km) and between Hong Hong and China (50 km, with 5.5 km of submersed tunnel). In the Baltic region works 4 long bridges: Oresund Bridge (7.85 km), Great Belt Bridge – Eastern (6.79 km), Great Belt – Western (6.61 km) and Oland Bridge (6.1 km). Oresund Bridge (Fig. 5) is connected with artificial island and underwater tunnel (Fig. 6).

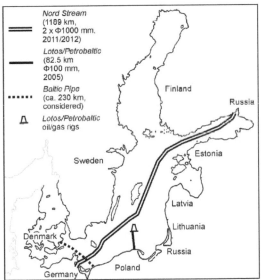

Figure 4. Present and considered gas pipelines in the Baltic Sea.

Interesting solution of the traffic connection through marine area is implemented in the Gulf of Finland so called Saint Petersburg *Flood* Prevention Facility Complex (former Leningrad Flood Barrier). Figure 7 shows the location of this construction, in which the tunnel beneath navigation canal is build.

This navigation pass can be closed by the huge flood doors.

Figure 5. Oresund Bridge (landscape impression).

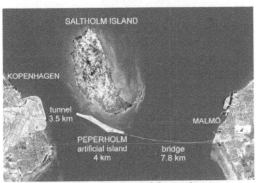

Figure 6. Components of the Oresund Connection.

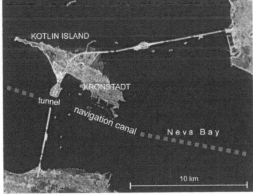

Figure 7. Saint Petersburg flood barrier.

2.6 *Coastal protection*

Encroachment of the sea into the land is a normal phenomenon in many points of the coastline in the world. Especially in the southern Baltic where land immerses itself successively even 0.2 mm/year (Uścinowicz 2011). The seashore is intensively defended in numerous places along the Polish coast. For example, in Jastrzębia Góra region (Figure 8), where gabion techniques is implemented, coastal protection cannot definitely stop the sea transgression – simply, coastal protection structures will sooner or later destroyed. Destroying processes of coastal structure can be weakened by detached breakwaters or underwater sills. Those underwater structures can be placed even several hundred meters from seashore line.

Figure 8. Example of poorly effective coastal protection: partly destroyed gabions at Jastrzębia Góra.

2.7 *Other technical activities*

Despite of above described human activities in the Polish Marine Area, other than navigation and coastal structures are gravel extraction (region of Słupsk Bank and Middle Southern Bank), undersea fiber-optics telecommunication cables, and military areas on the west of Ustka. Additionally, in the world scale, floating transloading terminals should also be classified as the place of technical activity influencing marine environment (CARGOTEC 2012).

3 DISCUSSION

3.1 *Navigational problems*

Marine technical investments are also discussed from point of view of navigation [Weintrit et al. 2012].

It is expected that the competition for marine space will be observed in the future - especially for the ocean shelf and for shallow seas (for example the Baltic Sea). Anyway, at the present, also traditional users of the sea (navigation, fishery) are forced to search the space for them. Therefore closely delimited shipping routes are formed, with the land continuous navigational assistance. High developed

current technologies allow reducing the risks caused by introduction of technical objects into marine areas. Only in emergency conditions - like sudden failure of equipment or extremely bad weather conditions - marine technical constructions can escalate serious hazards to navigation.

If coastal technical defense measures are considered, only constructions put forward (several hundreds meters) into the sea (wave breakers) can endanger small vessels (tourist yachts, motorboats, fishing boats).

3.2 *Environmental concerns*

Any human activity in the sea areas impacts the natural environment. The issue of hazards from navigation is known for several dozen decades already - International Convention for the Prevention of Pollution from Ships (MARPOL) regulates problems connected with negative influence of maritime transport on environment. However, the other than navigation activities are poorly analyzed in relation to the safety of the sea. Some adverse or beneficial effects can be predicted before the construction of the particular installation. Unfortunately some of negative effect may be unpredictable for a given sea area. Therefore, a thorough study in the design phase, construction phase and in operation phase are needed.

Every type of technical construction has its own specific impact on environment. For example, if HVDC system is considered, the first problem in the projecting phase was the route of he system. A lot of discussions affected final decision (Fig. 9). Figure 10 shows principle of operation of typical marine HVDC whereas Figure 11 presents distribution of the magnetic induction (horizontal component) above the cables (in the *SwePol Link* case). Fish profile voltage in the vicinity of electrode is shown in Fig. 12.

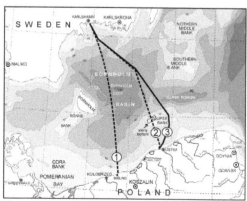

Figure 9. Evolution of projected route of the SwePol link [Andrulewicz at al. 2003]. The third is the final one – it bypasses chemical weapons deposits in the Bornholm Deep (1) and protected areas in the Słupsk Bank (2).

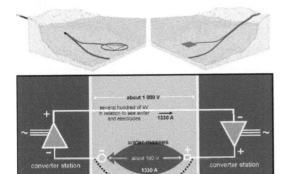

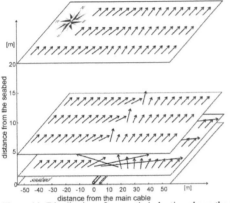

Figure 10. Scheme of marine HVDC connection

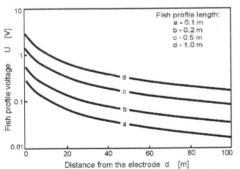

Figure 11. Distribution of magnetic induction above the HVDC cables (calculated by the author).

Figure 12. Fish profile voltage vs. distance from the electrode in the Baltic Sea waters (calculated by the author).

The question of environmental impact of marine technical construction is very broad and requires more intensive study (Otremba & Andrulewicz 2007).

REFERENCES

Andrulewicz E., Napierska D., Otremba Z., Environmental effects of installing and functioning of submarine HVDC transmission line SwePol Link: case study related to the Polish Marine Area of the Baltic Sea, Journal of Sea Research, 49, 337-345 (2003)
CARGOTEC 2012, Floating transfer terminal, http://www.cargotec.com/en-global/PS/Bulk-handling-equipment/Floating-transfer-terminals/Pages/default.aspx
Cieślak A. 2008. Development Strategy for Sea Areas, Min. of Regional Development. Proc. of Seminar on Development of Poland's Sea Areas", Warsaw (Poland), April, 2008.
Hajduk J.: Safety of Navigation and Spatial Planning at Sea. TransNav - International Journal on Marine Navigation and Safety of Sea Transportation, Vol. 3, No. 3, pp. 261-266, 2009
NOAA 2010, Types of Offshore Oil and Gas Structures http://oceanexplorer.noaa.gov/explorations/06mexico/background/oil/media/types_600.html
OffshorePoland 2012a, Total capacity of offshore wind farms in EU member states in the years 1991 – 2012 http://morskiefarmywiatrowe.pl/en/market-zone
OffshorePoland 2012b, The issued location decisions for offshore wind farm projects in Polish maritime area http://morskiefarmywiatrowe.pl/en/database/maps
Otremba Z., Andrulewicz E., Environmental Concerns Related to Existing and Planned Technical Installations in the Baltic Sea, Polish Journal of Environmental Studies, Vol. 17, No. 2, 173-179, 2008.
Uścinowicz 2011, Historia Bałtyku http://www.pgi.gov.pl/pl/ wody-podziemne-pomorza/387-historia-baltyku.html
Weintrit A., Neumann T., Formela K.: Some problems of the offshore wind farms in Poland, TransNav - International Journal on Marine Navigation and Safety of Sea Transportation, Vol. 6, No. 4, pp. 459-465, 2012

Chapter 5

Container Transport

Development of Container Transit from the Iranian South Ports with a Focus on the International North South Transport Corridor

M. Haghighi
College of Management, Tehran University, Iran

T. Hassangholi Pour
College of Management, Tehran University, Iran

H. Khodadad Hossani
Tarbiat Modares University, Tehran, Iran

H. Yousefi
Khoramshahr University of Marine Science and Technology, Iran

ABSTRACT: In this paper, we are going to investigate the growth of the Iranian maritime transport with a focus mainly on the North - South transit Corridor, review of dry ports in order to support the Iranian container terminals at the south ports in order to increase the volume of container storages at the terminals. For the purpose of improving the outcome of the ports operational management, it is recommended to concentrate on new strategies such as the utilize of transit corridors for development of the Iranian South Ports. The main part of this paper is dedicated to evaluate the role of container transit from the North - South transport corridor with a focus on the Iran North South transport corridor in order to improve Maritime Transport of Iran and increase the Iranian transit market share. Next part of this paper will be designated to consider the Iranian Southern Container Terminals at Khoramshahr, Imam Khomani, Busher, Bandar Abbas and Chabahar port which have suitable strategic position as transit base in the region. The next segment of this paper is dedicated to consider the growth of the container activities, container ships, container terminals logistic and services, advanced handling of containers at the hub container terminals, and rapid port operation.

1 INTRODUCTION

Although, cargo as container has been increased more than five times since 1990, therefore it caused that the world's fleet of container ships has developed about seven times. Apart from the significance of time and expenses on transit of containers, safety of containers or quality of container handling operation is vital issue at the container terminals.

It should be noted that in the last two decades, container transportation system has been faced under increasing development, in such a way that the rate of this development has reached to 7 or 9 percent in a year and it is predicted that this increase will have a rate of about 10 percent until 2020 while for other sea transportation means, the rate will be just 2 percent annually.[1] The development of dry ports has become possible owing to the increase in multi-modal transit of cargoes utilizing road, rail and sea.

2 GLOBAL CONTAINER TRADE

Table 1 shows part of the latest figures available on world container port traffic for 65 developing economies and underline Islamic republic of Iran with an annual percentage growth of container trade which changes between 2008- 2009 to 10.31 and between 2009-2010 to 17.50. It should be noted that Container trade in 2010 increased by 8 per cent on the Far East–Europe route, and by 10 per cent on the trans-Pacific Asia–North America route.

In 2010, the port of Shanghai for the first time took the title of the world's busiest container port from Singapore, with a throughput of 29.2 million TEUs. The 10 countries registering the highest growth were Ecuador (49.2 per cent), Djibouti (45.7 per cent), Namibia (44.7 per cent), Morocco (32.9 per cent), Jordan (15.8 per cent), Lebanon (15.4 per cent), the Syrian Arab Republic (12.2 per cent), Dominican Republic (11 per cent), the Islamic Republic of Iran (10.3 per cent) and Sudan (10.3 per cent).[2]

The country with the largest share of container throughput is China, with nine ports in the top 20. The Dominican Republic has been on the list of ports with double-digit growth for the last three years. The country with the largest share of container throughput continues to be China. [3]

Table 1. Container port traffic of the Islamic republic of Iran
Source: The Report of UNCTAD 2011

Country	2008	2009	Preliminary estimates for 2010	Percentage change 2008-2009	Percentage change 2010-2009
China	115 080 378	107 492 861	126 944 458	-6.58	19.58
Singapore	30 891 200	26 592 800	29 178 200	-13.91	9.73
China, Hong Kong SAR	24 494 229	21 040 096	23 532 000	-14.10	11.84
Republic of Korea	17 417 729	15 699 161	18 467 580	-9.87	17.78
Malaysia	16 024 820	15 691 296	17 975 796	-2.21	14.71
United Arab Emirates	14 768 177	14 425 038	15 196 323	-2.24	5.34
China, Taiwan Province of	12 971 224	11 352 097	12 302 111	-12.48	8.37
India	7 672 457	8 011 810	8 942 726	4.42	11.62
Indonesia	7 404 831	7 343 587	8 960 360	-2.18	23.70
Brazil	7 036 078	6 574 617	7 979 828	-6.11	21.37
Egypt	6 099 218	6 250 449	6 695 421	2.48	6.54
Thailand	6 726 297	5 897 856	6 648 530	-12.31	12.73
Viet Nam	4 393 990	4 548 598	5 474 452	10.17	13.00
Panama	5 129 499	3 537 112	5 906 744	-31.06	25.40
Turkey	5 218 316	4 521 713	5 508 974	-13.36	21.53
Saudi Arabia	4 652 022	4 430 678	5 313 541	-4.76	19.90
Philippines	4 471 428	4 306 733	5 048 669	-3.68	17.29
Oman	3 427 000	5 768 045	3 774 562	9.08	-0.17
South Africa	3 875 862	3 726 303	4 099 341	-3.86	9.40
Sri Lanka	3 687 496	3 464 297	4 000 000	-6.05	15.46
Mexico	3 312 713	2 874 287	3 705 608	-13.23	29.03
Chile	3 164 137	2 796 909	3 167 719	-11.64	13.12
Russian Federation	3 307 675	3 537 634	3 091 322	-25.31	32.34
Iran (Islamic Republic of)	2 600 230	2 206 475	2 600 527	10.31	17.50
Pakistan	1 936 001	2 058 056	2 151 058	6.12	4.52
Colombia	1 969 215	2 058 740	2 442 756	4.44	18.62
Jamaica	1 916 843	1 689 670	1 891 770	-11.81	11.96
Argentina	1 997 146	1 626 361	1 972 369	-18.57	21.27
Bahamas	1 702 000	1 297 000	1 126 000	-23.80	-13.26
Dominican Republic	1 138 421	1 263 496	1 302 601	10.96	9.43
Venezuela (Bolivarian Republic of)	1 325 104	1 298 717	1 228 364	-6.53	-0.84
Peru	1 206 326	1 230 849	1 533 926	-3.20	24.41
Morocco	919 360	1 222 306	2 066 430	32.92	68.46
Bangladesh	1 091 200	1 182 121	1 360 459	8.33	14.24
Ecuador	870 831	1 000 995	1 221 049	49.20	22.03
Lebanon	1 061 921	994 601	949 155	15.30	-4.57
Guatemala	937 642	966 328	1 012 369	-3.34	11.70
Costa Rica	1 004 971	875 497	1 013 463	-12.96	15.74
Kuwait	941 654	854 944	938 206	-11.19	4.00
Syrian Arab Republic	810 607	685 299	710 643	10.73	3.69
Côte d'Ivoire	713 625	677 329	704 119	-3.13	4.00
Jordan	582 515	674 525	619 000	15.80	0.67
Kenya	615 733	618 816	643 569	0.50	4.00

Table 2. Top ten container terminal and their throughput
Source: The Report of UNCTAD May 2011

Port name	2008	2009	Preliminary figure for 2010	Percentage change 2009-2008	Percentage change 2010-2009
Shanghai	27 980 000	25 002 000	29 069 000	-11	16
Singapore	29 918 200	26 490 800	28 490 800	-14	10
Hong Kong	24 494 229	31 640 096	23 532 000	-14	17
Shenzhen	21 413 888	13 250 100	23 526 700	-15	23
Busan	13 452 786	11 954 861	14 167 291	-11	18
Ningbo	11 226 000	10 502 800	13 144 000	-6	25
Guangzhou	11 001 300	11 190 000	12 550 000	2	12
Qingdao	10 320 000	10 260 000	12 012 000	-1	17
Dubai	11 827 299	11 124 082	11 600 000	-6	4
Rotterdam	10 800 000	9 743 200	11 146 804	-10	14
Tianjin	8 500 000	8 700 000	10 080 000	2	16
Kaohsiung	9 676 664	8 581 273	9 181 211	-11	7
Port Klang	7 973 579	7 309 779	8 870 000	-8	21
Antwerp	8 662 891	7 309 638	8 468 475	-16	16
Hamburg	9 737 000	7 007 704	7 900 000	-28	13
Los Angeles	7 849 985	6 748 994	7 831 902	-14	16
Tanjung Pelepas	5 600 000	6 000 000	6 530 000	7	9
Long Beach	6 487 816	5 067 597	6 263 399	-22	24
Xiamen	5 034 600	4 680 355	5 820 000	-7	24
New York/New Jersey	5 265 053	4 561 831	5 292 005	-13	16
Total top 20	247 021 180	220 000 801	254 387 802	-11	16

Table 2 shows the world's 20 leading container ports for 2008–2010. This list includes 14 ports from developing economies, all of which are in Asia; the remaining 6 ports are from developed countries, 3 of which are located in Europe and 3 in North America.[2] The majority of the ports listed remained in the same position for the third consecutive year, although the ports further down the league were subject to considerable shifting of fortunes and jostling for position. The top five ports all retained their respective positions in 2010, with Shanghai retaining its lead as the world's busiest container port, followed by Singapore, Hong Kong, Shenzhen and Bussan as shown in table.2. The gap between Shanghai and Singapore shortened as it shows that Singapore was in first step in 2008 and 2009, the modification of the figures for the both terminals in 2010 is 638,200 TEUs, and in2009 is 864,400 TEUs.

The resumption of manufacturing activity and global trade in containerized goods led to a recovery of demand for liner shipping services in early 2010. In 2009, however, the market was particularly bad for container shipping, as demand dropped by 9 per cent while supply grew by 5.1 per cent (Fig.2), the difference between these two figures being a staggering 14.1 percentage points. For the first time since 2005, demand is now forecast to grow faster than supply (in 2010).

A market segment of particular interest to many developing countries is containerized trade in refrigerated cargo, such as fruit, vegetables, meat and fish. Until the mid-1990s, the majority of this trade was transported in specialized reefer vessels. Since then, the entire growth in this market has been taken over by container shipping, installing slots for reefer containers on new container ships

At the beginning of 2010, the capacity to carry reefer cargo in containers stood at 2,898 million cubic feet, which was 9.5 times greater than the capacity on specialized reefer ships. The export of refrigerated cargo by container benefits from the global liner shipping networks and better door-to-door transport services. At the same time, it obliges ports and exporters to invest in the necessary equipment. Over the last decade, exporters have benefited from the increased competition between containerized and specialized reefer transport providers. As the reefer fleet is getting older and vessels are being phased out, this market segment will become almost fully containerized.[2]

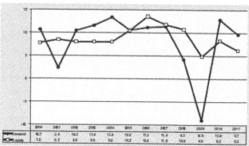

Figure 1. Supply and Demand of container shipping, 2000 – 2011, Source: The Report of UNCTAD 2011

3 IRANIAN CONTAINER TERMINALS OPERATION

Location of the Iranian container terminals are as follows: Khoramshahr, Imam Khomani, Bandar Abbas, Bushahr and Chabahar Port in South and Bandar Anzali, Noshahar and Amirabad Port in North of Iran. It should be noted that due to additional available capacity and a strong market, traffic at Bandar Abbas, Iran's main container terminal, has continued to increase. The port handled 2,231,200 TEU in 2010, an increase of 15% on the same period of 2009. The port is expecting to handle around 2.5M TEU for the year as a whole. Phase one of the port's second container terminal opened in February 2008, increasing overall capacity to 3.3M TEU per year and there are plans to double that in the next 36 months. Phase II of the new facility with another terminal operator became operational at 2012. Since 2010 a computerized system or automation system called TCTS 2010 system installed at Shahid Rajaee container terminal which is located a Bandar Abbas port. Based on the International regulations an online communication system can be carried out by port operator, custom, cargo receivers, shipping companies, and Transportation companies, etc.

4 ADVANCED EQUIPMENT REDUCE HANDLING TIME OF TRANSIT CONTAINERS AT THE TERMINALS

Container terminals are designated for the handling, storage, and possibly loading or unloading of cargo into or out of containers, and where containers can be picked up, dropped off, maintained, stored, or loaded or unloaded from one mode of transport to another (that is, vessel, truck, barge, or rail). Normally, a container terminal consists of different section such as POV (Parking Of Vehicles), Administration Building, Container yard, MY (Marshalling Yard) with inbound and outbound flow of containers in the terminal. It should be noted that the latest efficiency in container terminal automation provided by Zebra Enterprise Solutions is aimed at increasing container terminal capacity while improving port safety and security.[3] Designed to assist container terminal operators in the management of manned and automated port equipment, our container terminal automation solutions improve procedures and processes, as well as enhance container terminal equipment usage accuracy. Equipment management information such as maintenance schedules, equipment idle times, fuel levels and driver accountability of motorized and (non-motorized vehicles) and equipment can be tracked, monitored and managed in real-time. There have been a number of recent changes in the uses of advance technologies at Port container terminals that are designed to improve efficiency and productivity of operations. It is becoming common practice to see terminals operate with Optical Character Reader (OCR), Automatic Equipment Identification (AEI), Electronic Data Interchange (EDI), and other technologies such as cameras that are all designed to speed up the processing of containers through the terminal. In recent years, simulation has become as an useful tools in order to improve container terminal operation. Simulation can be distinguished as the following three groups: Strategically, operational, and tactical simulation. Strategically is applied to study of terminal layout and efficiency and costs of equipment, operational simulation is related to test different types of terminal logistics and optimization methods and finally, tactical simulation means integration of simulation systems into the terminal's operation systems.

5 DRY PORTS AS LOGISTIC POINT FOR CONTAINER TRANSIT OF THE TERMINALS

At first, it is better to understand the concept of a dry port. Mrs.Violeta Roso senior lecturer of Chalmers University in Sweden stated in this regard that "A "dry port" is defined as "an inland intermodal terminal directly connected to a seaport, with high capacity traffic modes, where customers can leave/collect their goods in intermodal loading units, as if directly at the seaport". And also, H.Yousefi (2011) expressed that A dry port is generally a rail terminal situated in an inland area with rail connections to one or more container seaports. The development of dry ports has become possible owing to the increase in multi-modal transit of goods utilizing road, rail and sea. This in turn has become increasingly common due to the spread of containerization which has facilitated the quick transfer of freight from sea to rail or from rail to road. So, Dry ports can therefore play an important part in ensuring the efficient transit of goods from a factory in their country of origin to a retail distribution point in the country of destination.[4]

The Persian Gulf has an area of approximately 240,000 km^2 and is very shallow, averaging just 50m-80m (1994; 1997), with only one opening– the Strait of Hormuz linking the Persian Gulf with the Arabian Sea. There are eight littoral Gulf States – Iran, Iraq, Kuwait, Saudi Arabia, Emirates, Bahrain, Qatar and Oman. The establishment of a shared place as dry port for all the above Gulf States will improve maritime transportation at the Persian Gulf. Based on IMO and WTO and the other relevant International regulations, it is necessary to consider the experiences of the container terminals operation at the Persian Gulf ports. It is useful for specifying the hub of container terminals at the Persian Gulf for further consideration. The capacity and operational methods of maritime transport, trade and transit of container at the terminals can be analyzed by using the Balanced Scorecard as an effective management tool which is used for improving container terminals activities at the Gulf. At the operational stages, relevant agencies including government departments directly involved in daily operations of dry ports are shown in the diagram below: In order to run a dry port efficiently, various agencies including government and nongovernment agencies with due authority are to be placed in the designated dry port premises.

6 DRY PORTS IN IRAN

6.1 Aprin central Terminal

Aprin is located at 21 kilometer of south western Tehran which is at the intersection of East-West & North-South railway junctions and is accessible to a number of highways. Aprin goods interchange central area is 100 hectares & construction of 110 storage houses & container Terminals are predicted in its area which their establishments are not completed. Before the Islamic Revolution in 1977, a region called Aprin was earmarked to the IRISL, which is linked to the railway. A dry port is a yard used to place containers or conventional bulk cargo, and which is usually connected to a seaport by rail or road and has services like, storage, consolidation, and maintenance of containers and customs clearance. They may be used for shipping, receiving and distribution centers designed to relieve the congestion in increasingly busy seaports, like an inland port. Aprin Terminal as a dry port has the potential to feed Tehran consuming Market & its surrounding industrial regions. At present, a storage house having the area of 9000 m2 is activated at Aprin center & its annual output is approximately 5000 TEU. Aprin Terminal which upon its formal activation commencement will be classified as off-shore dry ports, would have considerable interests; in case of coordinating with customs, the corresponding goods will be transported to the destination of Aprin center through railway after discharge & loading at sea ports of Shahid Rjaee in this context goods owners are able to directly refer to Aprin Center to receive their respective goods.

Aprin terminal which is equipped with Reach Stacker, Side lift, Lift truck, Bascule, container washing equipment 10 ton lift truck, the telescopic boom 35 ton crane, and a version of TPTS comprehensive & commence software is able to commence its activity as the first Iranian dry port. Recently, the Managing Director of Islamic Republic of Iran Shipping Lines stated that

Maragheh Special Economic Zone (SEZ) in Iran will turn into a dry port in the near future.

Figure 2. Three Intermodal Terminals in Iran
Source: Feasibility of establishment of "Dry Ports" in the developing countries—the case of Iran, Springer Science, Sep 2010.

Since Maragheh boasts railroad and is adjacent to Kurdistan province, the proposed terminal in Maragheh will expedite cargo transport and make it easier to provide services.

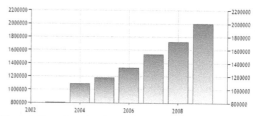

Figure 3. Iran, Container Port Traffic (TEU:20 Foot Equivalent Units)
Source: trading Economics.com

6.2 Shahid Motahhari terminal

This terminal is located at 36 km south of the city of Mashhad in the north-east of Iran. The site also has an access to the national highway. Its area is about 190 ha and there are required facilities to handle container transport needs. This terminal is managed by private sector (Domestic Various Related Statistics and Information [7].

6.3 Sarakhs terminal

This terminal is located at 170 km north-east of city of Mashhad in the northeast of Iran at Iran-Turkmenistan border. Its strategic position can be looked as an important gate to the markets of some land-locked countries. The site is also outfitted by required facilities (Domestic Various Related Statistics and Information [7].

6.4 *Sirjan terminal*

Sirjan is located at Kerman Province which is away 180 Km from the capital of the state, its distance from Banadr Abbas Port is 300 Km. Since Sirjan linked to Bandar Abbas port and Tehran through railway, therefore it has good location in order to support Shahid Rajaee Container terminal at Banadr Abbas Port.

7 THE INTERNATIONAL NORTH-SOUTH CORRIDOR

India has taken the lead in what it calls "kick-starting" an "international north-south corridor" from Iran to Russia via Turkmenistan and Kazakhstan to ensure a seamless connectivity to Central Asia. It should be noted that India wants this corridor to be operational by 2013. The North-South Transport Corridor is a term used to describe the ship, rail, and road route for moving freight from South Asia to Europe through Central Asia, the Caucasus, and Russia. [11] The route primarily involves moving goods from India via ship to Iran. From Iran, the freight moves by ship across the Caspian Sea or by truck or rail to Southern Russia. From there, the goods are transported by truck or rail along the Volga River through Moscow to Northern Europe. In 2001, Russia, Iran, and India signed an agreement to further develop the route. The Government of India had started this project with the view to enhance trade relations between South Asia and Central Asia. The primary step towards trade enhancement, was signing of Memorandum of understanding between Indian and Iran over the development of Chabahar port and transshipment facility at Banadr Abbas and Imam Khomani port. This "International North-South Transport Corridor" will have its starting point from Mumbai, and via transshipment the goods will reach Bandar Abbas Port in Iran, then a railway link will be established between Iran, Turkmenistan, Kazakhstan and it will finally reach Russia. India and Russia will strive to revive the North-South Transport Corridor (NSTC) through Iran that has failed to take off more than 10 years after the three countries signed an agreement to set up the trade route. India and Russia agreed to "enhance connectivity" through the NSTC including through China, the Embassy release said. The international North-South transport corridor (ITC) linking India, Iran and Russia is becoming increasingly popular, and is likely to become the "Silk Road "20th century." This trade route will cut by two-thirds the

time of cargo transportation from Indian Ocean areas and Persian gulf to Europe, and reduce the price of shipment of each cargo container by 400 dollars. By 2005 Russia, India and Iran plan to double cargo flows, bringing them to eight million tones.

ITC is to ensure the delivery of cargos from Indian Ocean and Persian Gulf regions to Iranian ports on the Caspian Sea, to be taken up by ferries to Russia's railroad terminals or, by river-marine vessels, through Russia's inland waters to countries of eastern and central Europe and Scandinavia. Distance of new route (Green line in the following figure) in comparison to the previous route is 40% shorter; it means that instead of 45-60 days it becomes 25-30 days.

Figure 4. Green line as new, short, and cheap route
Source: Internet, Site of Wikipedia

In addition, in the new route the price of shipment of each cargo container 30% cheaper than previous route.[11] The ITC Coordination Council such as representatives of Iran, India, and Russia, had invited delegations of Belarus and Kazakhstan, to attend. The latest meeting of the Coordinating Council in Tehran approved Belarus, Kazakhstan, Oman and Tajikistan as new ITC members. The applications from Syria, Azerbaijan and Armenia are currently under consideration. Tajikistan and Oman will join the ITC agreement shortly. The INSTC was expanded to include eleven new members, namely: Republic of Azerbaijan, Republic of Armenia, Republic of Kazakhstan, Kyrgyz Republic, Republic of Tajikistan, Republic of Turkey, Republic of Ukraine, Republic of Belarus, Oman, Syria, Bulgaria (Observed). The Astrakhan transport hub occupies an important location in the North-South ITC. It is a transit point for some three million tones of cargoes bound for Caspian Sea ports of Iran. An increase in cargo flows, which began after the emergence of the ITC, required the construction of additional port facilities in this region.

In near future, train ferries will take cargos to the port of Amir Abad in Iran. The construction of a container terminal in the port of Makhachkala began in 2002, and by the end of this year its capacity will reach 30520 TEU a year.

Figure 5. Astrakhan as main location of North South ITC
Source: Source: Internet, Site of Wikipedia

It is planned to bring the capacity of the terminal to 61000 TUE a year in the nearest future. At present, container shipment through the port of Makhachkala mostly proceeds along the Russia-Iran route.[11]

Figure 6. The location of Makhachkala port in Caspian Sea
Source: Source: Internet, Site of Wikipedia

At present, several operators work on the ITC transport market, who gained the necessary experience, made business contacts and now successfully fulfill cargo owners' orders. A healthy competition between them ensures a stable high level of transport services. The North-South ITC will have a tremendous economic, political and strategic significance this century. To bridge the two continents is the important mission of the North-South ITC.

8 CONCLUSION

Improvement of the Iranian container transit from the International transit corridor can be carried out by using the advanced equipment for handling operation at the container terminals; it causes to reduce the time vessels spent in the ports. It should be noted that the Competitive advantage of a container terminal at the Iranian ports is achieved by the integrated scheduling of various types of handling equipment with an aid of the useful Strategic Models at an automated container terminal. It can be observed that use of an optimum strategic planning at the container terminals may cause to develop the efficient scheduling of the equipment in order to increases the productivity of the container terminals. Apart from the significance of strategic planning on container transit, the other factor which has great influence on developing the Iranian south ports is the establishment of dry ports under the following three categories: a). Close to the Iranian container terminals, b). At the costal water area, c). Very far from the Iranian South Ports, in order to increase the capacity of Iranian transit ports.

REFERENCES

[1] R. S. Kaplan and D.P. Norton, "The balanced scorecard: measures that drive performance," Harvard Business Review, vol.70, no.1, pp.71-79, 1992.
[2] Review of Maritime Transport, (2010) Report by the UNCTAD secretariat, Chapter 5, UNITED NATIONS, New York and Geneva.
[3] Trade and Development report, (2011) UNCTAD, New York and Geneva
[4] H.Yousefi, et al (2011), Development of the Iranian Maritime Transport: A Focus on Dry Ports and the Iranian Container Terminals Operation, Proceeding IMLA19 Conference, Opatija, Corasia.
[5] H.Yousefi, et al (2011), Balanced Scorecard: A Tool for Measuring Competitive Advantage of Ports with Focus on Container Terminals, International Journal of Trade, Economics and Finance, Vol. 2, No. 6.
[6] I. Vacca, M. Bierlaire, M. Salani, (2007) Optimization at Container Terminals: Status, Trends and Perspectives. 7 Th Swiss Transport Research Conf .Monte Verita /Ascona , September 12-14.
[7] Andrius Jarzemskis and Aidas Vasilis Vailiauskas, (2007) "Research on Dry Port concept as Intermodal Node" by Vilnius Gediminas Technical University.
[8] L. Henesey, P. Davidsson, J. A. Persson, (2006) Agent Based Simulation Architecture for Evaluating Operational Policies in Transshipping Containers. Multiagent System Technologies. LNAI, Vol. 4196, Springer, pp. 73-85.
[9] FDT (2007) Feasibility study on the network operation of hinterland hubs (Dry Port Concept) to improve and modernize ports' connections to the hinterland and to improve networking, Integrating Logistics Centre Networks in The Baltic Sea Region [Project], January.
[10] Ehsan Dadvar & S. R. Seyedalizadeh Ganji &Mohammad Tanzifi, (2010), Feasibility of establishment of "Dry Ports" in the developing countries—the case of Iran, Springer Science+Business Media

[11] "Export Practice and Management" by Alan Branch, Thomson Learning, (2006).

[12] North – South Transport Corridor, the Internet Site of Wikipedia.

[13] Kaplan, Robert S., and David P. Norton, (1996) "Using the Balanced Scorecard as a Strategic Management System" Harvard Business Review 74, no. 1.

[14] Lester, Tom. "Measure for Measure: The Balanced Scorecard Remains a Widely Used Management Tool, but Great Care Must Be Taken to Select Appropriate and Relevant Metrics." The Financial Times, 6 October 2004.

[15] Lewy, Claude, and Lex Du Mee.(1998) "The Ten Commandments of Balanced Scorecard Implementation." Management Control and Accounting,.

[16] McCunn, Paul. (1998)"The Balanced Scorecard…the Eleventh Commandment." Management Accounting 76, no. 11.

[17] van de Vliet, Anita. (1997) "The New Balancing Act." Management Today, July 1997, 78.

[18] Williams, Kathy. "What Constitutes a Successful Balanced Scorecard?" Strategic Finance 86, no. 5 (2004).

[19] Domestic various related statistics & information (2006–2008).

[20] UNESCAP (2008) Logistics sector development, planning models for enterprises & logistics clusters. New York.

[24] Roso V (2008) Factors influencing implementation of a dry port. Int J Phys Distrib Logist Manag 38(10).

[21] Roso V (2007) Evaluation of the Dry Port concept from an environmental perspective: a note. Transp Res Part D 523–527.

[22] Roso, V., (2009). The dry port concept. Doctoral Thesis, Department of Logistics and Transportation, Chalmers University of Technology, Goteborg, Sweden.

[23] UNCTAD, (2009), ASYCUDA Home Site. Available at: http://www.asycuda.org [Accessed: 12 December 2009].

Green Waterborne Container Logistics for Ports

U. Malchow

Hochschule Bremen – University of Applied Sciences, Bremen, Germany

ABSTRACT: With the Port Feeder Barge a new type of harbour vessel has been designed – firstly for the operation within the port of Hamburg. Other major and even minor ports could benefit from the operation of this innovative type of vessel as well as it improves the efficiency and at the same time reduces the ecological footprint of intra-port container logistics. It can even facilitate container handling at places which are not suited at all for container operation yet – be it for reasons of shallow water or insufficient handling equipment.

1 THE 'PORT FEEDER BARGE' CONCEPT

Figure 1. Port Feeder Barge (artist impression)

The internationally patented Port Feeder Barge is a self-propelled container pontoon with a capacity of 168 TEU (completely stowed on the weather deck), equipped with its own state-of-the-art container crane mounted on a high column (Fig. 1). The crane is equipped with an automatic spreader, extendable from 20ft to 45ft, including a turning device. A telescopic over height frame is also carried on board. The barge is of double-ended con-figuration, intended to make it extremely flexible in connection with the sideward mounted crane. Due to the wide beam of the vessel no operational restrictions (stability) for the crane shall occur. The crane has a capacity of 40 tons under the spreader, at an outreach of 27 m (maximum outreach: 29 m). The unique vessel is equipped with 2 electrically driven rudder propellers at each end in order to achieve excellent manoeuvrability and the same speed in both directions. Hence the vessel can easily turn on the spot. While half of the containers are secured by cell guides, the other half is not, enabling the vessel to carry containers in excess of 40ft length as well as any over-dimensional boxes or break bulk cargo. 14 reefer plugs allow for the overnight stowage of electrically driven temperature controlled containers.

Table 1. Port Feeder Barge - Main Data

Type:	self propelled, self sustained, double-ended container barge
Length o.a.:	63.90 m
Beam o.a.:	21.20 m
Height to main deck:	4.80 m
Max. draft (as harbour vessel):	3.10 m
Deadweight (as harbour vessel):	2,500 t
Gross tonnage:	approx. 2,000 BRZ
Power generation:	diesel/gas electric
Propulsion:	2 x 2 electrical rudder propeller of 4 x 280 kW
Speed:	7 knots at 3.1 m draft
Class:	GL ✠ 100 A5 K20 Barge, equipped for the carriage of containers, Solas II-2, Rule 19 ✠ MC Aut
Capacity:	168 TEU (thereof 50% in cellguides), 14 reefer plugs
Crane:	LIEBHERR CBW 49(39)/27(29) Litronic (49 t at 27 m outreach)
Spreader:	automatic, telescopic, 6 flippers, turning device, overheight frame
Accommodation:	6 persons (in single cabins)

The vessel shall fulfil the highest environmental standards. A diesel- or gas-electric engine plant with very low emissions has been chosen to supply the power either for propulsion or crane operation. The vessel can be operated by a minimum crew of 3 whereas in total 6 persons can be accommodated in single cabins.

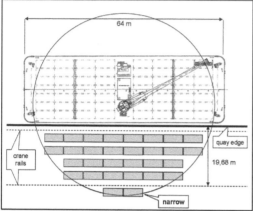

Figure 2. Turning cycle of crane

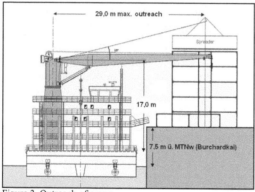

Figure 3. Outreach of crane

The key element of the worldwide unique Port Feeder Barge concept is its own full scale heavy duty container crane. All its mechanical components have been especially designed for continuous operation – unlike standard shipboard cranes, which are designed for operation only every few weeks when the vessel is in port. Due to its nature the load cycle requirements of the Port Feeder Barge are even higher than for many quayside cranes, which has significant consequences on the layout of the mechanical components.

When berthed, the Port Feeder Barge is able, without being shifted along the quay, to load or discharge 84 TEU in three layers between the rails of typical quayside gantry cranes (Fig. 2). This is more than sufficient, with a total loading capacity of 168 TEU. That is why the full outreach of the crane

is not always needed. Berthing the vessel with the crane on the opposite side of the quay (Fig. 3 and 4) would speed up crane operation as the turning time of the outrigger is minimised. The height of the crane column is sufficient to serve high quays in open tidewater ports even at low tide while stacking the containers in several layers (or to serve even deep sea vessels directly, Fig. 7). Due to its short length of 64 m the Port Feeder Barge needs only a small gap between two deep sea vessels for self sustained operation (Fig. 4).

The operation of the Port Feeder Barge is not limited to inside seaports. As the hull is classified according to Germanischer Lloyd's class notification for seagoing vessels the operation in (sheltered) open waters off the coast is also possible which opens some interesting opportunities for additional employment.

The design of the vessel has been developed by PORT FEEDER BARGE GmbH[14] in close co-operation with Wärtsilä Ship Design Germany GmbH, both of Hamburg.

Figure 4. Port Feeder Barge is working independently from quayside equipment at a deep sea terminal requiring only a small gap between two deep sea vessels

Within the port of Hamburg the Port Feeder Barge shall ply between all the major waterfront container facilities, including a dedicated berth to meet with the inland waterway vessels.

2 BUSINESS FIELDS

2.1 Intra-port haulage

The Port Feeder Barge shall serve as a 'floating truck' in the course of its daily round voyage

[14] PORT FEEDER BARGE GmbH,
Grosse Elbstrasse 38, 22767 Hamburg / Germany,
T: +49 (0)40 / 401 6767 1, E: info@portfeederbarge.de,
www.portfeederbarge.de

throughout the port, i.e. shuttling containers between various container facilities. Hence container trucking within the port can be substantially reduced [1].

Figures on port's internal box movements which are predominantly carried on the road are generally difficult to obtain. It is estimated that in 2011 within the port of Hamburg approx. 300,000 containers, i.e. approx. 85% of the anticipated entire intra-port volume, have been carried by truck (which is corresponding to approx. 450,000 TEU).[15, 16] The remaining was carried on the water by ordinary barges. The reason for the poor share of barge transport is very simple: conventional inland barges or pontoons employed in intra-port container transportation are dependent from the huge quayside gantry cranes for loading/discharging. However one move by gantry is already exceeding the costs of the entire trucking. Naturally two moves are needed and the barge has to be paid as well. Hence in most cases intra-port barging of standard containers is not competitive unless the liftings by the quayside gantries are subsidised by the terminals.

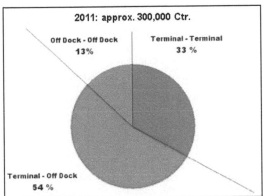

2011: approx. 300,000 Ctr.

Off Dock - Off Dock
13%

Terminal - Terminal
33 %

Terminal - Off Dock
54 %

Figure 5. Split of intra-port container haulage by truck within the port of Hamburg

According to industry sources $^{1}/_{3}$ of road haulage is of terminal-to-terminal nature while more than half is between a terminal and an off dock facility (depot, packing station, repair shop etc.) of which many have their own water access (Fig. 5).[16] Hence the present cargo potential for the Port Feeder Barge is estimated to roughly 150,000 containers p.a. (corresponding to approx. 225,000 TEU). Approx. 50% (!) of all intra-port container trucking has to pass the notoriously congested Koehlbrand bridge, which is one of Hamburg's landmarks and the major link between the eastern and western part of the port (Fig. 6). Hence the Port Feeder Barge would offer a viable and more environmentally friendly alternative compared to trucking.

[15] In Hamburg the average TEU ratio is approx. 1.5 TEU/box
[16] Extrapolated information provided by CTD GmbH (HHLA), November 2009

Figure 6. Typical view on Hamburg's Koehlbrand bridge linking the port's eastern and western part (photo: dpa)

2.2 Feeder operation

In multi terminal ports common feeder services have to accept and deliver containers from/to all facilities where deep sea vessels are berthing. For this reason the feeder vessels have to call at all terminals within the port – sometimes even if only a few boxes have to be handled. E.g. in Hamburg each feeder vessel has to call in average at 4 different facilities (incl. waiting berths) [2] [3]. That is why the feeder lines are already a major customer of the road hauliers. Otherwise the number of berth shiftings within the port would have been even higher.

From the terminal's perspective all vessels with less than approx. 100 boxes to handle are critical with respect to profitability. However, in Hamburg almost $^{2}/_{3}$ of all terminal calls by feeder vessels are below that figure! Smooth and efficient feeder operation is essential for the port's economic well-being as its entire container throughput relies to more than $^{1}/_{3}$ on transhipment.

As the feeders are already big customers of the trucking companies for intra-port haulage the Port Feeder Barge can also replace trucking for collecting and distributing their containers. The Port Feeder Barge will offer a more competitive service than the trucks can do, especially for lots of many or over-dimensional boxes (flats with overwidth/-height). Hence the Port Feeder Barge can be used by the feeders more intensively than the trucks at present enabling the feeders to concentrate on the major terminals only, thus reducing the number of berth shiftings, reducing their time in port and related costs, increasing terminal and berth efficiency as well as improving safety (collisions). In 2011 two feeder vessels ran already into each other while manoeuvring with one even in danger to sink within the port.

2.3 Inland navigation

Inland navigation is facing a dilemma as far as the hinterland transport of containers to and from seaports is concerned. On the one hand there is a common understanding that its share in hinterland transport has to be substantially increased – for capacity and environmental reasons. On the other hand in sea ports inland waterway vessels have to berth at the facilities which are tailor made for the biggest container vessels sailing on the seven seas (with a capacity of 14,000 TEU and even more). Hence the efficiency of the big gantry cranes is rather low when serving the small vessels. Not surprisingly but most disadvantageous inland navigation enjoys the last priority when it comes to berth allocation.

Figure 6. The Port Feeder Barge is serving an inland waterway vessel midstream (artist impression)

Inland barges suffer more than feeder vessels as they have to call at even more facilities. E.g. Rotterdam has approx. 30 terminals and depots which are frequently served by inland container barges. The average number of terminal calls per vessel is about 10 whereas in about 50 % of the calls only less than 6 containers are handled [4]! This kind of inefficient and not coordinated 'terminal hopping' is very time consuming and each delay at a single terminal results in incredible accumulated waiting time during the entire port stay. Not surprisingly only $^1/_3$ of the time in port is used for productive loading/unloading [5].

In Hamburg where inland navigation has still a share of less than 2% in hinterland container transport the inefficient operation has been identified as one of the major reasons for such small share [6]. Some Dutch and German studies regarding the problems of transhipment procedures between inland navigation and deep sea vessels have been already published [4] [5] [6]. One common result is that container handling for inland navigation and deep sea vessels should be separated from each other. In other words: Inland vessels should not call at the deep sea facilities any more.

It is claimed that dedicated inland waterway berths have to be introduced at deep sea terminals. However most terminals do not have any shallow draught waterfront left where such berths could be meaningfully arranged. Transforming existing valu-

able deep sea quays to exclusive inland navigation berths with dedicated (smaller) gantry cranes does not pay off for the terminals as such a measure would reduce their core revenue earning capacity.

The erection of a central and dedicated inland navigation terminal within a port, where all inland barges call only once, has also been proposed to spare the inland barges their inefficient 'terminal hopping'. However this would burden the most environmentally friendly mode of hinterland transport with the costs of two further quayside crane moves and one additional transport within the port (either on the water or even by truck). The opposite of more waterborne container hinterland transport would be achieved. Hence increasing the share of inland navigation in hinterland transport of containers is really facing a dilemma in many major container ports which can be solved by introducing Port Feeder Barges.

The Port Feeder Barge could act as a dedicated 'floating terminal' for inland navigation. During its daily round voyage throughout the port the vessel is collecting and distributing the containers also for inland navigation. Once a day, the Port Feeder Barge will call at a dedicated berth to meet with the inland barges where the containers shall be exchanged ship-to-ship by the vessel's own gear, independently from any terminal equipment (virtual terminal call). Not even a quay is required but the transhipment operation can take place somewhere midstream at the dolphins (Fig. 6).

Such kind of operation will strengthen the competitiveness of inland navigation and contributes to increase the share of the most environmentally friendly mode of hinterland transport. Employing one or more Port Feeder Barges as a 'floating terminal' is less costly and much quicker and easier to realise than the erection of comparable quay based facilities (not to mention that less parties have to be involved for approval).

3 FURTHER APPLICATIONS

3.1 Emergency response

The Port Feeder Barge can also help to keep consequences of maritime averages at a minimum. When container vessels are grounded in coastal zones they mostly have to be lightered very quickly to set them afloat again in order to avoid further damage to the vessel, the environment and in extreme cases to sustain even the accessibility of a port at all. However it has honestly to be conceded that most container ports are not prepared for such situation and do not have suitable floating cranes (if any) available to quickly lighter big container vessels.

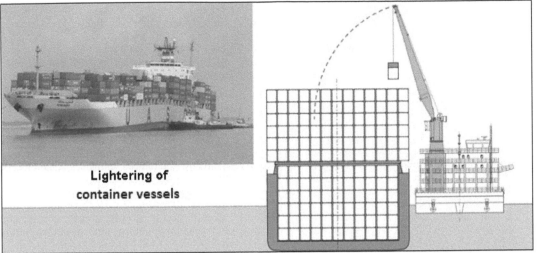

Lightering of
container vessels

Figure 7. Grounded panamax container vessel on Schelde river in 2005 and how it could have been quickly lightered

Unlike some other heavy floating equipment, the Port Feeder Barge can navigate in very shallow waters due to its light ship draught of only 1.2 m. Despite its small size the Port Feeder Barge can quickly lighter grounded container vessels with even more than panamax beam by working from both sides (Fig. 7).

3.2 *Hong Kong style midstream operation*

In Hong Kong approx. 10% of the huge port's container throughput, still relies on floating units serving deep sea vessels directly while laying at anchor (Fig. 8)! These traditional but unique midstream barges are equipped with their own cargo gear, but the handling method is far from being sophisticated. The A-frame derricks have a single beam just controlled by wires and are not even fitted with a spreader, but instead rely only on steel wires being fitted manually to the comer castings of the containers. In fact this is cargo handling technology from the 1950s and complies hardly with international port labour safety standards. Such midstream barges are only operating in Hong Kong (except a few in Angola and Vietnam). In average 4 fatal accidents are officially reported each year. Quite apart from the health and safety issues, they are not self propelled (not even pushed but towed). Port Feeder Barges would significantly improve such operation with regard to safety, efficiency, speed, flexibility and accessible ship sizes.

Figure 8. Typical midstream operation in Hong Kong

At other places throughout the world where terminal facilities are insufficient or congested or water depth is limited the introduction of advanced midstream operation by Port Feeder Barges could be a viable alternative to the construction of costly terminal facilities.

3.3 *Floating crane*

With a capacity of 49 t under the hook (40 t under the spreader) the Port Feeder Barge can also be employed as a flexible floating crane with sufficient deck space for any kind of cargo other than containers.

4 URBAN ISSUES

Operating Port Feeder Barges is also beneficially affecting urban issues as ports can supersede with heavy land based investments to improve their intermodal connectivity for inland navigation. With respect to investment, availability of land reserves, construction approval, flexibility and not to forget environmental and townscape issues a 'floating terminal' is much smarter than any land based facility.

5 OPTION: LNG AS FUEL

All costly measures to be taken to keep the exhaust emissions of the diesel-electric engine plant at an envisaged minimum (e.g. exhaust scrubbers, urea injection, filters etc.) could be saved when choosing LNG as fuel. The Port Feeder Barge would be an ideal demonstrator for LNG as ship fuel:
– As a harbour vessel it does not rely on a network of bunkering stations. Only one facility is sufficient. At the initial stage the vessel could even be supplied out of a LNG tank truck (a standard 50 m³ truck load is sufficient for approx. 14 days of operation).
– Due to its pontoon type there is plenty of void space below the weather deck. Hence the accommodation of the voluminous LNG tanks would not be a problem at all which is not the case with all other types of harbour vessels. Up to approx. 500 m³ of tank capacity could be theoretically installed which is by far more than sufficient (Fig. 9).

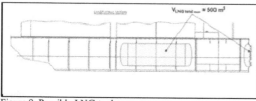

Figure 9. Possible LNG tank arrangement

Another option would be to use ISO LNG tank containers for fuel storage. This would allow for cost efficient intermodal supply chain solutions. However the legal framework has to be checked very carefully as different safety rules apply for LNG stored as cargo in containers or as fuel in shipboard tanks.

Bunkering of LNG requires certainly some additional safety measures. However in Norway where numerous coastal ferries are fuelled with LNG in the meantime the bunkering procedures have become a daily routine without any incident for the time being.

Beside the logistic and environmental advantages the Port Feeder Barge concept would provide LNG as fuel would further boost its green quality.

6 CONCLUSION

As there is no doubt that container volumes will certainly continue to increase ports have to prepare to ease already experienced bottleneck situations and to reduce the environmental impact of container transhipment procedures.

The Port Feeder Barge concept is a 'green logistic innovation' for sea ports (whose inherent beneficial effects to the environment can even be further increased by using LNG as fuel) that helps to…
– shift container trucking within sea ports from road to waterway,
– ease feeder and transhipment operation within multi terminal ports,
– improve the intermodal connectivity of inland navigation within sea ports,
– be prepared for grounded container vessels.
At places with insufficient or congested terminal facilities and/or shallow water restrictions the Port Feeder Barge could facilitate the handling of deep sea container vessels at all.

REFERENCES

[1] Malchow, U., "Port Feeder Barge – Intermodaler Anschluß für die Binnenschifffahrt" (conference presentation), 2. DVWG-Binnenschifffahrtsforum, Duisburg, 4. September 2007
[2] Nordsee-Zeitung, June 28, 2008
[3] Deutsche Verkehrs-Zeitung, August 16,2008
[4] Konings, R., "Opportunities to improve container barge handling in the port of Rotterdam from a transport network perspective", Journal of Transport Geography, Vol. 15 (2007), Elsevier Ltd.
[5] Konings, R., "Smart collection and distribution of containers on barges in the port of Rotterdam", NECTAR cluster meeting, 12-13 November 2004, Lugano
[6] Beyer, H. and Pistol, B. "Konzeptstudie zur Verlagerung vom Lkw auf Binnenschiffe und zur Stärkung der Hinterlandverkehre", Uniconsult GmbH, Hamburg, 14. Januar 2009

The Concept of Modernization Works Related to the Capability of Handling E Class Container Vessels in the Port Gdynia

K. Formela & A. Kaizer
Gdynia Maritime University, Gdynia, Poland

ABSTRACT: In the following article, the authors decided to present a concept of modernization works, which need to be done in order to enable the handling of E class container vessels in the port of Gdynia. This concept is based on the data, available to public usage, covering development plans and investments to be conducted in the Port of Gdynia in the nearest future. Modernization works will enable further development of two container terminals: Baltic Container Terminal (BCT) and Gdynia Container Terminal (GCT) located in the Port of Gdynia.

1 INTRODUCTION

Seaports are very important elements of transportation networks. Although the localization of any port has significant influence on its development, proper operation and comfortable access to marine terminals are directly reflected in the volume of freights handled in the specified area. The current development of containerization results in a huge increase in popularity of the transportation of unit cargo and significantly influences spatial layouts of both existing and newly designed seaports. The annual increase in shares of this branch in the global transport causes other terminals, especially those concentrated on conventional general cargo, to note a significant drop in transshipment. This tendency results in the number of containers transshipped being the main determinant of the development of a specified port.

2 THE PORT OF GDYNIA AND E CLASS CONTAINER VESSEL

2.1 *Characteristics*

The port of Gdynia is a merchant seaport on the Gulf of Gdańsk in Pomeranian Voivodeship, its coordinates being 54°32' N, 18°34' E. It is the third Polish seaport in the amount of transshipped cargo (after Gdańsk and Szczecin-Świnoujście port complex). The port of Gdynia is a universal port, specialized mainly in handling general cargo

vessels. Port localization is characterized by very convenient navigation conditions. The Hel Peninsula is a natural protection for vessels anchored at the roadstead and the entrance to the port, which is 150 m wide and 14 m deep (Admiralty Sailing Directions, Zarząd Portu Gdynia 2013), makes the port easily accessible. The port working 24 hours a day, in a 3 shifts system. It is non-tidal, accessible throughout the whole year because it never freezes up in winter. The length of the wharf is 17700 meters, of which over 11000 is designated as transshipment operations area. The overall premises of the object is 755,4 hectare, of which 508 ha is on land. According to port regulations, pilot services are compulsory for vessels bigger than 60 metres in length and tug assistance is compulsory for vessels over 90 metres in length or over 70 metres if they carry dangerous cargo. The greatest load capacity of the port of Gdynia is containerized cargo, handled mainly by two container terminals, located at the west port, that is Baltic Container Terminal and Gdynia Container Terminal. Freights in the form of conventional general cargo are handled by Baltic General Cargo Terminal. Freights handled by the following terminals are bulk freights: Baltic Grain Terminal, Maritime Bulk Terminal Gdynia Ltd, Baltic Bulk Terminal Ltd, Westway Terminal Poland and Petrolinvest. The ferry slip that operates vessels from Stena Line, which provides services on the route between Gdynia and Karlskrona, is located at the Helsinki II wharf. The general group of companies operating in the area of the port of

Gdynia is presented in Figure 1, together with their locations.

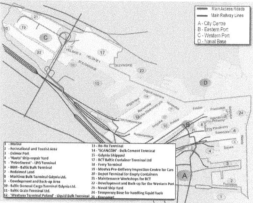

Figure 1. Main companies operating in the area of the port of Gdynia. [2]

2.2 *Development possibilities*

According to the developmental strategy of the port of Gdynia, emphasis is put on modernization of port infrastructure, which allows for handling even bigger vessels. The strong development of containerization and steadily growing volume of unit cargo freights provides ample developmental chances for general cargo terminals. However, capabilities to handle the huge ocean container vessels are fairly limited at Gdynia terminals. Up to December 2012 the biggest container ship ever handled in the port of Gdynia was a vessel of Mediterranean Shipping Company – a container ship MSC Krystal, which called at the port of Gdynia on 18th January 2012, has 277,3 meters in length and 40 meters in width. It was possible for such a big vessel to enter the Baltic Container Terminal thanks to deepening the port up to -13,5 m.

Modern trends in operating container terminals put strong emphasis on directly handling ocean vessels, creating container hubs and giving up feeder shipping. That is why the port of Gdynia aims also at creating appropriate standards that will fulfill the demands of ship owners.

A significant developmental problem in Gdynia is limitation to handle bigger vessels, which is caused by the depth of the fairway and basins in the port. There are also sizeable difficulties with manoeuvring vessels which are longer than 350 m. Due to spatial layout of the port of Gdynia, a big developmental chance (through handling E class container ships) is the capability to increase the parameters of the turning-basin no. 2, which would be possible by removing the Gościnne wharf located at the Naval Shipyard, as well as dredging the port canal and the wharfs near container terminals up 16,0 m. Of course this kind of modernization

changes must comply with the regulations and requirements set by approved organizations before deployment (PIANC, 1985 and 1987; Safe Waterways; McCartney, 2005)

2.3 *E class container vessel*

E class container vessels (Figure 2) are a series of akin ships build in 2006-2008 by Odense Steel Shipyard for Danish A.P. Moller – Maersk Group, known also as Maersk. These vessels are characterized by L=397,0 m length and B=56,0 m. width of the hull, as well as the maximum draught of T_{max}=16 m. In the world ranking of the biggest container vessels they are only overtaken by CMA CGM Marco Polo, which is a leader in everything but the hull size (Length Over All=396,0 m., Beam=53,6 m.). E class container deadweight is 15 000 TEU, compared to 16 000 TEU of Marco Polo. This situation is going to change in the near future, as Maersk company ordered Triple E class container vessels in Daewoo Shipyard, which will be the biggest vessels of this type in the world. The new type of vessels will be a breakthrough in the vessels' construction, thanks to significant reduction of fuel consumption and a serious reduction of CO_2 emission, in order to protect the natural environment. The new vessels' hull will have LOA = 400m. in length and B = 59m. in width. The new hull shape will allow for larger cargo space with a minimal possible increase in size, compared to other, existing class E Maersk vessels.

Figure 2. Model of E class container vessel in Transas Navi-Trainier 5000 Professional Simulator.

3 THE CONCEPT OF MODERNIZATION WORKS IN THE PORT OF GDYNIA

The term "modernization works" includes the proposed changes in existing navigational and technical infrastructure in the form of: turning-basins (localization and size) and wharfs, as well as the deepening of the port canal. This concept was created on the basis of the data available to public usage about developmental plans and investments that will be conducted in the nearest future in Gdynia Port (Zarząd Morskiego Portu Gdynia S.A. 2013; Rynek Infrastruktury, 2013; Puls Biznesu, 2013). Deepwater Container Terminal in Gdańsk, which operates in neighborhood, handles, among others, class E container vessels. Because of that, enabling Gdynia Port to handle these vessels should be interpreted as a development opportunity for Baltic Container Terminal and Gdynia Container Terminal.

3.1 Proposed changes of existing navigational and technical infrastructure

To enable handling of E class container vessels, the authors propose changes in the present, shown in the Figure 3, navigational and technical infrastructure in the form of:

- the removal of the Gościnne wharf. Development opportunities for the terminals in Gdynia Port are significantly dependent on improving manoeuvring conditions for container vessels. To do that, the only discernible project is the purchase and then demolishing the Gościnne wharf located on the premises of the Naval Shipyard.
- relocating and resizing the existing turning-basin No. 2 The removal of the Gościnne wharf will enable relocating the turning-basin deeper in the port canal and increasing maximum its diameter from d=385 m. up to d=485 m. This will significantly improve the possibilities of safe conducting manoeuvrs for vessels longer than 350m.

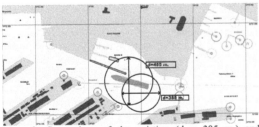

Figure 3. Comparison of the existing (d = 385 m.) and proposed (d = 485 m.) turning circle after the removal of the Gościnne wharf.

- relocating the floating dock. Considering the scope of aforementioned modernization works, the floating dock moored presently at Dock IX significantly limits the possibilities of relocating and resizing of the newly planned turning-basin. The authors proposed locating it in the north part of Dock IX, as presented in Figure 4. This solution will allow for using the available space more efficiently.

Figure 4. The proposed changes concerning relocating and resizing of the existing turning-basin No.2, as well as relocating the floating dock after removing the Gościnne wharf.

- Dredging the port canal up to 16 m. Such depth should provide sufficient under keel clearance during maneuvers of E class vessels, with the draught of T=15 m. It should however be mentioned that the reserve under keel clearance will depend, among other things, on speed of the vessel during maneuvers in the canal (settling effect influence) and the present hydro-meteorological conditions (e.g. water level, waves). Before the dredging works start a project should be prepared. It should take into account the predicted amount of spoils (excess material), the amount of time required for the work, the kind and accessibility of required equipment, as well as the possible need to strengthen the wharfs. In the case of dredging near Norweskie and Słowackie wharfs (No. 11 and No. 14 on Figure 1), which is box placed wharf type, the possibility of the need to strengthen the construction to keep the proper stability conditions needs to be taken into account.

The Figure 5, presented below, depicts the simulation area of Gdynia Port, including the present navigational and technical infrastructure. On the upper part is chart of a part of the Port. On the lower part, a visualization, including the highlighted, present turning-basin d=385 m. in diameter and a model of E class container vessel L=397 m. i B=56 m. length and width of the hull.

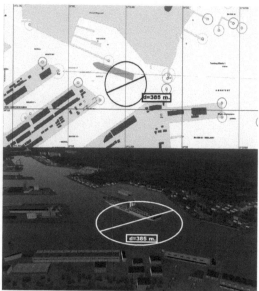

Figure 5. The chart (upper part) and visualization (lower part) of existing navigational and technical infrastructure in Gdynia Port.

A new simulation area of Gdynia Port, together with visualization, was made specifically for this article in order to present simulating method which is one of the methods of ship's turning-basins designing (Kornacki, 2007 and 2011). All aforementioned proposed changes were included in it. This area is presented in the Figure 6 and 7.

Figure 6. The visualization of the New simulation area of Gdynia Port.

Figure 7. The chart of the New simulation area of Gdynia Port.

For the initial assessment of the topic discussed the concept of modernization works in Gdynia Port in order to handle E class container vessels, a real-time simulation of entering and manoeuvring of the said class vessels in Gdynia Port was conducted. The simulation was run at the laboratory of Faculty of Navigation at Gdynia Maritime University. Navi-Trainer 5000 Professional navigation and manoeuvring simulator, ECDIS Navi-Sailor 4000 electronic maps and systems simulator and Model Wizard (v. 5.0) application were used for it. In the simulation area it is possible to generate a variety of hydro-meteorological conditions (e.g. wind, waves, current) that influence the vessel manoeuvring in this area.

The simulation was conducted in good weather conditions (wind direction: NE, wind speed: 6 m/s; wave height h=0,4 m) and with two tow boats assistance, both with azimuth thrusters (tractive force of 50 t.). Figure 8 presents the trajectory.

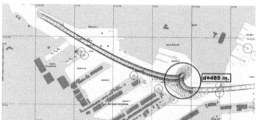

Figure 8. The visualization of the New simulation area of Gdynia Port.

The trajectory of movement presented in the picture above with white colour records the position and outline of the vessel in Δt=30 s. intervals. The analysis of recordings of the simulation allows for later assessment of not only vessel's position in relation to land/other navigational obstacles, but also assessment of settings and indications of navigational and manoeuvring devices, at any given moment of the simulation.

4 CONCLUSIONS

After analyzing the recording of the trajectory of movement a conclusion may be drawn that aforementioned manoeuvre is safe and the presented concept of modernization works is justified. However, it is crucial to remember that the aim of the above simulation was only to present the possibilities and to initially assess the planned works and show the method which can be used to optimizing sizes of turning-basins. For thorough assessment of the navigational risk connected with manoeuvring a vessel of this size in the port all possible threats should be taken into account in order to conduct the planned modernization works.

What is more, a series of simulations in comparable conditions should be conducted, that will support the mathematical, statistical and stability calculations, as well as theoretical assumptions, good practice and experience.

Fast development of containerization makes transport of cargo by containers very popular and relative easy. Annual growth observed in the global transport makes that containerization changing existing transport schemes. Finding new opportunities and new development solutions are chance to follow the trends that determine the modern maritime transport. Concept of modernization works related to the capability of handling E class container vessels in the port Gdynia allow to increase in the number of containers handled in Gdynia container terminals.

REFERENCES:

A Guide and in Port Approaches, 3rd Ed, ICS, 1999 r.
A Guide to good practice on port Marine Operation - prepared in conjunction with the Port Marine Safety Code, 2009 r.
Admiralty Sailing Directions, Baltic Pilot, Volume I, NP18, edition 15/2010 r.
Briggs M.J. &Borgman L.E. & Bratteland E.2003. Probability assessment for deep-draft navigation channel design. International Journal for Coastal Harbour and Offshore Engineering, Coastal Engineering 48, Elsevier.
Coastal Engineering Manual – Part 5, August 2008, publication no EM1110-2-1100
Dziennik Ustaw z 1998 r poz. 98,101.645 Rozporządzenie Ministra Transportu i Gospodarki Morskiej z dnia 01.06.1998 r. w sprawie warunków technicznych jakim powinny odpowiadać morskie budowle hydrotechniczne i ich usytuowanie
Dziennik Ustaw z 2006 r poz. 06.156.1118 Ustawa, Prawo Budowlanc (tekst jednolity)
EM 1110-2-1100, Coastal Engineering Manual - Part V

Kornacki, J. & Galor, W. 2007. Gdynia: Analysis of ships turns manoeuvres in port water area. TransNav'07.
Kornacki, J. 2007. The analysis of the methods of ship's turning-basins designing. Szczecin: XII International Scientific and Technical Conference on Marin Traffic Engineering, Maritime University of Szczecin.
Kornacki, J. 2011. Simulating Method of Ship's Turning-basins Desgning. Gdynia. International Journal on Marine Navigation and Safety of Sea Transportation.
McCartney B. 2005. Ship Channel Design and Operation. American Society of Civil Engineers.
PIANC, 1985. Underkeel Clearance for Large Ships In Maritime Fairways with Hard Battom. International Navigation Association
PIANC, 1997. Approach Channcls – A Guide for Design, International Navigation Association
Rozporządzenie Ministra Gospodarki Morskiej z 23.10.2006 r. w sprawie warunków technicznych użytkowania oraz szczegółowego zakresu kontroli morskich budowli hydrotechnicznych
Safe Waterways. A Users Guide to the Design, Maintenance and Safe Use of Waterway, Part 1, Canada
Zarządzenie nr. 12 Dyrektora Urzędu Morskiego w Gdyni z dnia 14 czerwca 2005 r. Przepisy portowe. Tekst ujednolicony wraz z późniejszymi poprawkami.
Bussines Portal Puls Biznesu www.pb.pl Website dated on 23.02.2013:
http://www.pb.pl/2555444,56153,port-gdynia-wyda-ponad-451-mln-zl-na-inwestycje-w-latach-2012-2014
Gdynia City gdynia.naszemiasto.pl Website dated on 23.02.2013:
http://gdynia.naszemiasto.pl/artykul/1248883,gdynia-zaloga-stoczni-marynarki-wojennej-przeciwna,id,t.html
Rynek Infrastruktury portal. Website dated on 23.02.2013:
http://www.rynekinfrastruktury.pl/artykul/81/1/rozbior-stoczni-marynarki-wojennej-nauta-chce-przejac-firme-a-port-w-gdyni-nabrzeze.html
Zarząd Morskiego Portu Gdynia S.A. Port Gdynia. Website dated on 23.02.2013:
http://www.port.gdynia.pl/pl/oporcie/daneportu
http://www.port.gdynia.pl/pl/oporcie/strategia-rozwoju/64-strategia-rozwoju-portu?start=3
http://www.port.gdynia.pl/podsumowanie2011/prezentacja_Z MPG_podsumowanie_2011.pdf

Container Transport Capacity at the Port of Koper, Including a Brief Description of Studies Necessary Prior to Expansion

M. Perkovic, E. Twrdy & M. Batista
University of Ljubljana, Faculty of Maritime Studies and Transport, Slovenia

L. Gucma
Maritime University of Szczecin, Poland

ABSTRACT: The vision of the Port of Koper is to become the main container terminal in the northern Adriatic. The northern Adriatic ports of Koper, Rijeka, Trieste, Venice and Ravenna operate in a relatively closed system in which the market and customers are limited and therefore the ports are forced to co-operate while at the same time they compete with each other. In addition they are located in three countries, with different transport policies and plans of development. The Port of Koper is Slovenia's only port, and therefore extremely important, as it contributes significantly to the Slovenian GDP. The state of Slovenia is the largest shareholder and the future development of the port depends on decisions made by the Ministry of Infrastructure. The increase in container throughput in the Port of Koper requires a reconstruction and extension of the current container terminal as an absolute priority. Regarding economic sustainability the extension must be in line with the estimated growth of traffic as well as with the exploitation of present and future terminal capacities. The occasional expansion projects must fulfil environmental and safety requirements. For large container vessels (LOA more than 330 m) calling at the Port of Koper the safety of the berthing and departure conditions have to be simulated under various metocean conditions. At the same time manoeuvres should not be intrusive – expected propeller wash or bottom wash phenomena must be analysed. When large powerful container vessels are manoeuvring in shallow water bottom wash is expected and because sediments at the port are quite contaminated with mercury some negative environmental influence is expected. The most important expected investment in the container terminal is therefore extending (enlarging) and deepening the berth. The paper will present statistics and methods supporting container terminal enlargement and a safety and environmental assessment derived from the use of a ship handling simulator.

1 INTRODUCTION

The ports of Koper, Trieste, Venice, Ravenna and Rijeka are located in the northern part of the Adriatic Sea, which penetrates deep into the middle of the European continent, providing the cheapest maritime route from the Far East, via Suez, to Europe. Large commercial and industrial hubs like Vienna, Munich and Milan are just a few hours' drive away. In the last twenty years the total container traffic in the northern Adriatic ports has increased almost exponentially, on average 7% per year, though the rate has varied among ports (Fig. 1), (NAPA). The fastest growth of container traffic was recorded at the Port of Koper, at an average of 14% per year, in Venice the growth was constant, while at Ravenna the traffic barely increased at all. The minimum throughput was and remains at the Port of Rijeka, which lost a great deal of traffic due

to the state of war in Croatia; since about 2003 the increase in Rijeka's container throughput has been more in line with that of Koper, Trieste, and Venice.

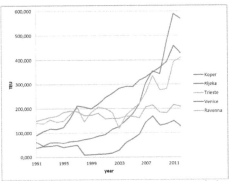

Figure 1. Container throughput in NAPA ports during 1991-2012 (1000 TEUs), (*NAPA*)

2008 and 2009 – the worst years of the global economic and financial crisis – offer some interesting results. In Venice during this period, throughput kept steadily increasing by 5% per year; the other four ports experienced a decrease averaging 15%. The largest drop in traffic was recorded in Trieste, a decrease of more than 58,000 TEUs (17.5%), though by percentage Rijeka fared worse, declining at the rate of 22.5% (38,000 TEUs less). We performed the shift-share analysis proposed by Notteboom (2007). In this analysis we include absolute growth of container traffic ABSGR and the share effect among ports. The calculation is based on the following formulas

$$ABSGR_k = TEU_{k1} - TEU_{k0} \qquad (1)$$

$$SHARE_k = \left(\frac{\sum_i TEU_{i1}}{\sum_i TEU_{i0}} \right) \qquad (2)$$

Table 1. Absolute growth of container throughput and total shift of containers for NA ports (1991-2011)

Period	Koper	Rijeka	Trieste	Venezia	Ravena
	Absolute growth TEU				
1991 1995	-3,758	2,000	8,200	32,300	34,600
1995 1999	19,821	-29,866	35,163	76,703	-11,595
1999 2003	48,033	18,164	-66,765	83,864	-13,045
2003 2007	179,411	116,742	147,465	45,845	46,220
2007 2011	283,666	5,637	127,323	128,851	8,756
1991 2011	527,173	112,677	251,386	367,563	64,936
	Containers shift TEU				
1991 1995	-11,851	-4,065	-12,307	16,801	11,422
1995 1999	10,121	-34,235	10,592	51,003	-37,481
1999 2003	38,067	16,572	-84,646	59,706	-29,698
2003 2007	68,289	71,020	37,144	-124,752	-51,700
2007 2011	125,743	-50,263	8,739	-16,309	-67,910
1991 2011	263,476	13,175	-62,571	-7,743	-206,337

Results of the calculation are displayed in Table 1, which shows that the port of Koper has by far experienced the largest absolute growth and shift in the region. All other ports have oscillations. The most unpleasant situation is at the port of Ravenna.

2 HOW TO HANDLE MORE CONTAINERS IN THE PORT OF KOPER

The trend in the shipping market is an increase in orders of ships of over 7500 TEUs, which is why the Port of Koper began to further develop the infrastructure in its container terminal. Significant financial investments in the extension of the container shore, expansion of storing space and the purchase of specialized transport equipment paid off during the increase in transport (despite the global crisis) in the year 2010. The quantity of transported containers has reaching enviable numbers. This very success, though, has at the same time created a

problem. The growth of container throughput in the Port of Koper is at the limit of the capacity for the existing container terminal. Therefore, it is necessary to start construction on a new container terminal and reconstruction and extension of the current container terminal. The extension is in line with the estimated growth of traffic as well as with the exploitation of present and future terminal capacities. New projects and potential investments are important steps for the development of the Port of Koper, enhancing its performance and increasing its market share. The figure below (Fig. 2) shows the enlargement plan of the Port of Koper. A new pier, 3, is foreseen as an additional container terminal, while the existing container terminal shall be extended to accommodate one more berth (berth 7D). At present a large container vessel can call at the container terminal (berth 7C) with a limited draft of 11.6 m.

Figure 2. Existing and extended piers (Perkovic et al. 2012b)

To extend the pier and to determine the appropriate channel depth, deterministic and semi-probabilistic methods for designing a channel were applied. The minimum width and shape of the channel must be appropriate for safe calling and departure of characteristic container vessels presented at wind conditions up to 5 knots. As a result of the extension of pier 1 the entrance into basin 1 will narrow, which can affect the safety of approach for the largest cruisers (LOA up to 347m, draft 14.0m) calling at berths 1 and 2.

Figure 3. Basin 1 - fully loaded berths; at top extended container terminal (berth 7D)

The extended plan with a fully loaded berth is presented in figure 3. The initial step was to analyse aspects relative to safety of an approaching cruiser while the extended container terminal is occupied by a large container vessel – figure 4a, shows the approaching trajectory and measurement lines of safe margins. Figure 4b shows the results of the first attempt at designing an entrance to a channel dredged to -15 m and the trajectory of a large container vessel entering basin 1. The designer hoped to make the channel as short as possible to minimise dredging costs, which is why the designed entrance was steep and narrow. Such an approach was also chosen because of limited amount of landfill capacities. Even brief simulation using a full mission ship handling simulator (Transas NTPro 5000, version 5.25) (Transas 2012) running with previously chosen container vessel model - clearly shows that such a channel is not an adequately safe approach for large container vessels. Based on those initial simulations further research work was ordered.

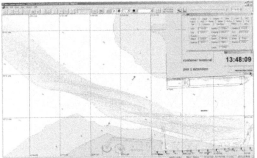

Figure 4a. Basin 1 – approaching trajectories of cruising vessel

Figure 4b. Basin 1 – approaching trajectories of container vessel (initial test)

2.1 Determining nominal channel width by the determinist method

The fundamental criterion for defining and dimensioning elements forming a navigation channel or a harbour basin is safety in manoeuvring and operations carried out within them (Puertos del Estado 2007). The criteria for the geometric layout

definition of the following navigation channels and harbour basins: fairways, harbour entrances, manoeuvring areas, anchorages, mooring areas, buoy systems, basins and quays is based on knowing the spaces occupied by vessels, which depends on: a) the vessel and the factors affecting its movements, b) the water level and factors affecting its variability. The main references for defining those factors are ROM 3.1 *"The Recommendations for the Design of the Maritime Configuration of Ports, Approach Channels and Harbour Basins"* (Puertos del Estado, 2007) and PIANC *"Permanent International Association of Navigation Congresses"* (PIANC 1997). The key parameters in approach channel design according to PIANC and ROM are alignment, traffic flow, depth, and width. They are all interrelated to a certain extent, especially depth and width. Factors included in determination of the channel width include: vessel manoeuvrability (oo), ship speed (a), prevailing cross wind (b), prevailing cross current (c), prevailing longitudinal current (d), significant wave height (e), aids to navigation (f), bottom surface (g), depth of waterway (h), cargo hazard level (i), width for bank clearance (j). The minimum channel width designed for the analyzed container vessel turned out to be 162.64 meters for wind conditions 4-6 according to the Beaufort scale (*Table 2*). As a particular (gusty) katabatic wind is present in that area - manoeuvres should not be allowed at wind stronger than 5 according to the Beaufort scale. That limit was confirmed by the simulation (semi-probabilistic) method described in the next paragraph. The effectiveness of such simulations depends on the simulator capabilities to properly represent maneuvering characteristics and factors influencing ship behavior (Kobylinski 2011).

Table 2. PIANC approach – factors determining minimum channel width

Basic manoeuvring lane width and additional widths		Factors for multiplying vessel beam (B=42.8)	
00	vessel manoeuvrability (poor)	1.8	
Additional Widths for Straight Channel Sections		wind <4^{0}Bf	wind 4-6^{0}Bf
a	ship speed (slow, less than 5 knots)	0.0	0.0
b	prevailing cross wind	0.1	0.5
c	prevailing cross current (low, 0.2 – 0.5 knots)	0.1	0.2
d	prevailing longitudinal current (low)	0.0	0.1
e	significant wave height	0.0	0.1
f	aids to navigation (moderate with poor visibility)	0.3	0.3
g	bottom surface (smooth and soft, < 1.5T)	0.1	0.1
h	depth of waterway (h/T) 1.25–1.5	0.4	0.4
i	cargo hazard level (low to medium)	0.2	0.2
j	width for bank clearance	0.1	0.1
Sum		3.1	3.8
The bottom width of the waterway (channel)		132.68 m	162. 64 m

2.2 Determining nominal channel width through the semi-probabilistic method

Channel geometric design in this procedure is mainly based on statistically analysing the areas swept by vessels in the different manoeuvres considered, which, should a sufficient number of manoeuvre repetitions be available, will enable the resulting design to be associated to the risk present in each case (Brigsa et al. 2003, Solari et al. 2010). This method was applied on the basis of real simulator studies. The simulations were performed in different meteorological conditions. Under every type of condition adequate numbers of trials were executed by human navigators. After the simulations, each trial was processed statistically in order to obtain the probability density function of ships' maximum distances from the centre of the waterway and the accident probability calculation in the given conditions. Finally, a safe water area was plotted with consideration of previously set up admissible risk level. The navigational risk R is defined as:

$$R = P \cdot C \tag{3}$$

where: P - probability of accident, C – consequences. The risk is expressed usually in monetary values over a given period of time (one year in this kind of analysis). The vessel can safely navigate only in such an area where each point satisfies the depth requirement. If such case exists, the area is referred to as the safe navigable area. The vessel carrying out a manoeuvre in a navigable area sweeps a certain area determined by the subsequent positions of the vessel. The parameters of that area have a random character and depend on a number of factors. Therefore, for fairways and harbour entrances the navigational safety condition can be transformed to this form (Gucma 2013).

$$D_m\left(t\right) \ge d^s_{ijkm} \tag{4}$$

where:

$D_m\left(t\right)$ – breadth of the navigable area at the m-th point of the fairway at the moment t, for which the safe depth condition is satisfied: $h(x, y, t)$ $T(x, y, t) + (x, y, t)$;

d^s_{ijkm} - breadth of the safe manoeuvring area at the m-th point of the fairway for the i-th vessel, performing the j-th manoeuvre in k-th navigational conditions.

$h(x, y, t)$ – area depth at a point with the coordinates (x, y) at the moment t,

$T(x, y, t)$ - vessel's draft at a point with the coordinates (x, y) at the moment t,

$\Delta(x, y, t)$ – under-keel clearance at a point with the coordinates (x, y) at the moment t.

$$d^s_{ijkm} = f\left(d_{ijkm}\right) \tag{5}$$

where:

d_{ijkm} – swept path of the i-th vessel performing the j-th manoeuvre in k-th navigational conditions for the m-th point of the waterway.

The layout of a swept path is presented in figure 5

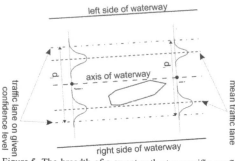

Figure 5. The breadth of a swept path at a specific confidence level at points (i) and ($i+1$) of the fairway.

2.3 Simulations and results

First it was necessary to build the planned, enlarged port area based on precise bathymetry. The sailing area was created using Transas application Model Wizard (Transas 2011). Highly precise bathymetry (*Figure 6*) (spatial resolution 1m x 1m) was inserted and the projected manoeuvring area was quickly created. Figure 7 is a screenshot from the ship handling simulator NTPro5000 (Transas 2012).

Figure 6. Modified channel - created with Model Wizard sw.

Figure 7. Fully loaded container vessel (111626 DWT) approaching terminal with assistance provided by pilot and two tugs – wind 5 knots from 060°

44 simulations were executed in various metocean conditions. Manoeuvres were processed according to the model previously described. The resulting safe waterway area at a 0.95% confidence level is presented with figure 8 (green colour). Such a confidence level is used most frequently for the design of the waterways.

In more critical solutions the level 99% could be considered. In port basins, however, the ship's speed is slow enough to significantly reduce the consequences of accidents, which explains the tolerance of 0.95% as a starting point for more serious considerations and risk analyses.

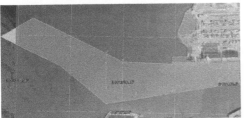

Figure 8. Final layout – the green sector shows the minimum required width based on statistical analyses and a 95% confidence level at various metocean conditions.

3 BOTTOM WASH

Among the many environmental issues concerning transport, one that seems to be largely overlooked is that of re-sedimentation, the effect of maritime vessels on the sea bottom - particularly, of course, in and near ports. The Gulf of Trieste is a semi-enclosed gulf in the north-eastern part of the Adriatic Sea, a shallow water area with an average depth of 16 m and a maximum depth of 25 m. This shallow area is subject to special pollution consideration related to bottom wash phenomena. There is a high mercury concentration (*In the town of Idrija, Slovenia, the world's second largest mercury mine was active for 500 years and an estimated 37,000 tons of mercury has in consequence dispersed throughout the environment*) in the subaquatic sediment which rises into the sea column while ships are manoeuvring. This sediment cloud (smaller particles) is then moved by currents for several hours before re-sedimenting, which has a nefarious effect on the aquatic food chain. The process of bottom wash is basically a function of the size, type and speed of propeller, vessel speed, sub-propeller clearance and sediment conditions (Gucma & Jankowski 2007). It is obvious that the process is dynamic; continuously changing vessel position results in variable bathymetry and vessel/tug propulsion. This process can be simulated and compared with actual manoeuvring results where telegraph recording data is collected together with vessel dynamics.

3.1 Model and some results

As a vessel moves, the propeller produces an underwater jet of water. This turbulent jet is known as propeller wash, or bottom wash (or propwash). If this jet reaches the bottom, it can contribute to re-suspension or movement of bottom particles. Velocity distribution behind the propeller is, for fully developed turbulent flow, given by (Albertson et al. 1950):

$$\frac{v_x}{v_0} = \frac{1}{2\xi} \exp\left(1 - \frac{\rho^2}{2\xi^2}\right) \quad \left(\xi > \frac{1}{2}\right) \tag{6}$$

where

$$\xi \equiv \frac{C_1 X}{D_0} \quad \rho \equiv \frac{r}{D_0} \quad \left(r^r = z^2 + y^2\right) \tag{7}$$

and v_0 is initial velocity, D_0 propeller diameter, C_1 empirical constant and x, y, z are coordinates. The maximal velocity at a given ρ is obtained from the condition

$$\frac{d}{d\xi}\left(\frac{v_x}{v_0}\right) = -\frac{\xi^2 - \rho^2}{2\xi^2} \exp\left(-\frac{\rho^2}{2\xi^2}\right) = 0 \tag{8}$$

so

$$\xi = \rho$$

and maximal velocity is

$$\frac{v_{x,max}}{v_0} = \frac{1}{2\rho} \exp\left(-\frac{1}{2}\right) \tag{9}$$

At the bottom we have $\rho = \dfrac{h}{D_0}$ therefore

$$\frac{v_{b,max}}{v_0} = \frac{e^{-\frac{1}{2}}}{2\frac{h}{D_0}} \approx \frac{\sim 0.303}{\frac{h}{D_0}} \tag{10}$$

In Propeller Wash Study *(Moffatt & Nichol, 2005)* the maximal bottom velocity is given by

$$\frac{v_{b,max}}{v_0} = \frac{\alpha}{\frac{h}{D_0}} \tag{11}$$

where α is 0.22 for open propellers and 0.3 for ducted propellers.

The simulated berthing procedure described in figure 7 was this time analysed for the purpose of bottom wash calculation. Ship position, dynamics and tug forces were recorded with a time resolution of one second (*1 Hz*). Data were stored and used for the bottom wash model where velocity streams are calculated for the sea bottom level.

Figure 9 shows propeller jet streams at the sea bottom for the approaching manoeuvre of the analysed container carrier vessel. Wherever bottom velocity streams exceed 0.5 m/s some re-sedimentation is expected. Further modelling must be done to calculate the total amount of sediment transport divided further into bed-load, suspended-load and wash load, analysed separately for approaching and departure manoeuvres.

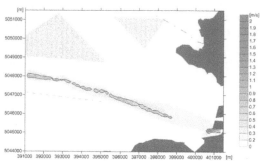

Figure 9. Bottom velocity streams for approaching Container carrier

At any rate the next two figures demonstrate that there will be no major increase of re-sedimentation for large container vessels calling at the Port of Koper. Installation power of main engine will increase by 10%, but when analysing bottom wash at zero speed (when the vessel is on stop and start to accelerate, maximum wash is expected) with telegraph command ordered to "*Slow Ahead*" propulsion power is equal 2.803 kW for larger container carrier compared with 2.545 kW for existing vessel. The main hull and propulsion particulars are:

Studied Container ship		Existing Container ship
Displacement	= 132.540 t	120.000 t (estimated)
Engine power	= 60.950 kW	54.853 kW
Service speed	= 22.8 knt	25.0 kt
Length o.a.	= 346.98 m	318.20 m
Breadth m.	= 42.80 m	42.80 m
Draft	= 14.00 m	14.00 m (limited to 11,6 m)

Figure 10 shows the axial and vertical velocity streams, where the left edge of the image represents the water surface, while the right edge is the sea bottom margin. The image shows the velocity streams of the studied vessel where the shaft line is -9.9 m under the sea surface and 11.4 m above the sea bottom while the existing vessel has 2.4 meters smaller draft (limited vessel draft of 11.6 m comparing to 14 m draft after the dredging). The studied sea depth is 21.3 m. Figure 11a and 11b show the bottom velocity streams in the axial direction. The main difference in bottom velocity streams between existing (Fig 11b) and larger container carriers (Fig 11a) is mostly due to the increase of the vessel draft. Again such increase is minor; maximum speed at bottom will increase only by approximately 0.2 m/s.

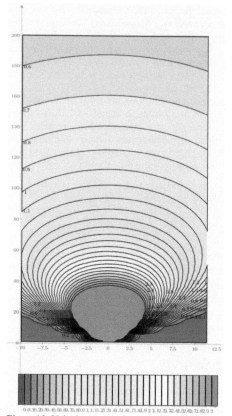

Figure 10: Velocity streams for planed container carriers

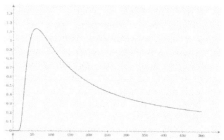

Figure 11a. Bottom velocity streams for planned container carriers

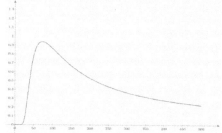

Figure 11b. Bottom velocity streams for existing container carriers

212

4 CONCLUSION

The Port of Koper intends to increase their cargo operations from the current 16-18 million tons to 30-40 million tons in five to ten years, doubling the cargo capacity, so more or less doubling the potential number of vessels calling. Containers continue to increase their share of shipping visits; Ro-Ro's, another intrusive vessel type, continues to accommodate an increasing traffic in automobiles; and passenger vessels are growing in size while ports strive to expand to accommodate this increase.

For each alteration at the precise point where the land meets the sea at a port, a number of considerations are likely to arise. The two concerns discussed here are safety and potential environmental harm. Not for the first time, we demonstrated that ship handling simulators can help reconstruct real domain thrust conditions in a variety of circumstances. A number of careful simulations were necessary to determine the best, that is, safest and most cost-efficient, means for expanding a berth and channel, the extent of dredging required, and the best approach for large vessels.

The environmental factor covered here is one that does not seem to attract much research as of yet – the effect of vessel manoeuvres in and near ports in regard to bottom wash and re-sedimentation. The effects of current shipping trends on the sea bed must be understood with a long term view to eliminating environmental damage, in this case particularly as it may affect cross-border sedimentation.

It is thus far unclear whether the maritime transport business will reach a period of something like stasis, when ships are of optimal size for each type of cargo, when ports have reached optimal or maximal capacity, and, perhaps most important of all, when all negative effects on the environment have been eliminated. Until then, perhaps every change must attract careful scientific scrutiny, so that the potential harmful effects of growth in wealth are mitigated.

NOTE

Part of the paper is the result of work performed with national ARRS project titled *"Influence of circulation and maritime traffic on sediment transport in wide open bays"* number L2-4147 (D)

REFERENCES

Albertson, N., Dal Y., Jensen, R., Rouse, H. 1950. Diffusion of Submerged Jets, *ASCE Paper* No. 2409

Briggsa, M.J., Borgmanb, L.E., Brattelandc, E. 2003. Probability assessment for deep-draft navigation channel design, *Coastal Engineering* 48: 29–50.

Gucma S., Jankowski S. 2007. Depth Optimization of Designed New Ferry Berth. TransNav - International Journal on Marine Navigation and Safety of Sea Transportation, Vol. 1, No. 4: 407-411

Gucma, L. (Ed.). 2013. LNG terminals design and operation. Navigational safety aspects. *Marine Traffic Engineering*. Szczecin: 1-268

Kobylinski, L.K. 2011. Capabilities of Ship Handling Simulators to Simulate Shallow Water, Bank and Canal Effects. TransNav - International Journal on Marine Navigation and Safety of Sea Transportation, Vol. 5, No. 2: 247-252

Moffatt & Nichols. 2005. Propeller Wash Study. Kitimat LNG Import Terminal, MN Project No. 5499.

NAPA, North Adriatic Ports Association, http://www.portsofnapa.com/about-napa

Notteboom, T. 1997. Concentration and load centre development in the European container port system, *Journal of Transport Geography*, Vol 5: 99-115.

Perkovic, M., Batista, M., Malacic, V., Zagar, D., Jankowski, S., Gucma, L., Gucma, M., Rostopshin, D. 2012a. Bottom Wash Assessment Using a Full Mission Ship Handling Simulator, MARSIM: 10, Singapore

Perkovic, M., Gucma, L., Przywarty, M., Gucma, M., Petelin, S., Vidmar, P. 2012b. Nautical Risk Assessment for LNG Operations at the Port of Koper. *Strojniški vestnik - Journal of Mechanical Engineering*, 58 vol 10: 607-613

PIANC. 1997. Permanent International Association of Navigation Congresses, Approach channels: *a guide for design, final report of the joint working group PIANC and IAPH, in cooperation with IMPA and IALA*, Supplement to Bulletin No. 95.

Puertos del Estado (Ed.). 2007. ROM 3.1-99. *Recommendations for the Design of the Maritime Configuration of Ports, Channels and Harbour Basins*, ISBN 84-88975-39-2

Solari, S., Moñino, A., Baquerizo, A., Losada, M. 2010. Simulation model for harbour verification and management, *Coastal Engineering*, No 32: 1-11

Transas. 2011. Model Wizard technical manual, Transas ltd, http://www.transas.com/products/simulators/sim_products/navigational/components/tools/

Transas. 2012. NTPro 5000 technical manual, NTPRO – Navigational simulator, Transas ltd http://www.transas.com/products/simulators/sim_products/navigational/

Chapter 6

Intermodal Transport

Overview of Intermodal Liner Passenger Connections within Croatian Seaports

V. Stupalo, N. Jolić & M. Bukljaš Skočibušić
Faculty of Transport and Traffic Sciences, University of Zagreb, Zagreb, Croatia

ABSTRACT: Republic of Croatia is a coastal country with coastline length of 5,835.3 km, with around 350 ports and small harbours located alongside the coast. Out of 718 islands around 50 of them are permanently inhabited. Enhancement of island accessibility in Croatia is necessary for better facilitation of tourism flows to/from and within tourist destinations as well as to ensure sustainability of Croatian islands. Sea connections and the quality of land connections are important for passengers and drivers arriving to and leaving the port. In order to increase the number of people using public transport it is important to make it more attractive by simplifying the transportation, e.g. by offering passengers one ticket for the entire trip even when using several transport modes, and by assuring integrated schedules of different transport modes to provide quick, qualitative, reliable and flexible transport service. This paper gives an overview of the intermodal liner passenger connections within Croatian seaports and the level of service that they provide.

1 INTRODUCTION

Republic of Croatia is a maritime country with 718 islands and 467 islets and reefs (Klemenčić 1997), of which 50 islands are inhabited all year around. Of the total Croatian surface area 56,594 km^2 (64.56%) is land area, while the surface of coastal sea occupies an area of 31,067 km2 (35.44%). The total length of the coast is 5,835.3 km, which is 1.4 m coastline per capita. Of the total coastline length 1,777.3 km (30.46%) is mainland, while 4,058 km (69.54%) island coastal length (CBS 2011a). Along the total coastline approximately 350 ports and harbours is located (TCD 2010).

Croatian port are divided into ports open to public traffic and port for specific purpose, which may be open to international and / or domestic traffic. Ports open to public traffic by the size and importance for the Republic of Croatia are divided into three groups: 1. ports of special (international) economic interests for the Republic of Croatia, 2. ports of county importance, and 3. ports of local importance.

For the purpose of this paper, the state lines connecting the Croatia's mainland and its islands were analysed and port of call of this lines in respect of their intermodal connections availability. The data was gathered by the authors through sailing schedule of year 2012 and 2013, disseminated by the

Coastal liner traffic agency, while the data on intermodal connections was gathered through project Adrimob, but due to the difference between identified ports missing data was gathered by authors. Only Croatian state lines that connects Croatian mainland with its islands, which are defined by the Decision of Government of Republic of Croatia, were analysed together with the ports of call of identified lines. International passenger maritime lines were not analysed.

The first part of the paper gives an introduction on the organisation of the coastal maritime public liner transport service in Croatia. In the following chapter the definition of intermodal passenger transport is given as well as the key issues of intermodal passenger transport, and the current situation of usage of different modes of transport for one journey in Croatia is overviewed as well as the level of the services provided in the port and availability of different modes of transport in identified ports.

2 PUBLIC STATE LINES

Coastal maritime public liner transport service in Croatia, between Croatia's mainland and its islands, is regulated and organised by Coastal liner traffic agency, agency under the auspice of Croatian

Ministry of maritime affairs, transport and infrastructure (hereinafter: Ministry). Main scope of the work of Coastal liner traffic agency, established by the Act on regular and occasional coastal maritime transport services (OG 33/06, 38/09, 87/09, 18/11), consist of awarding of concessions for performing public transport services in coastal maritime liner transport, and regulations of the terms and conditions of the provision of services, i.e. regulates general terms of public service compensation and subsidy arrangement for the provision of these services, invites public tenders for provision of services in maritime public liner transport, ordinance on the requirements imposed on ships and ship operators performing those services, concludes concession contracts, and continuously monitors implementation of concluded concession contracts, and the whole Act.

Maritime public state lines are defined by the Decision of Government of Republic of Croatia,[17] on the proposal of the Ministry, where three types of lines are defined, according to the type of the ship performing the service: ferry, high-speed and conventional passenger (classic) ship lines. For all of these lines the decision stipulates vessel capacity, route and minimum frequency. Minimum frequency, i.e. number of return trips per week, is determined depending on the time period of the year, whereby three different time periods are defined: 1. off season, 2. pre-season and post-season, and 3. high season. Total 56 lines are defined, 27 ferry lines, 16 high-speed lines and 13 conventional passenger shipping lines. Two conventional passenger shipping lines, although listed and defined by the Decision on state lines, have not been included in the timetable due to certain technical problems.[18] Most of the lines are operated during the whole year; only six lines[19] in total (three ferry lines, two high-speed lines and one conventional ship line) are operated seasonally. Also, although Adriatic Croatia is divided into seven counties, most of the lines operate within the ports located within the same county; only seven lines[20]

are operated within the ports located within two to max three counties. Duration of the trip also varies from a line to a line, from trip duration of only five minutes[21], to a trip that last almost a whole day[22].

Out of 54 current lines only five ferry lines are profitable[23], while others are not financially viable, due to the low capacity utilization and seasonal dimension, and are subsidized by the state, in year 2011 the state has allocated 383 mill HRK.[24]

3 INTERMODAL PASSENGER TRANSPORT

For the scope of this paper the following definitions, provided by the European research projects and studies, of intermodal passenger transport were used as guidance. First definition is of more technical dimension and defines intermodal transport as 'A route of an individual passenger or goods unit consisting of a combined chain from origin to destination involving at least two different modes (excluding walk for passengers)' (SORT-IT 1999), while other definition is more oriented on quality of service offered to the user: 'Passenger intermodality is a policy and planning principle that aims to provide a passenger, using different modes of transport in a combined trip chain, with a seamless journey' (Towards Passenger Intermodality in the EU 2004, LINK 2007-2010, KITE 2007-2008).

Although the first definition excludes walking as one of the transport modes of intermodal travel, it is wrong to dismiss walking wholly. Also, the study Towards Passenger Intermodality in the EU, recognize walking as one of the important mode combinations in long-distance intermodal passenger transport, where the following mode combinations are profiled: (Towards Passenger Intermodality in the EU 2004)

1 (High-speed) Rail /air
2 Urban public transport / long-distance train (coach, ferry)
3 Car / long-distance train (coach, ferry)
4 Cycling / long-distance train (coach, ferry)
5 Walking / long-distance train (coach, ferry)
6 Cross-border transport.

In that study walking was analysed as transfer from home to urban public transport stops, from interchanges to physical destinations and the aspects of walking around interchanges, i.e. as one of door-to-door journey segments (Towards Passenger Intermodality in the EU 2004).

[17] Decision on defining state public transport lines in coastal maritime line transport of 5 December 2008, and amendments to the Decision of 4 March 2010 and 20 October 2011 (OG 10/11)

[18] For line 409 Zadar-(Ošljak)-Preko the intention to offer a concession will be published after the ferry transport will have been moved to Gaženica port while line 506 Prvić Luka-Prvić Šepurine-Vodice waits for the decision on granting concessions, which was set aside by the Ministry in year 2009.

[19] That are the following lines: Ferry line 101 Rijeka-Split-Stari Grad-Korčula-Sobra-Dubrovnik, Ferry line 604a Vela Luka – Lastovo, Ferry line 641 Drvenik-Dominče, High-speed line 9141 Pula-(Unije)-Mali Lošinj-(Ilovik)-Zadar, High-speed line 9603a Split-(Milna)-Hvar and Conventional ship line 616 Trogir-(Slatine)-Split

[20] That are the following lines: Ferry line 101 Rijeka-Split-Stari Grad-Korčula-Sobra-Dubrovnik, Ferry line 337 Stinica-Mišnjak, Ferry line 604 (Lastovo) – Vela Luka - Split, Ferry line 641 Drvenik-Dominče, High-speed line 9141 Pula-(Unije)-Mali Lošinj-(Ilovik)-Zadar, High-speed line 9309 Novalja-Rab-Rijeka and High-speed line 9604 Lasto-vo-Vela Luka-Hvar-Split.

[21] Conventional ship line 501 Brodarica-Krapanj, acording to the timetable lasts only 5 min.
[22] Seasonal ferry line 101 Rijeka-Split-Stari Grad-Sobra-Dubrovnik, according to the timetable lasts 23h and 30min.
[23] These lines are: No 332 Valbiska-Merag, No 334 Brestova-Porozina, No 335 Prizna-Žigljen, No 337 Stinica-Mišnjak, and No 631 Split-Supetar.
[24] http://www.mppi.hr/default.aspx?id=9327, 01.10.2012

Thus, walking (short-distance) should be included in the integrated intermodal system as a mean of integration of two transport modes (air, rail, road, waterborne), ensuring that the walking information are integrated part of real-time door-to-door information systems (both pre-trip and on-trip), and by ensuring adequate pedestrian routes.

The information about the multimodal journey can be provided by the online multimodal journey planners, such as TransDirect (UK), 9292 (NL), Reiseauskunft (DE), etc.[25], providing door-to-door travel information and ticketing services. Most of the intermodal journey planners are on regional or national scale, while intermodal journey planner on a European scale is still not available (LINK 2007-2010). Information provided by the multimodal journey planner the to the user varies, as well as the selection of transport means (train, airplane, ship, bus, tram, bicycle, walking, taxi, private car, etc.), possibilities to calculate and compare travel cost and travel time depending on the transport mean, selection of the "greener" mode of transport, to the availability of maps etc.

To provide conditions for the establishment of harmonised EU multimodal travel information service, development of binding measures (requirements, procedures, etc.). is included in ITS Directive (Directive 2010/40/EU). The railway sector has already in 2011 adopted a set of common standards - TAP TSI[26] (Regulation (EU) No 454/2011), according to which all computerised travel information and reservation systems have to be adopted. The same standards will be provided as interfaces for the inclusion of other modes (COM 898 2011).

To provide a seamless journey the following core elements/issues, identified in the project Towards passenger intermodality in the EU, have to be integrated to produce passenger intermodality: (Towards Passenger Intermodality in the EU 2004)
1 Networks and interchanges (availability of infrastructure of a certain quality, e.g. physical interconnection and interoperability, and integration of the different transport services, e.g. integrated timetables, average waiting time, etc.)
2 Intermodal services (integrated ticketing and tariffs, handling of baggage, pre-trip and on-trip information)
3 Promotion of intermodality (at individual level, site level and general public level).

Optimisation of intermodal transport and market integration are set as one of the goals of 2011 White paper (COM 144 2011), promoting collective transport and multimodal journeys as an easy and reliable alternative to the private transport through insurance of uniform access conditions for passengers, insurance of service quality at a basic level, and through better mobility planning. New rules on passenger rights, one of the objectives of White Paper 2001 (COM 370 2001), based on three keystones: 1. non-discrimination, 2. accurate, timely and accessible information, and 3. immediate and proportionate assistance, as a first step in the EU passenger transport policy moving from a purely modal approach to a more intermodal vision have been adopted.[27]

4 INTERMODAL LINER CONNECTIONS

Passenger liner transport services on state lines in the Republic of Croatia, according to the available sailing schedules, is at the moment carried out by 13 different marine companies: Jadrolinija Rijeka, Linijska nacionalna plovidba d.d. (Inc.) Split, G&V Line d.o.o. (Ltd.) Dubrovnik, U.T.O. Kapetan Luka Krilo Jesenice, Mediteranska plovidba d.d. (Inc.) Korčula, Rapska Plovidba d.d. (Inc.) Rab, Miatrade d.o.o. (Ltd.) Zadar, T.U.O. Mankul Zadar, PRZ 'Vrgada' Vrgada, Bura line & Offshore Slatine, Nautički centar Komiža, d.o.o. (Ltd.), City of Šibenik-Gradski parking d.o.o. (Ltd.) and Trade 'Trn' from Ždrelac. Most of the lines, 37 in total, are operated by Croatia's largest passenger shipping company, Jadrolinija, which is state owned.

For the passenger planning a trip all sailing schedules at state level are available at the web page of Coastal liner agency, also sailing schedules, and in many cases also the ticket price is available on the web pages of the service providers. Unfortunately not all service operates have web pages, also only two companies provide at least partial on-line ticket service.[28]

On the state level there is no available multimodal journey planner. Currently, multimodal journey ticket is provided by only a few bus carriers who through the price of one ticket (bus ticket) includes transportation by ferry, and these services are in most cases available only during the season.

---

[25] For more see: http://ec.europa.eu/transport/its/multimodal-planners/examples-of-existing-national-journey-planners/index_en.htm, 09.01.2013.
[26] TAP TSI - Technical Specifications for Interoperability for Telematic Applications for Passenger, defined by the ERA (European Railway Agency)

[27] For more information see: COM 898 (2011) A European vision for Passengers: Communication on Passenger Rights in all transport modes. Relevant EU Regulations in waterborne transport: 1. Regulation (EU) No 1177/2010 of the European Parliament and of the Council of 24 November 2010 concerning the rights of passengers when travelling by sea and inland waterway and amending Regulation (EC) No 2006/2004, and 2. Regulation (EC) No 392/2009 of the European Parliament and of the Council of 23 April 2009 on the liability of carriers of passengers by sea in the event of accidents.
[28] One operator offers service of purchasing ticket on-line but only for the coastal (currently there is no lines available) and international lines while other operator offers service of on-line purchase of the ticket, by e-mail correspondence, for only one line, and the ticket has to be picked up at the information desk.

Also time schedules, between different transport modes (air, rail, road, waterborne) are not synchronized. For some ports, e.g. port of Stari Grad on the island of Hvar, there are bus connections connecting the port with different island places, buses are located near the ferry and are waiting for the passenger to disembark from the ferry, so the passenger can continue its journey, but there is no available service that will enable a passenger to book or prepay its ticket for the ferry and for the bus, thus ensuring his place.

To increase the system efficiency and service accessibility, in March 2012 establishment of the *Maritime coastal liner traffic information system* was launched by the Ministry. Operational implementation of the overall system is under the responsibility of Coastal liner agency. The system main purpose is to reduce expenses and to improve efficiency of the maritime passenger liner traffic monitoring system.[29] Prerequisite of the system implementation is creation of legislation that will follow informatiozation process, one of the current plans is to expand the scope of the work of Coastal liner agency by assumption of booking and ticketing service.

To attract more passengers and to better utilize shipping capacity, it is necessary to explore current market and possibilities of multimodal passenger transport introduction. Introduction of multimodal passenger transport is one of the strategic determinations set by the Coastal liner agency.

5 CROATIAN PORTS

Current state passenger coastal lines connects 94 different ports alongside Croatian coast, these ports are of different meaning (state, county or local importance).

Depending on the ports meaning (state, county or local importance) state-port relation varies, so each port focuses its development plans accordingly to this relation, also, the level of the service differs between these ports. Ports of state importance are under the jurisdiction of state port authority while ports of county and local importance are under the jurisdiction of county port authority. As before mentioned, Adriatic Croatia is divided into seven counties, but more than half of the ports, 48 ports, are located within two counties, Zadar County and Split-Dalmatia County. Within County of Zadar 27 different ports are connected with state coastal lines, within County of Split-Dalmatia there are 21 ports, within County of Dubrovnik-Neretva 16, within County of Primorje-Gorski kotar 15, County of

Sibenik-Knin has nine ports, County of Lika-Senj four, while County of Istria has only two ports.

Out of 94 ports 21 ports are located alongside mainland coastline, while 73 ports are located alongside island coastline. Island ports are of county or local importance and are under the jurisdiction of county port authorities, most of the islands ports are located within or close to the town centre, and so the level of the service provided within or in close vicinity to the port area varies according to the size and the offer of the town. These 73 ports are situated on 43 different islands. Number of inhabitants (CBS 2001) varies from the one island to another, from 17,860 (island of Krk) to 8 (island of Vele Srakane) inhabitants per island, also varies the island surface area (CBS 2011a) from 405.78 km^2 (Isl. Krk and Isl. Cres) to less than 6 km^2.

Of 94 ports, in year 2010, eight ports had an annual traffic of more than one million of passengers, excluding cruise passengers, see Table 1.

As shown in Table 1, the busiest Croatian passenger ports are port of Split and port of Zadar. Passenger port of Split and passenger port of Zadar are both situated in the city centre, both cities are one of the most populated cities alongside Croatian mainland coastline, Split with 178,102 and Zadar with 75,062 inhabitants (CBS 2011b).

Table 1. Maritime transport of passengers (excluding cruise passengers) in '000

Port/Year	2008	2009	2010
Split	3.725	3.558	3.523
Zadar	2.227	2.227	2.144
Preko (Isl. Ugljan)	1.606	1.563	1.495
Korčula (Isl. Korčula)	1.618	1.583	1.474
Cres (Isl. Cres)	1.465	1.559	1.416
Jablanac*	1.349	1.539	1.402
Supetar (Isl. Brač)	1.509	1.532	1.383
Dubrovnik – port of Gruž	879	1.037	1.151

* Note: Jablanac port was port of the call for ferry line 337 but from 2012 this line is departing from Stinica (Stinice was one of the port identified)
Source: Eurostat 2013

These two ports are well connected with maritime passenger public coastal lines connecting Croatian mainland with its islands; Split is connected with 14 different lines while Zadar is connected with 13 different lines.[30] Both ports are equipped with necessary infrastructure and suprastructure to accommodate all types of passenger ships (cruise, ferry, catamaran, etc.), and near the port area connections with other modes of transport is

[29] For more information see: http://www.mppi.hr/default.aspx ?id=8827 (01.08.2012.)

[30] Port of Rijeka, also located within the city centre of one of the populated maritime cities (128,624 inhabitants, Census 2011), due to the lack of liner connections, only three, in year 2010, according to Eurostat data, achieved the traffic of only 185,000 passengers.

available (railway[31], bus[32], taxi, rent-a-car[33], highway[34], airport[35], etc.) As before mentioned, both ports are located within the city centre, which ensures additional comfort services to be available within or near the port.[36]

For other ports main infrastructure needed for organisation of integrated passenger transport is there, only the service is not, see Table 2.

Table 2. No of maritime ports with multimodal connections in Croatia

Distance from the passenger terminal to inland connections	TYPE OF CONNECTION				Rent a -	
	Rail	Bus	Taxi	bike	car	boat
0 - 500 m	1	35	37	18	15	23
500 m -1 km	1	7	2	11	5	9
> 1 km	3	3	1	3	2	8
TOTAL	5	45	40	32	22	40

Source: Compiled by authors. Data source: ADRIMOB project, and Port authorities

6 CONCLUSION

As stated at the beginning, Croatia is a maritime country, but also a tourist country. Ensuring availability of the adequate transport connection that connects mainland with the islands and islands mutually is essential to ensure sustainability of the island and for the development of wider tourist offer. Change of transport, by changing transport mode, or even change of transport between the same modes of transport, represent a discomfort for the passenger, especially for the passenger of impaired mobility, passengers traveling with heavy baggage and for the passengers traveling for the first time especially if they don't know the language needed. To reduce passenger perception of transport breakings (change of transport) it is necessary to ensure adequate interchanges where timetables of different transport modes will be synchronized to ensure reliable, fast, flexible and comfort transfer. Information provided and infrastructure ensured, to the passenger, should be user friendly.

Currently, multimodal journey ticket in Croatia is provided by only a few bus carriers who through the price of one ticket (bus ticket) includes transportation by ferry, and these services are in most cases available only during the season. Also timetables, between different transport modes (air, rail, road, waterborne) are not synchronized. This implies that passenger have to wait in line and buy a ticket for each part of its journey, and due to the unsynchronized timetables and lack of the information this often represents a big problem to the passenger. To tackle this problem in March 2012 establishment of the *Maritime coastal liner traffic information system* was started, creation of legislation as a prerequisite of the system implementation is in process.

With 54 currently operated state maritime passenger lines 94 ports are connected. Most of these ports are of local or county importance, connected with at least one to four different lines. Only eight ports have annual traffic greater than one mill of passengers, while only two ports have more than ten different line connections. Out of 94 ports in less than 15 ports multimodal journey ticket is offered through the price of one bus ticket that includes transportation by ferry. For many ports main infrastructure needed for organisation of integrated passenger transport is there, only the service is not.

To attract more passengers and to better utilize shipping capacity, it is necessary to explore current market and possibilities of multimodal passenger transport introduction. Introduction of multimodal passenger transport is one of the strategic determinations set by the Croatian Coastal liner agency.

REFERENCES

Commission Regulation (EU) No 454/2011 of 5 May 2011 on the technical specification for interoperability relating to the subsystem 'telematics applications for passenger services' of the trans-European rail system. Official Journal of the European Union L123/11

COM 370. 2001. White Paper. European transport policy for 2010: time to decide. Brussels: Commission of the European Communities

COM 898. 2011 Communication from the Commission to the European Parliament and the Council. A European vision for Passengers: Communication on Passenger Rights in all transport modes. Brussels, European Commision

COM 144. 2011. White Paper. Roadmap to a Single European Transport Area – Towards a competitive and resource efficient transport system. Brussels: European Commission

Croatian bureau of statistics, CBS. 2001. The Census of Population, Households and Dwellings. Zagreb: Croatian bureau of statistics

Croatian bureau of statistics, CBS. 2011a. Statistical yearbook. Zagreb: Croatian bureau of statistics

Croatian bureau of statistics, CBS. 2011b. Population in major cities and municipalities, The Census of population, households and dwellings. Zagreb: Croatian bureau of statistics

---

[31] Railway station in Split is located less than 500 m from the port, while Zadar railway station is located within less than 2 km from the port area.
[32] Bus station in Split and in Zadar is located within walking distance from the railway station. There are national and international bus lines available.
[33] Taxi station and rent-a-car offices are for both ports situated within walking distance from the port area.
[34] Port of Split is around 18 km away from the highway, while port of Zadar is around 11 km away from the highway.
[35] International airport is 18 km away from port of Split, while 16 km away from the port of Zadar.
[36] Additional services implies: restaurant, bars, shopping centre, banks, tourism info points, hotels, travel agency, internet facilities, toilets, etc.

Decision on defining state public transport lines in coastal maritime transport. 2008. Zagreb: Government of the Republic of Croatia, Class 342-01/08-01/05; Ref.no: 5030116-08-1

Decision on Amendments of the Decision on defining state public transport lines in coastal maritime transport. 2010. Zagreb: Government of the Republic of Croatia, Class 342-01/10-01/01; Ref.no: 5030116-10-1

Directive 2010/40/EU of the European Parliament and of the Council of 7 July 2010 on the framework for the deployment of Intelligent Transport Systems in the field of road transport and for interfaces with other modes of transport. Official Journal of the European Union L207/1

Eurostat. 2013. Maritime transport-passengers-annual data-all ports-by direction. Available at: http://epp.eurostat.ec.europa.eu

Government of the Republic of Croatia. Decision on Amendments of the Decision on defining state public transport lines in coastal maritime transport. 2011. Zagreb: Official Gazette 10/2011

KITE. 2007-2008. Knowledge Base on Intermodal Passenger Travel, Available at: http://www.kite-project.eu/kite/cms/

Klemenčić, M. (ed) 1997 Atlas Europe. Zagreb: Leksikografski zavod Miroslav Krleža

LINK. 2007-2010. Intermodal passenger transport in Europe. Passenger intermodality from A-Z. The European Forum for Intermodal Passenger Travel. DG MOVE

Project ADRIMOB. 2011. Sustainable coast MOBility in the ADRIatic area. Available at: http://adrimob-ipa.racine.ra.it/

Regulation (EU) No 1177/2010 of the European Parliament and of the Council of 24 November 2010 concerning the rights of passengers when travelling by sea and inland waterway and amending Regulation (EC) No 2006/2004

Regulation (EC) No 392/2009 of the European Parliament and of the Council of 23 April 2009 on the liability of carriers of passengers by sea in the event of accidents.

SORT-IT. 1999. Strategic Organisation and Regulation in Transport. Available at: http://www.transport-research.info/Upload/Documents/200310/sortit.pdf

Towards Passenger Intermodality in the EU. 2004. Report 1 Analysis of the Key Issues for Passenger Intermodality, Dortmund: DG TREN

Transport and communications department, TCD. 2010. Maritime, river and pipeline transport. Zagreb: Croatian chamber of economy

http://www.mppi.hr/default.aspx?id=9327, 01.10.2012

http://ec.europa.eu/transport/its/multimodal-planners/examples-of-existing-national-journey-planners/index_en.htm, 09.01.2013.

http://www.mppi.hr/default.aspx?id=8827, 01.08.2012.

http://www.agencija-zolpp.hr/Brodskelinije/tabid/1267/Default.aspx, 09.01.2013

Concept of Cargo Security Assurance in an Intermodal Transportation

T. Eglynas, S. Jakovlev & M. Bogdevičius
Vilniaus Gedimino Technikos University, Vilnius, Lithuania
Klaipeda University, Computer Engineering Department, Klaipeda, Lithuania

R. Didžiokas, A. Andziulis & T. Lenkauskas
Klaipeda University, Mechatronics Science Institute, Klaipeda, Lithuania

ABSTRACT: Intermodal transport can be defined as all means of transport that are required for transportation of cargo or container from point A point B. Intermodal transportation object (in this case the container) is exposed to a variety of forces that can deform the container and cargo. The main problems are security of container and cargo in intermodal freight transportation in sea port territory. In order to reduce the risk of transportation there is a need to add intelligent systems that allow the partial control of risk and its factors to existing structures. In order to ensure greater cargo security level it is necessary to install an additional smart system into existing infrastructure, and improve the container tracking and cargo delivering reporting system. This article presents the concept which uses RFID technology for information about container status transfer. Remote data transmission system provides information for operators and automatic control systems. Developed smart container crane control algorithms based on RFID system provides information about the cargo. It also provides suggestions for establishing the intermodal transportation of cargo in port territory depending on the container contents. Combining RFID technology and specific control algorithms helps to avoid cargo damage while warehousing and loading goods. In the future, a system concept that simulates realistic data transfer management processes will be developed.

1 GENERAL INSTRUCTIONS

In recent years, scientists are increasingly developing solutions to quay cranes problems. One of the most researched problems of recent years is the container crane and its control algorithms optimization [1-5]. Many different container transportation algorithms that solve specific problems have been already created, but there is still a need for complex algorithms targeted not only at individual handling process, but also at full of ship-to-shore-to-ship algorithm efficiency improvement. One of the most common new quay cranes control algorithms development [6] is associated with container swinging. Cases of this problem can be caused by many different sources [7-11]. Sometimes it is caused by swinging of the crane mechanisms such as an engine, or the uneven surface of the rails for conveying equipment, but also it can occur in cases where the swing cause is not known. This problem can be caused by a complex set of forces operating to the external area of the container, and even from inside the container.

Poorly designed algorithm for lifting mechanism, and all other external forces can cause intrusive actions that affect the container and its cargo [12-14]. This paper is an attempt to find quay crane parameters that can be optimized and controlled in order to reduce risks.

The container is main object used in intermodal transportation [15-17], and one of the most popular of its means of transportation is a vessel. Every year, in intermodal terminals millions of different types of containers are transshipped. One of the main objects used for container shipping in container terminals are large quay cranes. It can be said that the container transfer process is one of the most important intermodal transportation elements, it is necessary to ensure a high level of security both physically and as well from the hardware and software aspect. Wrong operator actions or inappropriate algorithm can not only damage the containers or cargo, but also it can cause hurt the terminal personnel or damage quay crane and other transport objects (container lifts, trucks and infrastructure). If security rules are breached, catastrophic accident can happen, such as collapse of or quay crane boom.

In recent years, scientists have been trying to solve the problem of container swinging [8-10, 17].

Container hoist and container crane are connected with steel cables. Container can start to swing because of different container weights, wind gusts and control algorithms. As a result, more time is required for the container shipment from point A to point B. Moreover, due to unstable accelerations there emerges a probability that the container can be damaged in the collision with the quay crane structures. In case of strong winds, operation of the quay cranes is often suspended. Currently existing technical and software solutions are not completely efficient to avoid all of these problems [10-11, 18].

2 SYSTEM DESIGN

2.1 *Problem of container safety and risk management*

Growing cargo traffic flows demand to develop innovative systems. These systems improve efficiency and security of transportation process. Containerized cargo flows have also changed and much attention is paid to its service and security. During intermodal transportation container is subjected to external forces and weather conditions, which can deform the container and/or its cargo. The seaport equipment also can be damaged. In order to ensure the cargo and port facility security it is necessary to complement existing systems with intelligent transportation technologies to partially control the risks.

The main problem is handling process control accuracy and security. New intelligent transport systems and their control algorithms are developed. In order to achieve better result in handling process it is necessary to analyze the existing systems to create more efficient control algorithms that can ensure the container and its cargo security. One of ways to improve intelligent transportation systems is supplementing them by additional information about external factors, such as changes in container transportation, the external forces acting on the container (container and contents status) and other useful information.

2.2 *Proposed wireless sensor network solution for risk management*

In order to collect information about the container and its cargo it is necessary to install wireless sensor network (WSN) which acquires all the necessary information about the cargo and status of intermodal transportation. This system could use RFID technology to securely acquire data from containers. Data collected using wireless sensor network subsystem [9] will help in estimating the current state of the cargo and container. RFID technology will to read and transfer data to freight company

databases and handling equipment in container terminal such as quay cranes, storage cranes, transportation machinery and other intermodal terminal equipment.

In the following figure (Figure 1) data exchange concept has been provided, which is very relevant to the crane control algorithms optimization. In order to implement different algorithms to variations in loading time, it is necessary to know a great number of external factors that may at first glance seem insignificant. Therefore, it is important to get the correct information about the cargo. It is well known that the container filling, loading variations and the type of content exposed to external natural factors (wind, precipitation or human activities) affect his swing during reloading. It can be the reason why most off existing standard algorithms for handling process do not work properly. Current algorithms are designed for fast handling and use the same handling parameters for all kinds of container cargo.

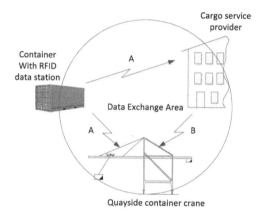

Figure 1. RFID data exchange conceptual model (A – data from container station, B – data from cargo service provider, for terminal handling systems (e.g. for quayside container crane).

Container with RFID based technology – uses data collection station. It collects all the data about cargo transportation total time, force fluctuations and weather conditions. When container reaches one of the intermodal transportation points, all the data is transmitted to freight company administration, which reports to the cargo owner. In the graph data about container and cargo status are marked with an "A". These data is scanned by quay crane before loading process is started. Data transfer is marked with the letter "B", it is the operator's direct influence to the loading process. For example, if the quay crane operator system chooses an algorithm that is ready for the load group, but the operator having assessed the condition of the cargo in the ship decides that the cargo has a higher degree of risk, he can offer to use "softer" algorithm in order to avoid the consequences which may result in "rough" shipment. This procedure will increase probability of secure shipment. In order to ensure the data safety [18], the RFID system can be activated only when the container is located in the data exchange area (container terminal territory).

3 CONCEPT OF THE CONTROL ALGORITHM FOR QUAYSIDE CRANE

The first step for creation of the control algorithm concept is to collect the relevant data and to make evaluation of its impact to the loading process [19]. The next step is very important and it consists of the evaluation and determination of the appropriate quay crane performance characteristics for each type of cargo. To achieve this, a theoretical diagram of the generating principle control crane algorithm was created (Figure 2).

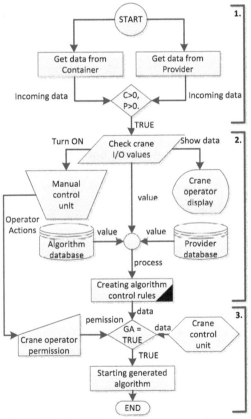

Figure 2. Specific transportation algorithm generation system (C – data from container (container status), P – data from cargo service provider, GA – data for generating algorithm).

In order to properly control algorithm for each type of goods, data of container and its cargo has to be evaluated. RFID technology on intermodal

containers (Figure 1.) will provide the opportunity to obtain the necessary real time information about the cargo. Also, in order to ensure the security of information transmission and to avoid situations where malicious attempts to damage the cargo for a variety of reasons could emerge, the statistical data will be collected on cargo and terminal service. This identification and evaluation process algorithm comprises the first part. The following is the data evaluation process for evaluating the RFID information received and the information about the quay crane. The crane controller evaluates available data of cargo, crane capabilities and the current position of the crane hoist after which authorizes and approves carriage control algorithm. The crane control system collects data from the database and terminal algorithms database, as well as an evaluation of the lifting mechanism position, starts to generate an algorithm mostly suited for current situation. The whole algorithm generated by the system is proposed to the crane operator which can approve or deny the system generated algorithm version. This process (generated verification algorithm) is carried out in the third part of the algorithm.

4 FUTURE WORKS

In order to solve this problem the intermodal terminal with a quay crane model has been created. It will be used for further stress management algorithms research, avoiding the additional funds loss which would be a result of interruption to the real system. The quay crane prototype is shown below (Figure 3).

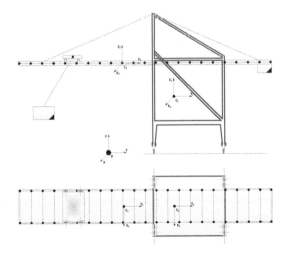

Figure 3. Quayside crane prototype.

The transport system model will help to address problems associated to the process control accuracy and security. Quay crane and intermodal container is one of the most important objects in a dynamically changing environment of the intelligent transportation systems. It is planned to make a crane dynamic model calculations that allow assessment of nodes which are also affected by the swinging motion, which increases the risk of damage to the cargo.

5 CONCLUSION

Completed system design for future studies will help to evaluate container safety, security and risk management processes that can be put in place thus enabling the risk control. This paper reviews only quay cranes operations with containers. For effective proposed system work it is necessary to develop and install smart algorithms based process shipping optimizations for other port facilities, such as elevators, storage equipment and auxiliary cranes.

System model designing has helped to clarify the characteristics which have to be improved in further research. It has also been found that for correct assessment of the external forces affecting the crane swinging it is necessary to create a dynamic crane model and to estimate the crane vibrations. Data transmission security in the port area also should be assessed in order to avoid intentional cases of disturbing overall system performance, by transmitting false information about the container and cargo in it.

ACKNOWLEDGMENTS

The authors are grateful of the project "JRTC Extension in Area of Development of Distributed Real-Time Signal Processing and Control Systems"(Subsidy Contract No: LV-LT/1.1./LLIV-215/2012/Cross-border DISCOS/), also of the project "Lietuvos jūrinio sektoriaus technologijų ir aplinkos tyrimų plėtra" (project code No. VP1-3.1-ŠMM-08-K-01-019, funding and administration contact No. VP1-3.1-ŠMM-08-K-01-019) for the financial support for writing and publishing this paper.

REFERENCES

[1] S.-L. Chao and Y.-J. Lin, "Evaluating advanced quay cranes in container terminals," Transportation Research Part E: Logistics and Transportation Review, vol. 47, no. 4, pp. 432–445, Jul. 2011.
[2] H. Schaub, "Rate-based ship-mounted crane payload pendulation control system," Control Engineering Practice, vol. 16, no. 1, pp. 132–145, Jan. 2008.
[3] C. Chang, and K. Chiang, "The nonlinear 3-D crane control with an intelligent operating method," pp. 2917–2921, 2008.
[4] L. I. Zhi-jun, L. I. Zhen, L. I. Hui-jie, Z. Ye, S. Cheng, and S. U. N. Yan, "Research of Coordinated Control Method of Hybrid Power Crane System," pp. 1093–1097, 2012.
[5] Y. Tanaka, Y. Konishi, N. Araki, T. Sato, and H. Ishigaki, "Development of High Speed Controller of Container Crane by Binary Input Using Mixed Logical Dynamical System," 2009 Fourth International Conference on Innovative Computing, Information and Control (ICICIC), pp. 181–184, Dec. 2009.
[6] D. Chang, Z. Jiang, W. Yan, and J. He, "Integrating berth allocation and quay crane assignments," Transportation Research Part E: Logistics and Transportation Review, vol. 46, no. 6, pp. 975–990, Nov. 2010.
[7] Y. Fang, P. Wang, and X. Zhang, "A Motion Planning-Based Adaptive Control Method for an Underactuated Crane System," vol. 20, no. 1, pp. 241–248, 2012.
[8] L. Cao and L. Liu, "Adaptive Fuzzy Sliding Mode Method-Based Position and Anti-swing Control for Overhead Cranes," 2011 Third International Conference on Measuring Technology and Mechatronics Automation, no. 2, pp. 335–338, Jan. 2011.
[9] H. Yoshihara, N. Fujioka, and H. Kasahara, "A new vision-sensorless anti-sway control system for container cranes," 38th IAS Annual Meeting on Conference Record of the Industry Applications Conference, 2003., vol. 1, pp. 262–269, 2003.
[10] H. Sano, K. Ohishi, T. Kaneko, and H. Mine, "Anti-sway crane control based on dual state observer with sensor-delay correction," 2010 11th IEEE International Workshop on Advanced Motion Control (AMC), pp. 679–684, Mar. 2010.
[11] Y. Tanaka, Y. Konishi, N. Araki, and H. Ishigaki, "Control of container crane by binary input using Mixed Logical Dynamical system," 2008 International Conference on Control, Automation and Systems, vol. 2, pp. 13–17, Oct. 2008.
[12] D. Oguamanam, "Dynamics and modelling of mega quayside container cranes," pp. 193–198, 2006.
[13] M. Georgijević, "Container Terminals in River Ports", pp. 199–204, 2006.
[14] T. G. Crainic and K. H. Kim, "Intermodal Transportation," vol. 14, no. 06, pp. 467–537, 2007.
[15] V. Roso, J. Woxenius, and K. Lumsden, "The dry port concept: connecting container seaports with the hinterland," Journal of Transport Geography, vol. 17, no. 5, pp. 338–345, Sep. 2009.
[16] S. Nundrakwang, T. Benjanarasuth, J. Ngamwiwit, and N. Komine, "Multivariable control of overhead crane system by CRA method," 2008 SICE Annual Conference, vol. 1, no. 3, pp. 3278–3282, Aug. 2008.
[17] H. Kawai, Y. Choi, Y. B. Kim, and Y. Kubota, "Position measurement of container crane spreader using an image sensor system for anti-sway controllers," 2008 International Conference on Control, Automation and Systems, pp. 683–686, Oct. 2008.
[18] F. Longo, "Design and integration of the containers inspection activities in the container terminal operations," International Journal of Production Economics, vol. 125, no. 2, pp. 272–283, Jun. 2010.
[19] A. Kusminska-Fijalkowska and Z. Lukasik, "The Land Trans-Shipping Terminal In Processes Flow Stream Individuals Intermodal Transportion," International Journal on Marine Navigation and Safety of Sea Transportation, vol. 5, no. 3, 2011.

Chapter 7

Propulsion and Mechanical Engineering

Diagnostic and Measurement System for Marine Engines'

A. Charchalis
Faculty of Marine Engineering, Gdynia Maritime University, Poland

ABSTRACT: Modern way of machines' exploitation, due to their high level of constructional complication, requires certain level of supervising. That supervising is generally reduced to detection of pre-failure states and evaluation of machines' single elements or components condition. In the frame of development of the research capacity of the Mechanical Faculty of Maritime Academy in Gdynia, has been developed the Exploitation Decision Aid System for marine engines exploitation, based on existing test bed with the marine diesel engine Sulzer AL 25/30. Modernization of the engine, significantly extended research and measurement capacity, what has resulted with improvement of quality, extension of the span, and acceleration of carried out research and development works in the domain of safety of exploitation and diagnostics of marine power plants. Above stated investments enables also an extension of the range of research and expertise related to engines' failures and exhaust gases emission pollution, in relation to broad spectrum of implemented fuels. The goal has been achieved in the way of the test equipment modernization including: effective pressure sensors, high pressure fuel sensors, monitoring and visualization of the engine systems' parameters, electronic indictors adopted to continuous operation at all cylinders in the same time, and high class decision aid computer equipment.

1 MARINE DIESEL ENGINE SULZER AL 25/30 TEST BED

Diagnostic system for evaluation of exploitation attributes of marine diesel engines consist of the diesel engine driving electric generator and top class operating station enabling monitoring and recording of working parameters. The operating station enables also remote control of the engine and auxiliaries. Diagnostic system consist of:

- three cylinders, four stroke diesel engine type 3AL 25/30 Cegielski Sulzer with power rate of 396 kW, with turbocharger type VTR 160 Brown Boveri
- synchronous, self - excitation electro generator GD8-500-50, 500 kVA
- operating station EMOS
- electronic indicator with Kistler combustion sensors
- electric switchgear
- fan coolers
- fuel tanks with fuel distribution system and the centrifugal.

1.1 Diesel engine

The marine diesel engine type 3 AL 25/30, is four stroke non-reversible self-ignition, turbocharged engine. The engine was manufactured by HCP Cegielski in Poznań, under licence of Sulzer.

Figure 1. Marine diesel type 3AL 25/30

Main technical particulars of the engine:
- type   - 3AL 25/30
- no. of cylinders   - 3
- bore [mm]   - 250
- stroke [mm]   - 300
- swept capacity [cm^3]   - 14726
- power rate [kW]   - 408
- rotational speed [rpm]   - 750
- compression ratio   - 1:13

1.2 *Operator station EMOS*

Operator station EMOS is dedicated to current control, visualization and archive of the working parameters of the engine Sulzer 3AL 25/30.

The station is equipped with personal computer with two displays (19" and 40") and operator board, consisting the set of control lights and elements of engine's systems work control. The station is also equipped with safety devices system, and system of auxiliary mechanism monitoring and steering, both are governed by PLC programmable Logic Controller) Schneider Modicon.

PLC Schneider Modicon is based on 4 basic processors with Modbus communication, and 3 processors for integration of 2 within 3 communication lanes (CANopen, Ethernet and Modbus) each. Every processor has a port USB mini-B, which is the programming port and also connecting port for graphic panels Schneider Electric. The system Modicon M340 is build up basing on the board enabling configuration of full spectrum of amplifiers, processors and in/out modules with "hot swap" function what means broken element exchange without switching off the system. In/out modules are: analog, digital (64 channels) and fast counters.

All parameters controlled by operator station are available for outer recorders.

The operator station fulfils tasks as follow:
- operator access to all controlled working parameters
- constant display of alarms list with alarm on, alarm off and acknowledge time,
- acknowledgement of appearing alarms using the keyboard or the mouse,
- possibility of setting four alarm threshold levels for analog signals,
- possibility of setting time delays of alarm signals,
- changing of configuration of measurement channels, selection of measurement ranges and calibration,
- constant archiving of data and simple mode of files outlook,
- recording of trends of analog data and trends of changes based on records history,
- Data export to outer receivers for subsequent analysis and processing,
- Three access levels for different operators,
- Printing of the rapports and data sets,
- Independent work of two monitors enabling display of two pictures in the same time.

Figure 2. Operator station EMOS

In the case of occurring the alarm, EMOS system initiates visual alarm in the form of a blinking light and a horn acoustic signal. The alarm initiation delay can be set on individual determined for every alarm channel.

For measurement of variable pressure in cylinders and high pressure fuel pipes, electronic indicator Unitest has been used. It is six – way indicator with piezoelectric sensors of combustion pressure Kistler type 6353A24, light pipe sensors of injection pressure Optrand AutoPSI-S-2000 and impulse head type MOC. It enables pressure measurements with discretion 0.5° of crankshaft angle.

Kistler sensors are connected by dedicated adapters, enabling measurement of combustion pressure before indicators cocks. Solution like that lets avoid errors of value run due to interference of the cocks. In this case, automatic recording of indicator charts *on-line* mode is possible.

Electronic indicator Unitest 2008 can be placed in a group of Mean Indicated Pressure (MIP) calculators. That device is a stationary indicator dedicated to measurement, digital recording and visualization of the runs of combustion pressure and fuel injection pressure in domain of crankshaft angle. The most important elements of the indicator are: pressure sensors, injection sensors, crankshaft angle sensor, signal amplifiers, analog-digital transducers and personal computer.

The indicator has been equipped with special program enabling measurement and visualization of pressure runs. Example of a window with combustion and injection pressure charts is presented on fig. 3.

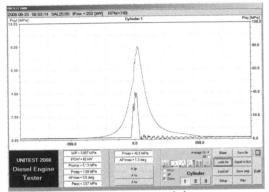

Figure 3. Electronic indicator program window

The indicator program includes broad spectrum of options for easy analysis gathered runs of combustion and fuel injection. Extension of selected parts of a picture is possible, and run of functions can be related to mean values of all cylinders and values of reference. Apart from graphic analysis of runs, automatically following parameters of combustion and injection are determined:
– indicated power of the engine;
– mean indicated pressure;
– peak of combustion pressure;
– angle of combustion pressure peak;
– expansion pressure (at angle $36°$ after TDC);
– peak injection pressure;
– angle of injection pressure peak.

Above stated elements are as follow laboratory: Technical Diagnostics, Tribology, Surface Engineering.

Those three laboratories, which equipment was funded by Ministry of Science and High Education are expected to enable realization of advanced research programs and contracted research in diapason of technical diagnostic, technical security engineering, analysis of mechanisms reliability, tribology and surface engineering.

2 THE TECHNICAL DIAGNOSTIC LABORATORY

The Technical Diagnostic Laboratory consist of equipment listed below:
– Vibration Analyzer PULSE by Brüel & Kjaer,
– Acoustic Emission Set by Vallen System,
– Analyzer/Recorder of working process by Sefram Instrumens & Systems,
– Mobile Gas Analyzer by Testo,
– Industrial Video endoscope XLG3 by Everest,
– Thermo vision Camera by NEC Avio Co.,Ltd,

2.1 Vibroacoustics

Vibration signals are carrying much information about technical condition of a machine and are a base for utilization in signals' monitoring systems as a condition trends factor of a machine. Spectral analysis of signals enables an identification of a failure type. Vibration signals monitoring is useful also for evaluation of bearing nodes, condition of shafts, and frictional couplings, including gears meshing and blades arrangements into rotary machines.

The vibration analyzer is the 6. channel recorder type 3050-A-60, the module LAN-XI 51,2 kHz (CCLD, V) Brüel & Kjaer. The set includes also the acoustic calibrator 4231 and the calibration's exciter 4294. The set consist also the tachometer probe MM360, set of microphones 4189-A-021 and the accelerometer 4515-B. Measurements and analysis are carried out using computer program PULSE time (FFT analysis program, harmonic analysis, signals' recorder). All is governed by the central station. The range of output voltage for typical accelerometer/microphone with build-in amplifier CCLD is 120 dB for broad band 10 Hz – 51 kHz, and 160 dB for narrow band 6 kHz. Maximum peak voltage is 10 V, and linearity ± 0,03 dB in the range of 120 dB. Data processing in the analyzer is 24 bit mode. Registered frequencies band is DC – 51 kHz.

2.2 Oil spectral analysis

The spectrometer is analyzing traces of radicals coming from: oil additives, wear processes and outer contamination. Comparison of results with previous ones and permitted limits enables observation of the normal mechanical wear process or early detection of potential damage, at its early stage. Moreover, enables evaluation of oil condition in reference to content of additives. It concerns mostly synthetic oils.

The spectrometer measures contents of radicals dissolved or floating particles in mineral or synthetic products, using the method of a rotational disc electrode (RDE). Basic configuration of the spectrometer enables detection of 22 radicals, ie. : Al, Ba, B, Cd, Ca, Cr, Cu, Fe, Pb, Mg, Mn, Mo, Ni, P, K, Si, Ag, Na, Sn, Ti, V, Zn.

The spectrometer range can be extended, what let detection additional radicals: Sb, Bi, As, In, Co, Zr, W, Sr, Li, Ce and detection of radicals In cooling liquids and water.

2.3 Video endoscope research

Video endoscope Everest XL G3 enables evaluation of technical condition of internal spaces, for example marine engines and machines, permanent and mobile pressure tanks, pipelines and masts, with possibility of dimensional evaluation of defects, visualization at LCD display and video recording. 3D phase measurement enables inspection and measurement of defects by only one lens, what eliminate necessity of its replacing by measurement lens. It lets scanning and measurement in 3 dimensions every detected discontinuity. Phase measurement analyzes available in observation zone (105° surface, and creates 3 dimensional movable model. Working probe in the system XL G3 is exchangeable.

2.4 *Thermo vision research*

Thermo vision camera Thermo Gear G100 from Japanese manufacturer NEC-AVIO Co., Ltd. enables tracking processes related to changes of temperature or emission in time or related to differentiation of thermal pictures of selected individual objects. The camera gives to operator many possibilities if measurement. It has a temperature preview function for 5 random points of the picture, with possibility of setting up individual coefficients of emission for every point. The camera enables also maximum/minimum temperature at whole display or in selected area, the value of difference of temperatures between two selected points, or linear profile of temperature.

As the camera is equipped with the optical focus with resolution 2 000 000 pixels, also registration of optical picture is possible. Pictures can be presented separately, parallel (one next to one at the display) or in penetrating mode.

During analysis of the picture, one has to put attention at changes of mutual position of pictures in relation to the distance from observed object.

The camera enables broad implementation for diagnostic research of machines and mechanisms as well research of technologic or energetic process.

The camera is equipped with the detector with dimension 320x249 elements. Works in real time, with refreshing frequency 60Hz. It has thermal sensitivity at least 0,08 °C at ambient temperature 30 °C. The camera can register temperatures in diapason from -400 °C to 500 °C, divided to two sub- ranges: -400 °C to 120 °C and 0 °C to 500 °C with accuracy ±20 °C or ±2%.

2.5 *Acoustic emission measurement method*

The AE method relay on detection and analysis of acoustic signal, emitted by a material being under mechanical stress. Emitted elastic waves are a result of interval elastic energy release. Thus energy is a phenomenon related to physical process taking place in materials or at their surface. Processes accompanying by acoustic emission are plastic displacement, cracking, structural and phase changes, corrosion, leaking and fibers cracking in composite materials. Accurate analysis enables definition of sources and kind of acoustic emission.

The set for non-invasive (without disassembling or destroying) measurement of a wear level of machines elements being under stress, deformations or load e.g. the wear of injectors, pumps, hydraulic elements, stress state of a fuselage or a hull sheets, pipelines etc.

The AE measurement set consist of 4 channels signal recorder AMSY 6 and the measurement module ASIP-2/S by Vallen System. The system is equipped with pre amplifier with a frequency range 20kHz to 1MHz and amplification 34dB, and AE signal sensor with range 100 – 450 kHz. The set has the recording module, putting down 8 MB's data bunches for every channel and data registration and analysis program.

2.6 *Marine engines exhaust gas analysis*

The mobile set dedicated for marine engines exhaust gas analysis enables measurement of emission of exhaust gases' toxic substances of different kinds of internal combustion engines, stationary or locomotive.

The set consist of high quality exhaust gas analyzer 350 XL by TESTO, including a industrial probe with particles filter, a infrared sensor calibration system and a rigid case. The analyzer has the Germanischer Lloyd Certificate, giving legacy for tests on board ships, in accordance to MARPOL Convention Attachment VI. Moreover, the set is equipped with the integrated temperature and humidity sensor, and atmospheric pressure gauge. Sensors are connected by 16 channels digital - analog transducer with industrial computer with dedicated programs, as a recorder. The recorder lets simultaneously connect all gas sensors, ambient parameters gauges and additional 13 random physical values sensors having standard 0-10 V outputs. The recorder has built-in parallel port RS-232, for connection with the recorder of TESTO analyzer. In fig.7. is presented the set of Exhaust gas analyzer Testo 350XL, and in tab.1.,exhaust gas measurement range.

Table 1. Gas analyzer measurement range

Parameter	Range	Unit
oxygen - O_2	$0 - 21$	% Vol.
carbon monoxide - CO	$0 - 5000$	ppm
carbon dioxide - CO_2	$0 - 20$	% Vol.
nitric oxide - NO	$0 - 2500$	ppm
nitro dioxide - NO_2	$0 - 500$	ppm
sulphur dioxide - SO_2	$0 - 3000$	ppm
gas temperature at measurement point	$0 - 1000$	°C
dynamic pressure	do 20	kPa

3 CONCLUSION

Modernization of the engine and extension of diagnostic base, will enable carrying out research specified below:
- diagnostic research based on active experiment, leading to determination of diagnostic parameters' data base;
- diagnostic research of the engine's functional systems, especially turbo charging, fuel injection and piston-connecting rod ensemble;
- research related to the utilization of combustion pressure charts, high pressure fuel fluctuation analysis and acoustic emission, for marine engines diagnostics;
- research on influence of alternative fuels implementation at the engine exploitation parameters including exhaust gases composition and toxic;
- research on influence of mode of the engine exploitation at its elements condition, including elements after recovery treatment;
- research on possibility of utilization of data base information for automatic gathering of knowledge (inductive methods of machine learning and knowledge reveal methods).

REFERENCES

Charchalis A., 2007. Conditions of Drive and Diagnostic Measurements During Sea Tests Journal of Kones vol. 14/4, Warszawa

Pawletko R., Polanowski S., 2011. Influence of TDC determination methods on mean indicated pressure errors in marine diesel engines, Journal of KONES, Vol. 18, No. 2, p. 355-364, Warsaw

Polanowski S., Pawletko R., 2011. Acquisition of diagnostic information from the indicator diagrams of marine engines using the electronic indicators, Journal of KONES, Vol. 18, No.3, p. 359-366, Warsaw

Pawletko R., Polanowski S., 2010. Research of the influence of marine diesel engine Sulzer AL 25/30 load on the TDC position on the indication graph. Journal of Kones Powertrain and Transport, Vol. 17, No. 3, p. 361-368 , Warsaw

Polanowski S., Pawletko R., 2010. Application of multiple moving approximation with polynomials in curve smoothing, Journal of Kones Powertrain and Transport, Vol. 17, No. 2, p. 395-402, Warsaw

Propulsion and Mechanical Engineering
Maritime Transport & Shipping – Marine Navigation and Safety of Sea Transportation – Weintrit & Neumann (Eds)

Develop a Condition Based Maintenance Model for a Vessel's Main Propulsion System and Related Subsystems

M. Anantharaman & N. Lawrence
Australian Maritime College (Utas), Tasmania, Australia

ABSTRACT: Merchant shipping has undergone a great transformation over the past three decades. The shipping market is highly competitive, which coupled with high crewing and fuel costs, leads to high operational costs. One of the paramount factor involved in vessel operation is the Maintenance cost and there is a dire need to keep this cost to a minimum. Fortunately the earlier policy of repair only maintenance in commercial shipping has been done away with, and was replaced by the policy of preventive maintenance. Planned Maintenance System was introduced by ship management companies in the early 90's. Planned Maintenance offered benefits over the repair only policy, but has its own demerits. Many a time machinery equipment is opened up for routine maintenance after a specified time interval, irrespective of the need. This could lead to potential failures, which is explained by the fact that preventive maintenance resulted in meddling of a well set piece of machinery equipment, leading to its subsequent failure. This is where Condition based maintenance or CBM steps into prominence. CBM monitors the health of the machinery equipment, analyses the condition and helps you in decision making. Accordingly a ship's engineer may decide to stop the running machinery equipment, open and overhaul the same, else postpone the overhaul for a later safe date.

1 INTRODUCTION TO PLANNED MAINTENANCE

Commercial shipping in the modern world is highly competitive, which coupled with high crewing and fuel costs, leads to high operational costs. One of the paramount factor involved in vessel operation is the Maintenance cost and there is a dire need to keep this cost to a minimum. Fortunately the earlier policy of repair only maintenance in commercial shipping has been done away with, and was replaced by the policy of preventive maintenance. Planned Maintenance System was introduced by ship management companies in the early 90's.

2 WHY CONDITION BASED MAINTENANCE

Planned Maintenance offered benefits over the repair only policy, but has its own demerits. Many a time machinery equipment is opened up for routine maintenance after a specified time interval, irrespective of the need. This could lead to potential failures, which is explained by the fact that preventive maintenance resulted in meddling of a

well set piece of machinery equipment, leading to its subsequent failure. The author would like to cite an incident experienced during his vast sailing career. A general cargo vessel was on her way from India to Europe. This passage involves the Suez Canal transit where large numbers of vessels go on a convoy. A few days before transiting the Suez canal the No.1 Steering gear motor was opened up for routine overhaul, as specified in the Planned Maintenance Schedule for the vessel. The motor was overhauled and reassembled. The vessel then entered the convoy and all went off well for an hour, under the guidance of the Suez Canal pilot who travels on board the ship during the transit. The pilot then ordered a helm movement and the vessel failed to steer as required. The reason was overload tripping of the overhauled motor, and investigations revealed errors during the reassembly of the motor. The vessel was then tied alongside the canal with the help of tugs, the motor had to be reopened again, and new bearings were fitted. Motor tried out and finally the vessel managed to transit the Suez Canal, though she had to be last on the convoy. resulting in considerable losses in terms of thousands of dollars to the company. This is where Condition Based

Maintenance or CBM steps into prominence. CBM monitors the health of the machinery equipment, analyses the condition and helps you in decision making. Accordingly a ship's engineer may decide to stop the running machinery equipment, open and overhaul the same, else postpone the overhaul for a later safe date.

3 ISM CODE AND MAINTENANCE

When it comes to operation of ships, all shipping companies need to abide by the ISM Code which is the International Safety Management Code, the purpose of this Code is to provide an international standard for the safe management and operation of ships and for pollution prevention. The above research will also address section 10.3 of the code which states that, 'The Company should identify equipment and technical systems the sudden operational failure of which may result in hazardous situations. The safety management system should provide for specific measures aimed at promoting the reliability of such equipment or systems. These measures should include the regular testing of stand-by arrangements and equipment or technical systems that are not in continuous use.'(ISM, 2002)

4 DIAGNOSIS AND PROGNOSIS

The concept of CBM for ship's machinery is still in its infancy. Reproducing a recent finding which says 'However, according to class records, only about 2% of thw world fleet is operating using CBM'.(MER,2012).Effective application of CBM techniques will result in large savings to the vessel owner / operator. A ship's machinery space is a large main propulsion system with several subsystems .All these systems have a fairly high degree of correlation; failure of any one subsystem could result in stoppage of the vessel, which is a highly undesirable event. CBM is a two sided coin with diagnosis on one side and prognosis being the other side. For an efficient vessel operation both the sides of the coin are vital.

Prognosis is an important element of the CBM program as it deals with the predication of failure faults. The above research should be useful to predict the occurrence and timing of a failure in a single subsystem (example ship's main propulsion and power generation system) or in several different subsystems (example ship's main propulsion and power generation system and control air system)

5 GROSS MAINTENANCE DEFECIENCIES

A few instances of major shipping disasters on account of gross maintenance deficiencies have been highlighted below. This information was gathered from leading reputable Marine Accident Investigation bodies in the commercial shipping world.

'The scavenge space inspection after the fire in number three unit, shortly after the first turbocharger failure, apparently revealed a high level of scavenge fouling. Similarly the condition of the scavenge spaces after the second turbocharger failure was poor around number two cylinder, albeit as a result of piston cooling oil leaking from the defective O-ring. Whether or not the condition of the scavenges led to a fire, which in turn caused the turbocharger failures, cannot be concluded with any certainty however their condition does indicate that the vessel's main engine maintenance regime in this respect may have been deficient.'(ATSB, 2006).

Figure 1. Condition of scavenge ports on Cylinder no. 1 of Main propulsion Engine Source based on ATSB (2006)

Figure 2. Condition of scavenge ports on Cylinder no. 2 of Main propulsion Engine Source based on ATSB (2006)

Figure 3. Condition of damaged main propulsion turbocharger rotor. Source based on ATSB (2006)

Apart from the failure of the auxiliary boiler, there were other examples where equipment did not work properly that were attributable to ineffective maintenance or equipment checks:
– Standby EGE circulation pump mechanical seal
– Automatic operation of soot blowers
– Fuel tank Quick Closure Valves
– CO2 drench pilot operating system
– Emergency diesel generator overheating
– Emergency fire pump suction

The maintenance system recorded that checks and planned maintenance were complete on all these items, and that there were no defects. While it is always possible for equipment not to work in an emergency, so many serious defects should not occur during the same incident. Neither the maintenance system nor any of the technical audits detected these latent defects, so the effectiveness of these systems must be questioned.'(MAIB, 2007)

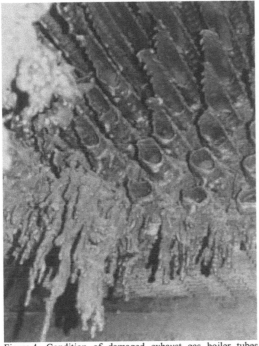

Figure 4. Condition of damaged exhaust gas boiler tubes. Source based on MAIB (2007)

6 RELIABILITY CENTERED MAINTENANCE

CBM leads to improved reliability of the machinery equipment and better inventory control of spares on board the vessel. This approach to maintenance has been advised to ship owners by leading classification societies.' By applying Reliability-centered maintenance (RCM) principles, maintenance is evaluated and applied in a rational

manner that provides the most value to a vessel's owner/operator. Accordingly improved equipment and system reliability on board ships and other marine structures can be expected by applying this philosophy'. (Robert, 2004)

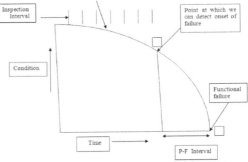

Figure 5. P- F Diagram Source ABS guidance notes 2004

ABS guidance notes on Reliability Centered Maintenance- 2004 highlights the P-F interval. 'If a potential failure is detected between Point P and Point F, it may be possible to take action to prevent the functional failure (or at least to minimise its effects). Tasks designed to detect potential failure are known as condition-monitoring tasks. See Figure 5 above.

7 RESEARCH PROPOSAL

My research proposal is to focus on main propulsion system and related subsystems (Figure 6) on commercial bulk carriers and naval vessels, collect data for these main and subsystems which will then be processed for statistical analysis and produce a reliable maintenance model for the vessel. To begin with, the research will be exploratory in nature, collecting data pertaining to real life examples and case studies published by reputed marine accident investigation bodies in world shipping mentioned above. I shall then start building up my theory developing from the knowledge gained in the exploratory process stage. This should eventually lead to development of a hypothesis which will be tested statically for a large sample size.

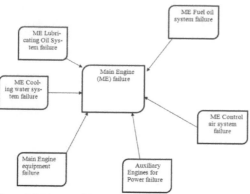

Figure 6. Main propulsion system and related subsystems for large vessels.

8 FAULT TREE ANALYSIS

I intend using FTA (Fault Tree Analysis) during development of the CBM (Condition based maintenance) model. FTA is a top down approach, which helps to identify basic event which can lead to the top undesirable event. I have included a few quotes from my earlier research work below: Looking at the fault tree one can easily recognise that a basic event of a distillate pump failure by itself can cause the top undesirable event to happen. Also past sailing experience of the author provides ample cases of Fresh water Generator failure on account of failure of the distillate pump. Hence the emphasis should be to avoid this basic event.'(Anantharamman, 2002)

'In conclusion the author would like to emphasise that recognising a Fault Tree proves to be a very reliable tool and healthy exercise for a Shipboard Engineer responsible for Operation, Running & Maintenance of ships. The benefits to cost ratio could be tremendous. Hence let us start planting this Tree for all shipboard activities and reap the benefits.'(Anantharaman, 2003).

9 SUMMARY AND CONCLUSION

We have seen in this paper that at times, just following a PMS regime on board vessels could lead to machinery failure, resulting in stoppage of a vessel at a critical juncture. In merchant shipping it is very important that we make a shift from PMS (Planned Maintenance System) to CBM (Condition based maintenance). The main propulsion plant of a vessel should be the focal point of CBM and we can work around the related subsystems. Fault Tree Analysis is one such approach wherein we can identify the basic event, failure of which could lead to a possible catastrophic failure of the plant. We need to look into the probability of failure of the system component using statistical tools for analysis.

A neuro-fuzzy modeling approach for CBM has been effectively utilised by researchers in merchant shipping and other shore based industries. (Kothamasu & Huang, 2007), (Xu et al, 2010).

This transformation could lead to a huge benefits to cost ratio at the same time ensuring safety and reliability.

REFERENCES

ISM(2002) 'International Safety Management (ISM) Code'. International Maritime Organisation (IMO).
MER, Operational Obstacle, MER,IMAREST Publications, November 2012, page 38.
ATSB, (2006). ATSB Transport Safety Investigation Report. Marine Occurrence Investigation No. 186 and 191 Final.
MAIB, (2007) Marine Accident Investigation Branch, Southampton, United Kingdom, Report 15/2007, July 2007
Robert M. Conachey, American Bureau of Shipping, Houston, USA
Presented at 2nd International ASRANet Colloquium, Barcelona, Spain, July 5-7, 2004
Anantharaman, Mohan, A Fault Tree Approach to Practical Shipboard Problem on a Fresh Water Generator – a Special Supplement, Maintenance Journal, Vol.15, No.4, pp.64-65, 2002
Anantharaman, Mohan, Fault Tree Analysis to Prepare Cargo Holds for Loading on Bulk Carriers, Maintenance Journal, Vol 16, No.3, pp 72 – 75, 2003
Ranganath Kothamasu & Samuel H.Huang, Adaptive Mamdani fuzzy model for condition-based maintenance, Journal Fuzzy Seta and Systems, Volume 24, December 2007, Pages 2715-2733 158 Issue
Q.Xu, X.Meng & N.Wang, Intelligent Evaluation System of Ship Management, International Journal on Marine Navigation and Safety of Sea Transportation Volume 4 Number 4 December 2010 479

Experimental Analysis of Podded Propulsor on Naval Vessel

M.P Abdul Ghani, O. Yaakob, N. Ismail, A.S.A Kader & A.F Ahmad Sabki
Marine Technology Centre, Universiti Teknologi Malaysia

P. Singaraveloo
Ship Clasification Malaysia

ABSTRACT: This paper describes the effect of pod propulsor attachment to the existing Naval Vessel hull form which was designed for conventional propulsor in aspects of resistance and motion characteristics. These investigations were carried out on a 3.0 m model by experimental works in the towing tank 120m x 4m x 2.5m at the Marine Technology Centre (MTC), Universiti Teknologi Malaysia (UTM). The basis ship chosen for this study is Sealift class type MPCSS (Multi Purpose Command Support Ship). In this study, the design for the new pod propulsor is based on a proven design and scaled down to suit this type of hullform accordingly. This paper describes the resistance comparison between bare and podded hulls in calm water as well in waves. The seakeeping test for hull with and without pod in regular waves at service speed of 16.8 knots were carried out at wavelength to model length ratio, Lw/Lm between 0.2 and 1.2. The outcomes from this experimental works on hull with and without pod were compared.

1 INTRODUCTION

Podded propulsion system is a new propulsion systems have been used for both commercial and naval ships. Propulsion pods are gondola shaped devices, hanging below the stern of a ship, which combine both the propulsive and the steering functions.

Pod propulsion offers attractive performance benefits over more conventional propulsion systems, especially in the areas of ship noise, hydrodynamic efficiency and fuel economy. The elimination of long shaftlines, support bearings, stern tubes, and other underwater protrusions typically with conventional system creates a smoother laminar flow over the hull and propeller

The first patent for a podded propulsion system was in 1826 by William Church and the first application was by John Ericson in 1836. The real application for this propulsion system in the past was applied to torpedoes. In Japan there are some vessels operating with podded propulsion system and the results from the application are good especially in reducing vibration level but rather complicated due to the conventional propulsion system using long shaft located between each other[5].

The podded propulsion system normally uses an electric motor driven by diesel electric drive. This propulsion drive has been used in icebreakers and other special purpose vessels. A pod consists of a motor located in a hydro-dynamically optimizes housing and stay attached to the hull. Well designed pods reduce resistance to motion by 5-10%. An optimally designed pod shape, positioning and angle in relation to ship's hull can increase propulsion efficiency up to 15% in comparison with an in hull propulsion system. Pods also decrease the vessel vibrations and noise levels and provide a more environmental friendly vessel to ship operators. Pods can be dismounted and serviced at sea, making dry docking for major propulsion repairs unnecessary.

Several model test series have been carried out to define a shape with optimal efficiency. CFD calculations have been made to investigate the flow and pressure pattern around the Pod. To reach a good propulsion efficiency, the underwater housing should be as small as possible.

2 BACKGROUND

Basically, a podded propulsion system consists of a fixed pitch propeller driven by an electric motor through a short shaft. The shaft and the motor are located inside a pod shell. The pod unit is connected to ship's hull through a strut and slewing bearing assembly. This assembly allows the entire pod unit to rotate and thus the thrust developed by the propeller can be directed anywhere over 360˚ relative to the ship.[4]

A small pod diameter or gondola diameter should be used to get a high total efficiency and to reduce the interaction effects between propeller and pod housing.

The pod diameter depends on the size of the electric motor inside the pod. The definition of the geometric parameters is shown in *Fig.* 1 and its proposed particulars as shown in Table 1.

The basis ship chosen for this study is Sealift Class Type MPCSS (Multi-Purpose Command Support Ship). The ship particulars and its body plan are shown in Table 2 and *Fig.* 2 respectively.

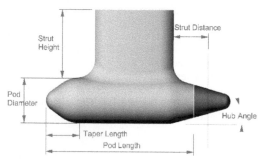

Figure 1. Pod Anatomy

Table 1. Proposed Pod Particulars

Parameter	Value
Propeller diameter, D m	3.887
Pod length, Lp, m	5.995
Pod diameter, Dp, m	2
Pod length ratio, Lp/D	1.542
Pod diameter ratio, Dp/D	0.514

Table 2. Ship Particulars

Parameter	Value
Length overall, LOA m	103.000
Length Between Perpendicular, LBP m	97.044
Breadth, m	15.000
Depth, m	11.000
Draught, m	4.409
Displacement, tonnes	4431.57
Speed (Operational), knots	16.8
Speed (Max), knots	19.98

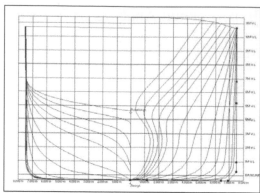

Figure 2. Sealift Class Type MPCSS of Naval Vessel

3 EXPERIMENTAL SET UP

The experiments have been carried out in the towing tank of the Marine Technology Centre (MTC), Univesiti Teknologi Malaysia (UTM). The dimensions of this tank are: length 120 meter, width 4.0 meter and water depth 2.5 meter. The maximum attainable speed of the towing carriage is 5 m/s with acceleration 1 m/s^2.

Table 3. Model Test Matrix for Resistance in Calm Water

Run	VS (knots)	Model Speed, Vm (m/s)	Fn
1	14	1.228	0.226
2	16	1.404	0.259
3	16.8	1.474	0.272
4	18	1.579	0.291
5	20	1.755	0.324

Table 4. Model Test Matrix for Resistance & Seakeeping in Regular Waves at Vm =1.474m/s

Fn	WAVE CHARACTERISTICS					
	Lw/Ls	Lw	Hw	Tw	ωw	Hw/Lw
0.272	0.5	1.5	0.015	0.980	6.02	1/100
0.272	0.6	1.8	0.018	1.074	5.85	1/100
0.272	0.8	2.4	0.024	1.240	5.07	1/100
0.272	1.0	3.0	0.030	1.386	4.53	1/100
0.272	1.2	3.6	0.036	1.518	4.14	1/100

4 RESULTS AND ANALYSIS

Most of the results of the measurements have been plotted based on the Froude number.

In general, podded hull has higher resistance value due to the additional wetted surface area. Based on the result obtained, at the design speed (16.8 knots), the total ship resistance value for hull with pod propulsor is higher than the hull without pod. Figures 3 and 4 shows the resistance for podded hull is higher than bare hull by differences about 22.5% but the differences between these two decreases with increasing of speed.

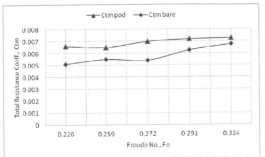
Figure 3. Ctm versus Fn

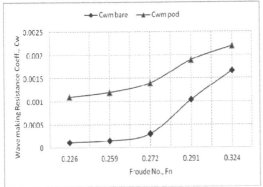
Figure 4. Cwm versus Fn

As shown in Figures 5 and 6, at the maximum (Lw/Lm=1.2) wave condition, the total ship resistance for podded hull is higher than for hull without pod. The difference between these two values is about 20%.

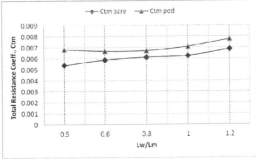
Figure 5. Ctm versus Lw/Lm at Fn=0.272

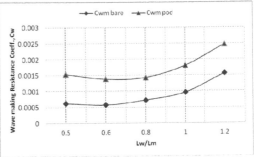
Figure 6. Cwm versus Lw/Lm at Fn=0.272

Figure 7 shows the pattern of pitch RAO for the model with and without pod are the same but there are small deviations in term of magnitude of the response whilst the values for model with pod are slightly higher hence the hull with pod produce higher pitching motion than than the hull without pod.

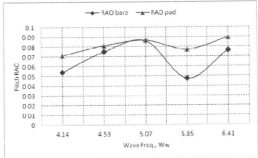
Figure 7. Response Amplitude Operator

5 CONCLUSIONS

From the above the following conclusions can be drawn:

It was found that the the hull with pod produce resistance 20% higher than the bare hull and also an increment about 22% higher in pitching response.

Generally pod technology has already made significant progress in the commercial shipbuilding industry. This new technology offers many unique advantages not offered by conventional electric propulsion systems. Pod propulsion is undoubtedly a viable option for future shipbuilding programs for the Malaysian navy.

ACKNOWLEDGEMENTS

The authors wish to express their sincere thanks to all staff of the Marine Technology Centre, UTM for their assistance in carried out this experimental works.

REFERENCES

Abdul Ghani, M.P, Mohd Yusop, M.Z and Islam, M.R, Design Study of Podded Propulsion System for Naval Ship, Proceedings of the 4th BSME-ASME International Conference on Thermal Engineering, 27-29 December 2008, Bangladesh

Marc Batsford, Pod Propulsion: A Viable Option for the Canadian Navy, Maritime Engineering Journal, Vol. 20, No. 2 Fall 2001/Winter 2002.

Cornelia Heinke, Hans-Jurgen Heinke, Investigations about the use of Podded Drives for Fast Ships, FAST 2003

Islam, M.F et al Numerical Study of Effects of Hub Taper Angle on the Performance of Propellers Designed for Podded Propulsion Systems, Marine Technology, 43, 1, pp. 1-10,2006

Modern Methods of the Selection of Diesel Engines Injector Nozzles Parameters

M. Idzior
University of Technology, Poznań, Poland

ABSTRACT: In the article one introduced the problems of the selection of diesel engines injector nozzles parameters and limitations of the pressure of the fuel injection. One talked over conditioning being with the stimulator of their systematical height. The methods of the selection guilty so to embrace, except the selection of constructional parameters , also the selection taking into account concurrent occurrences , for example the pressure and the speed of injected fuel or the stress distribution. One executed analyses of restrictive factors of the endeavour to the further lifting of maximum values of the pressure of the fuel injection. In the recapitulation one underlined important of the problem and his participation on the future development diesel engines.

1 THE INTRODUCTION

Is difficulty univocally to give methods of principle of the selection of parameters of injector nozzles, considering additionally their influence on impurities of combustion gases of the diesel engine. The methods of the selection, taking into account the environment protection, cannot assemble exclusively on parameters of the part executive, is what just the injector nozzle. It should make allowance for also the construction and parameters of cooperative parley and reasons and consequences of concurrent occurrences to the injection and the atomization of the fuel in the cylinder of the diesel engine.

The methods of the selection guilty so to embrace, except the selection of constructional parameters , also the selection taking into account concurrent occurrences , for example the pressure and the speed of injected fuel or the stress distribution.

Considering placings requirements and occurrences happening {reaching} in the correctly working nozzle, methods of the selection of parameters of constructional injector nozzles one can divide on two core groups:

1 methods of the selection:
- methods optimization, leaning on the theory of the optimization of the construction,
- methods simulatory, leaning on findings with the use of specialized computer programmes

(for example the stress distribution and thermal charges);

2 empirical methods of the selection:
- methods visualization leaning on the investigation (for example by means the instrument AVL Engine Video System) of the construction {the build} and the shape sprayed fuel,
- motor leaning methods on the practical investigation of sets of injector nozzles about accepted parameters with the regard of measurement of concerning issues of toxic relationships.

2 THE SELECTION OF PARAMETERS OF INJECTOR NOZZLES BY MEANS MATHEMATICAL OPTIMIZATION METHODS

2.1 *Preparing the new file with the correct template*

These methods consist in finding of best solution (in relation to the settled criterion) from the set of possibly (admissible) solutions. The conduct relies so on the research of the value of parameters for which is the realizing condition determining recorded mathematically the criterion of the examined occurrence, at the realization of recorded mathematically limitations.

The course of optimization problem one can divide on three stages:
- the acceptance of the criterion function and suitable groups of independent variables,
- the elaboration of the set of limitations,
- solution of optimization problem.

As the criterion function one can accept one from parameters of the work of the engine, for example the effective power N_e or the moment torque M_o. Answering to them independent variables will be each parameters of injector nozzles.

These parameters are treated independently, in the reality however the influence on the issue of impurities of combustion gases and the power or the moment generated by the engine have all combinations of occurrent sizes in at present the investigated injector. The only individual approach to every size gives the full possibility of the use of worked out technics and optimization algorithms.

The most of performance characteristics of the diesel engine can be approximated with the polynomial quadratic. Solution of the assignment must contain himself in the set of solutions admissible, appointed in our chance by admissible values of the issue of toxic relationships in combustion gases. This postulate assure only restrictive non-linear conditions, irregularity, for example HC_{tot} (parameters of the nozzle) $\leq HC_{totDOP}$.

Solution of optimization problem consists in the effective research the minimum of the criterion function in the admissible area traced by limitations - for example on the delimitation of the direction research of the point the minimum and on his qualification.

This type the approach to the theme of the selection of parameters of injector nozzles is, thanks to the quick development of computers and their possibilities counting, more and more often practical, from the regard even if on low costs of working out of optimum- solutions.

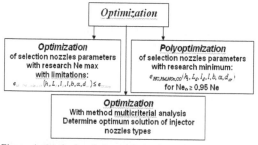

Figure 1. Methods of the optimization of the selection of parameters of injector nozzles

One cannot here however forget that is this typical theoretical approach and not always finding of best theoretical solution ties in with his immediate use in effect, because the progress in the development of computational methods considerably outdistances possibilities of their utilization in real model solutions.

Heaps of times on the hindrance to these best mathematical solutions stand up technological limitations or rememberings strength of materials limitations.

3 THE SELECTION OF PARAMETERS OF INJECTOR NOZZLES WITH THE REGARD OF THE SHAPE AND THE CONSTRUCTION OF THE STREAM

The spray pattern of the fuel, his construction, the quality of spraying - the drop diameter and their schedule chiefly decide about the degree of the entire and complete combustion in the cylinder of the diesel engine, and what himself with this binds and emissivities of this engine.

For the purpose of the graphic performance of the quality of spraying one prepares the characterization of the proportional participation drops of the fuel in the dependence from their diameter. These characterizations are often called in the literature unjustly a phantom of spraying; in reality are a thickness of the probability of drop diameters and can be prepared for different of their decisive sizes about spraying.

In at present produced engines self-igniting more and more are more often practical injectors with two spring which make possible the realization of the two-grade injection. How show research, the use of the two-grade injection and injector nozzles VCO lowers the issue of nitrous oxides and hydrocarbons in combustion gases [1].

Simultaneously research showed that such fuel injection conjointly with injector nozzles VCO unfavourably bore on the smokiness of combustion gases. Enlarging smokiness is especially visible at low engine loads.

4 THE SELECTION OF PARAMETERS OF INJECTOR NOZZLES WITH THE REGARD OF LIMITATIONS OF THE WEIGHT OF THE PREASURE OF THE INJECTION

Introduced to production engines are already provided into container parley Common Rail of the second generation, with enlarged pressure of the injection (160-180 MPa), in nearest years one foresees the enlargement of the pressure even to 220 MPa.

Nascent tensions at pressures 200 MPa are already too large for the persistence of some elements, first of all talked over injector nozzles. In spite that trunks of injector nozzles are executed steel chromic-nickel-tungstenic, about the large endurance on the extension, this however due tensions with high pressures of the injection can reach the border of the plasticity of given material. Such state of the load can as result of of the fatigue of material bring to the damage of the sprayer.

Calculations of the nozzle with eight holes [2] whose the section one showed on the fig. 2, laden with the pressure 200 MPa and with the pressure 300 MPa, so such, what appears at the destruction of the sprayer in some parley of the power supply, showed that greatest tensions came into being on the passage of the nest of the cone-shaped trunk into the well. They carry out for the first chance 710 MPa, while for second 1065 MPa. Large tensions come into being also at edges of intake- injection's openings and carry out properly 510 MPa and about 947 MPa. Itself bottom of the well is not strongly laden, because prevalent there tensions in the dependence from the pressure of the fuel hesitate from 80 to 125 MPa. From these calculations it results that the pressure of the injection carrying out 300 MPa seems greatest, possibly to the usage for the fuel injection in the diesel engine for the endurance and the persistence of injector nozzles. So high pressures demand usages of materials about greater than till now endurances on the extension, what doubtless increases costs of the realization of injector nozzles. Practical until quite lately universally to this end chromic-nickel-aluminium steels, chromic-nickel-tungstenic and similar, are taken place steels about the greater endurance, for example steels nickel- and other modern materials. Such are carbide-steels in which the participation of carbides, mostly TiC, carries out to 50%. Constantly these, after the heat-treatment, attain the hardness about 70 HRC, even in elevated operating temperatures and the large resistance on the erosion and cavitation, are however difficult in the tooling and expensive.

On injector nozzles of greater engines it begins to comply stellites. This are very hard alloys (W - 65%, Cr - 25%, In - 5%, C - 2% and V, Fe and other) about the very small linear expansion, what causes the very good dimension- stability in elevated temperatures, the high abrasion resistance and the resistance on aggressive fuels, the corrosion and the oxygenation.

Stellites have the high price, but are not as usual permanent and make possible the diminution of the mass of the nozzle, what for greater injector nozzles begins to be profitable economically.

The height of the pressure of the injection can be also limited by the compressibility and the stickiness of the fuel, and also rectifier valves master with the flow of the fuel which will have to quickly and unfailingly to work conditioned of enlarged loads.

The diesel fuel under the pressure 300 MPa diminishes her own volume about 15-20%. Such his wring at small doses of injected fuel can cause disturbances of the injection. Higher pressures cause also the height of the stickiness of oil, what favours to the formation large, badly burning up drops of the fuel.

5 CONCLUSION

Introduced problems of the selection of parameters of injector nozzles show as not as usual difficult is optimum- synchronizing of all parameters, so that they realize requirements placed to the present diesel engine, inclusive of with more and more sharper ecological requirements.

Perfecting of the injection's apparatus in today's engines of this type, with taking into account of norms of concerning issues of impurities, is one from most important criteria of the choice of best solutions of these decisive engines about the success of chosen constructions.

REFERENCES

[1] Idzior M.: Problemy doboru wtryskiwaczy silników o zapłonie samoczynnym zasilanych paliwami alternatywnymi. Wydawnictwo Naukowe Instytutu Technologii Eksploatacji –PIB, Radom 2012
[2] Idzior M.: Studium optymalizacji doboru parametrów rozpylaczy wtryskiwaczy silników o zapłonie samoczynnym w aspekcie właściwości użytkowych silnika. Wydawnictwo Politechniki Poznańskiej, Poznań 2004.
[3] Meyer S., Krause A., Krome D., Merker G.: Flexible Piezo Common-Rail-System with 3. Direct Needle Control. Motortechnische Zeitschrift nr 2, 2002
[4] Zbierski K.: Układy wtryskowe Common Rail. Łódź 2001.
[5] Materials of companies: BMW, Honda, Mercedes, Toyota, Volkswagen

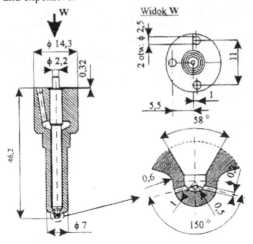

Figure 2. Injector nozzle with eight holes

Propulsion and Mechanical Engineering
Maritime Transport & Shipping – Marine Navigation and Safety of Sea Transportation – Weintrit & Neumann (Eds)

The Assessment of the Application of the CFD Package OpenFOAM to Simulating Flow Around the Propeller

T. Gornicz & J. Kulczyk
Wroclaw University of Technology, Institute of Machines Design and Operation, Poland

ABSTRACT: During the design process of the ship, a lot of attention is given to tune-up parameters of the propeller, which is one of the most important part of the propulsion system. Designing new generation, more effective and efficient propellers requires knowledge of hydrodynamic phenomena occurring in the propeller area. For this reason, a lot of time and funds is spent on the propeller research and development. Usually these are experimental tests in cavitation tunnels and towing tanks. However more and more research is carried out by CFD simulation. OpenFOAM software is open source CFD package, which becomes popular alternative to commercial codes. This paper attempts to answer the question: if OpenFOAM is suitable for such specific applications like accurate simulation of flow around the propeller? The paper presents the results of a CFD simulation of marine propeller created with OpenFOAM software. The obtained results were compared with the of the commercial CFD codes simulations and the experimental research.

1 INTRODUCTION

Designing propulsion system for new ship is responsible and important task. To do this properly it necessary to obtain knowledge about propellers characteristics. This is why so many different computational and experimental methods are now available. The experimental methods seem to be the most reliable. However they require a high technology facilities. Now a days they are being displaced by computer simulation methods. The research have confirmed (Carlton, 2007) that CFD calculations can be successfully applied to simulate complex flow (with separations, reattachments, huge pressure gradient, etc.) around rotating propeller. Now on market there are couple commercial CFD solvers (codes), designed for calculating complex flows, including propellers slip stream. Beside huge costs, their main impediment is closed source code, which prevents advanced users introducing modifications into program. In response to the need of fully customizable CFD solver, there have been started open source project called OpenFOAM. It was not designed for solving the problems of ship hydrodynamics, however it could be successfully used to determine propeller open water characteristic. What will be shown in next sections.

2 OBJECT OF STUDY

In research there was used a 5 blade propeller from "Potsdam propeller Test Case" program (Barkmann, 2011). Figure 1 shows the propeller geometry. Main reason of choosing this propeller was availability of experimental results. Main parameters of the construction are shown in table 1. Calculations were performed for constant propeller revolution n=10 [rps] and variable inflow velocity $V_A = 0.4$-4 [m/s].

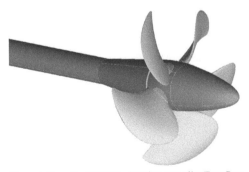

Figure 1. Propeller VP1304 - Potsdam propeller Test Case.

Table 1. Propeller main parameters.

Parameter name	Symbol	Value	Unit
Diameter	D	0.250	[m]
Pitch ratio (r/R=0.7)	$P_{0.7}/D$	1.635	[-]
Aspect ratio	A_E/A_0	0.77896	[-]
Hub ratio	d_H/D	0.300	[-]
Chord length (r/R=0.7)	$c_{0.7}$	0.10417	[m]
Number of blades	Z	5	[-]

3 CFD MODELS

3.1 *Computational domain*

The Propeller geometry was divided into 5 identical regions along symmetry axis, with respect to propeller's periodicity. Thus only one blade was modelled. Cyclic boundary walls were used to simulate periodicity of the flow. Outer boundary of computational domain was shaped in cylinder-like form. The main dimensions of computational domain were adjusted to propeller diameter (D) and they were defined as following: 1.5 D in upstream direction, 3.5 D in downstream direction, outer surface diameter - 2.5D. The geometry of the computational domain is presented on Figures 2 and 3.

Figure 2. Computational domain - global view.

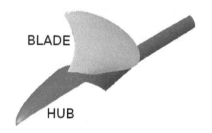

Figure 3 Computational domain - blade view.

Propeller geometry was simplified by closing the gaps between blades and a hub and between hub and shaft. This simplifications have negligible influence on propeller open water characteristic. Both simplification are shown on Figure 4. The computational domain were discretized in ICEM-CFD computer program. Unstructured tetrahedral element was chosen as a numerical mesh type. Advance size function was used to guarantee proper element size on high curvature blade surfaces. Geometry periodicity was taken into consideration during mesh generation process.

The OpenFOAM CFD toolbox contains its own grid generators. Authors of paper have not successfully built propeller mesh with provided tools. The main problem was unacceptable mesh quality. Mesh of the whole propeller (without periodicity), generated with OpenFOAM tools, was presented in (Klasson, 2011).

The toolbox provides also pack of numerical grid converters from external to OpenFOAM format. Program "fluent3DMeshToFoam" was used to import generated mesh into OpenFOAM environment.

Figure 4. Mesh on blade and hub. Closed gaps between: shaft and propeller, shaft and hub.

Result dependence from number of discrete elements were tested on two different meshes. Coarse mesh was about 900 thousands elements and detailed mesh was about 2 millions of elements. Both of the meshes were generated with caution to size of first layer elements. Non dimensional height of the first layer elements was set up in recommended range y+ = 30 - 300 (Wilcox, 2002). Numerical simulation of both size meshes gave similar quality of results. No significant differences were noted.

3.2 *Numerical methods*

Simulations were calculated with CFD toolbox OpenFOAM in version 2.1.1. Reference values were obtained from Ansys Fluent 13.0 system. The Ansys Fluent was validated many times and it was proven to be reliable tool in ship hydrodynamics applications (Jiyuan et al, 2008). In both calculation systems it was used the same numerical mesh.

3.2.1 *Ansys Fluent*

In Ansys Fluent system it was used pressure based, steady state calculation model. The propeller motion was simulated with rotating reference frame method. It was used SIMPLE solution algorithm and second order gradient schemes. Flow turbulence was calculated with "k-epsilon realizable" model. All parameters and numerical model details can be found in (Ansys, 2010). List of defined boundary conditions used in Ansys Fluent is shown in table 2.

Table 2. Boundary conditions types.

Patch name	Boundary condition
INLET	velocity inlet
OUTLET	outflow
BLADE+HUB	wall, rotating with reference frame
OUTER	slip wall
CYCLIC	periodic

3.2.2 *OpenFOAM*

In OpenFOAM CFD toolbox there was used two solution methods. Most of the calculation were made with "MRFSimple" solver. This is steady state, incompressible, multi reference frame solver, based on SIMPLE algorithm. Additionally to validate numerical simulation, for advance coefficient J=0.64, flow was also calculated with "PimpleDyMFoam". This is much more complicated incompressible, transient solver, based on PISO-SIMPLE merged algorithm. Additionally this solver can operate with dynamic moving meshes.

Flow turbulence was calculated with "k-omega SST" model.

It was used "linear" gradient schemes, and limited second order divergence schemes. Laplacian terms were treated with second order conservative schemes.

Computations periodicity was simulate with "Arbitrary Mesh Interface (AMI)", which is numerical method designed to operate with non-conformal patches. All detailed information about model and numerical methods can be found in official documentation (OpenFOAM, 2012). Boundary conditions defined in this case were listed in table 3.

Table 3. OpenFOAM boundary condition types.

Patch name	U	p	nut	k	omega
INLET	fixed Value	zero Gradient	calculated	Fixe Value	fixed Value
OUT-LET	inlet Outlet	fixed Value	calculated	inlet Outlet	inlet Outlet
BLADE, HUB	fixed Value	zero Gradient	nutk Wall Function	kqR Wall Function	omega Wall Function
OUTER	slip	slip	slip	slip	slip
CYCLIC	cyclic AMI	cyclic AMI	cyclic AMI	cyclic AMI	cyclic AMI

4 RESULTS

The "MRFSimple" solver from OpenFOAM CFD toolbox appeared to be reliable calculation tool for simulating flow around rotating propeller. The solver was stabile. Forces and moments stabilized after about 3000 iterations, which is comparable to commercial computational systems. Exemplary stabilization process of moment acting on a propeller blade is shown on Figure 5. Calculation residuals decreased to acceptable level, what is shown on Figure 6.

The "PimpleDyMFoam" solver generated similar results. Difference between "MRFSimple" and "PimpleDyMFoam" in values of forces acting on blade was about 2%. Only the time of calculation in case of transient solver was much longer (more than 10 times).

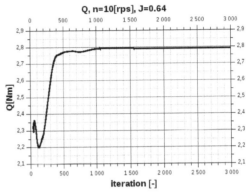

Figure 5. Stabilization of torque acting on a blade during solution process, J=0.64.

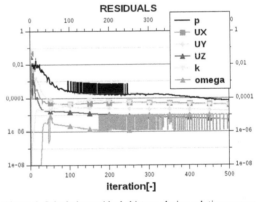

Figure 6. Calculation residuals history during solution process.

Propeller thrust coefficient characteristics is shown on Figure 7. Agreement between simulation and experiment results is very good. The average relative difference in results is 5%. However for large values of advance coefficient (J>1,5), when the thrust force is relatively small, the difference increase up to 58%. But this fact has no significant impact on shape of whole $K_T(J)$ characteristic. On Figure 7 there are also presented results from Ansys Fluent system. Their agreement to experimental results is very good as well.

The torque coefficient agreement between CFD and experimental results is very impressive. The comparison of the results is shown on Figure 8.

Average relative difference in that case is less than 4% and there is no increase in difference for large advance coefficient values. Figure 8 shows also the torque coefficient characteristic calculated with Ansys Fluent. This results are as good as others.

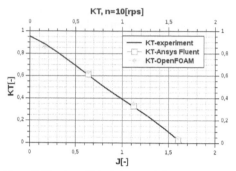

Figure 7. Thrust coefficient characteristics.

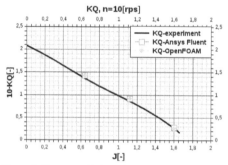

Figure 8. Torque coefficient characteristics.

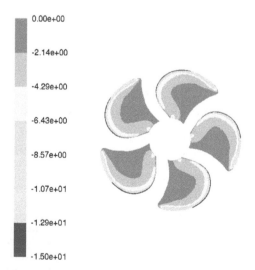

Figure 9. Pressure coefficient distribution on propeller blades (suction side) - Ansys Fluent.

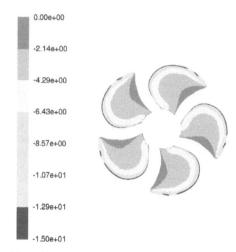

Figure 10. Pressure coefficient distribution on propeller blades (suction side) - OpenFOAM.

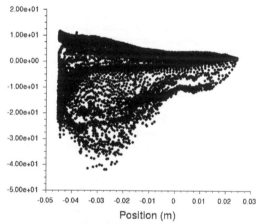

Figure 11. Pressure coefficient distribution on propeller blade in function X position - Ansys Fluent

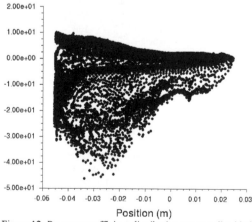

Figure 12. Pressure coefficient distribution on propeller blade in function X position - OpenFOAM

Figures 9 and 10 show the contours of pressure coefficients distribution on suction side. Figure 11 and 12 present values of pressure coefficient in function of X coordinate. Both sides (suction and pressure) have similar distribution for OpenFOAM and Ansys Fluent solver. The agreement is satisfactory. Relative differences are 9% for maximal and minimal values of calculated pressure coefficient.

5 CONCLUSIONS

Presented research confirms that OpenFOAM CFD toolbox can be successfully applied to evaluate propeller characteristic. The agreement between simulation and experimental results is very good. Flow and pressure fields calculated with OpenFOAM are not significantly different than those obtained from other commercial computational systems, like Ansys. OpenFOAM can be reasonable alternative for those who need flexible and not expensive CFD software.

ACKNOWLEDGMENTS

Calculations have been carried out in Wroclaw Centre for Networking and Supercomputing (http://www.wcss.pl), grant No 223.

REFERENCES

Carlton J, 2007, Marine Propellers and Propulsion, Elsevier
Klasson O.K. & Huuva T, 2011, Potsdam Propeller Test Case (PPTC), Second International Symposium on Marine Propulsors, Germany,
Wilcox D. C., 2002, Turbulence Modeling for CFD, DCW Industries inc.,
Jiyuan Tu & Guan Heng Yeoch & Chaoqun Liu, 2008, Computational Fluid Dynamics A practical Approach, Elsevier,
Ansys, 2010, ANSYS FLUENT User's Guide release 13, ANSYS, Inc
.Barkmann U. H., 2011, Open Water Tests with the Model Propeller VP1304, Schiffbau-Versuchsanstalt Potsdam GmbH, http://www.sva-potsdam.de/pptc.html
OpenFOAM, 2012, User Guide, OpenFOAM Foundation, http://www.openfoam.org/

On the Characteristics of the Propulsion Performance in the Actual Sea

J. Kayano & H. Yabuki
Tokyo University of Marine Science and Technology, Tokyo, Japan

N. Sasaki
National Maritime Research Institute, Mitaka, Japan

R. Hiwatashi
National Institute for Sea Training, Yokohama, Japan

ABSTRACT: The weather routing is widely adopted by oceangoing merchant ships in order to minimize the distance traveled and fuel consumption. The effectiveness of the weather routing mainly depends on the accuracy of weather forecast data and the ship's propulsion performance prediction in the actual sea. The authors performed full-scale experiments using a training ship in order to investigate the propulsion performance characteristics in the actual sea. This paper describes the results of the analysis on the Power Curves and Self Propulsion Factors under various weather and sea conditions.

1 INTRODUCTION

From the view point of earth environment protection, the shipping industry is required to develop and improve energy saving navigation technologies. The weather routing is one of the above technologies and it is often used for the navigation planning of oceangoing merchant ships.

The effectiveness of the weather routing on the energy saving mainly depends on the accuracy of the weather forecast data and that of the propulsion performance prediction in the actual sea where the effect of wind and wave on ship's motion exists. The weather forecast technology has been improved year by year and an easier method to obtain the worldwide accurate weather forecast data has been proposed (Yokoi 2010). On the other hand, the propulsion performance in the actual sea is usually predicted using the Self Propulsion Factors obtained by model tests due to the small amount of full-scale experiment data in the actual sea (Sasaki 2009, Tsujimoto 2000). In general, ship's speed in the actual sea is low compared to the speed measured at the speed trial in the still water. In order to improve the accuracy of a propulsion performance prediction, it is necessary to understand the effect of external disturbances such as wind and wave on the propulsion performance qualitatively.

From the above points of view, the authors conducted an experimental study using a training ship in order to investigate the characteristics of propulsion performance under various weather and sea conditions. In the study, the effects of winds and waves on the propulsion performance are analyzed separately according to the wind direction and wind force. The characteristics of Self Propulsion Factors in the actual sea are also examined. This paper describes the characteristics of a Power Curve (speed–BHP curve) and wake coefficient $(1 - w_t)$ in the actual sea compared with those obtained by model tests.

2 FULL-SCALE EXPERIMENTS

The test ship was a 6,720 G.T. training ship Ginga Maru and her principal particulars are shown in Table 1. A precise shaft horsepower meter with a thrust measurement function has been installed in the test ship.

Table 1. Principal particulars of the test ship

Hull	
Length: Lop (m)	105.00
Breadth: B (MLD, m)	17.90
Depth: D (MLD, m)	10.80
Cb	0.5186
Draft: d (m)	5.96
Main Engine	
Diesel	1 set
MCR	6,620 kW x 167 rpm
Propeller (CPP)	
Prop. Brade. No.	4
Prop. Dia.: Dp (m)	4.30
P.R. (Brade Angle)	0.9965 (24.4°)

The experiments were performed during her 2-month annual training cruises (from the middle of July to the middle of September) between Japan and Hawaiian Islands conducted years 2008, 2010 and 2011. The tracks of her 3 training cruises are shown in Figure 1. In the experiments, the propulsion performance data, engine operation data and navigation data were recorded automatically every 10 seconds using the Local Area Network System.

And, the following data were used for the analysis; main engine handle notch, propeller revolution, BHP (brake horse power), torque, thrust, ship's position, heading, speed, wind direction and wind velocity.

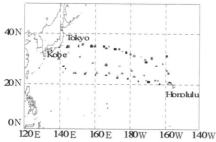

Figure 1. Track chart

In the propulsion performance analysis, the above raw data are processed according to the following procedure.

1. Identified first were steady parts of the raw data where the test ship proceeded at the steady condition continuously for two hours. The steady condition is judged by the main engine handle notch, propeller revolution, ship's heading, speed, wind direction and velocity.

2. The basic data are made by calculating average values of the steady parts mentioned above and these average values were divided into 5 wind directions relative to the ship's head and stern centerline as shown in Figure 2. In the figure, the word "Oblique Head Wind" indicates the port or starboard bow wind.

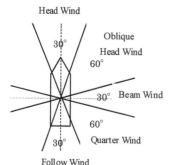

Figure 2. Definition of wind direction

3. For each of the wind directions, the mean value of basic data were calculated separately according to the Beaufort scale and were used for the analysis.

3 CHARACTERISTICS OF POWER CURVE

In order to investigate the propulsive characteristics of the test ship, the obtained data are compared with the Power Curve obtained by the power prediction using the model ship and the one obtained by the speed trial of the test ship as shown in Figure 3. The Power Curve obtained by the speed trial agrees well with that by the power prediction. The measured BHP values were transformed so that they were comparable with those obtained in the model test condition (displacement; 5,763 tons) using the 2/3 power rule based on the idea of the Admiralty Coefficient. The plotted data indicate mean values calculated according to the procedure described in the previous section. The obtained data were found to be in good order.

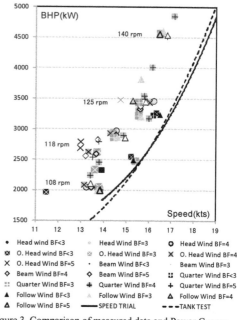

Figure 3. Comparison of measured data and Power Curves

Next, the authors examined the effect of wind on the propulsive characteristics for each of the wind directions shown in Figure 2. The comparison results of the Power Curves are shown in Figure 4 to 8. In the figures, Power Curves are shown for each of the Beaufort scale. On each of the wind directions, the Power Curves generally agree with those of speed trial and power prediction qualitatively.

When the test ship proceeds at steady BHP, her speed decreases as the wind force increases, and the degree of speed reduction decreases as the wind direction changes to afterward. The degree of speed reduction in the higher BHP region is greater than that in the lower BHP region. When the test ship proceeds under gentle sea condition (the range of the Beaufort scale less than 3), her speed is slow compared with that of speed trial and the two Power Curves are not in agreement. The same characteristics of the Power Curves as above were observed in the follow wind, and the effect of wave is considered to be one of the causes.

As the above propulsive characteristics are observed in the actual sea, it is important to predict a Power Curve taking into account the effect of wind direction, wind force and wave.

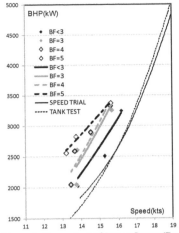

Figure 6. Comparison of Power Curves (Beam Wind)

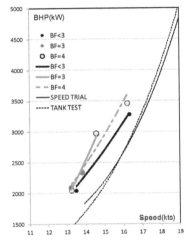

Figure 4. Comparison of Power Curves (Head Wind)

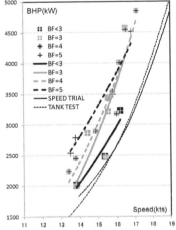

Figure 7. Comparison of Power Curves (Quarter Wind)

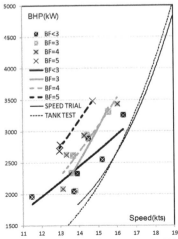

Figure 5. Comparison of Power Curves (Oblique Head Wind)

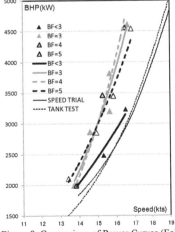

Figure 8. Comparison of Power Curves (Follow Wind)

255

4 CARACTERRISTICS OF SELF PROPULSION FACTORS

Although the effect of wind and wave is included in the propulsion performance data shown in the previous section, obtained Power Curves agree well with that obtained by the speed trial qualitatively. The authors examined the characteristics of Self Propulsion Factors using the data in the region of 16 knots where enough data were obtained.

4.1 Self Propulsion Factors prediction

When the actual value of thrust (T) and torque (Q) are measured at a ship, the propulsive efficiency (η_D) can be calculated using equation (1). Self Propulsion Factors such as wake coefficient ($1-w$) and relative rotating efficiency (η_R) can be also calculated using the propeller characteristic curve obtained by the propeller open test.

$$\eta_D = (T \cdot V_s)/DHP$$
$$DHP = BHP \cdot \eta_T \tag{1}$$

where V_s =speed; DHP=delivered horse power; BHP=brake horse power; and η_T=transmission efficiency

The procedure to obtain Self Propulsion Factors using the propeller open test results, speed, propeller revolution, BHP, thrust and torque is shown in Figure 9. When predicting the Self Propulsion Factors, it is necessary to measure the data accurately in an appropriate time interval and to know the effect of ship's condition such as displacement and trim on propulsion performance.

$Vs(kts), N(rpm), BHP(kW), T(kN), Q(kN \cdot m)$

⇩

$$\eta_D = \frac{T \cdot Vs}{2\pi(N/60)Q} \quad K_T = \frac{T}{\rho(N/60)^2 D^4}$$

$$K_Q = \frac{Q}{\rho(N/60)^2 D^5} \quad 1-w_T = \frac{(N/60) \cdot D \cdot J_T}{V_s \times 0.5144}$$

$$1-w_Q = \frac{(N/60) \cdot D \cdot J_Q}{V_s \times 0.5144} \quad \eta_R = \frac{K_{Q0}}{K_Q}$$

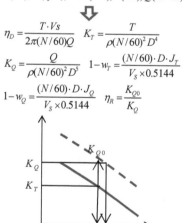

Figure 9. Calculation procedure to obtain Self Propulsive Factors

4.2 Effect of displacement

When discussing the effect of displacement on the propulsion performance, it is necessary to examine the effect of both mean draft and trim. The authors performed the analysis considering the effect of displacement alone due to scarce model test data on the trim.

The Power Curves of the test ship at full load (6,308 tons) and 75 % load (5,763 tons) obtained by the model tests are shown in Figure 10. Also shown in the figure are the Power Curves predicted under the same conditions using the propulsion performance prediction program "HOPE Light," which was developed by the National Maritime Research Institute Japan (Sasaki 2009). As the predicted values agree well with the observed values in the region of 16 knots, the HOPE Light can be used as a tool to discuss the characteristics of Power Curve in the waves under the above conditions.

From the Power Curves shown in Figure 10, the authors considered that the measured horse power value can be transformed into the horse power at the displacement in the model test using the equation (2).

$$BHP' = BHP \cdot (\frac{16}{V_s}) \cdot (\frac{5763}{Displacement})^{0.9} \tag{2}$$

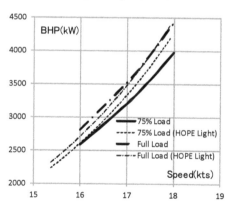

Figure 10. Power Curves obtained by model test and HOPE Light

4.3 Effect of wind and wave

The authors estimated the DHP under wind and wave disturbances using HOPE Light and the results are shown as the dotted line in Figure 11 together with measured DHP. Measured DHP are transformed into the value at the standard condition (displacement; 5,763 tons, speed; 16 knots) using the equation (2).

As shown in the figure, measured DHP data tend to increase in proportion to the increase of wind velocity and wave height, and this is mainly due to the increase of resistance by wind and wave.

However, on the distribution of measured data in the region of stronger wind and higher wave, regularity is difficult to be observed. Figure 12 shows the comparison results of DHP between measured and estimated. It seems that estimated DHP is lower than measured DHP in the region mentioned above.

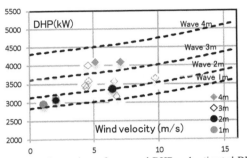

Figure 11. Comparison of measured DHP and estimated DHP for wind velocity

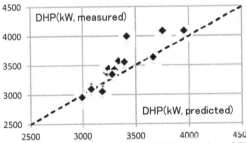

Figure 12. Comparison of measured DHP and estimated DHP by HOPE Light

In order to investigate the cause of the above results, the authors calculated propulsive efficiency (η_D) using measured BHP, thrust and speed according to the equation (1), and the obtained results are plotted against the propeller loading $(C_T = T/0.5\rho \cdot V_a^2(\pi D^2/4))$ as shown in Figure 13. The propulsive efficiency (η_D) decreases in proportion to the propeller loading. In the figure, estimated propulsive efficiency (η_{D0}) using the following simple equation that takes into account only the change of propeller efficiency (η_0) is also shown in dotted line.

$$\eta_{D0} = \eta_H(const) \cdot \eta_R(const) \cdot \eta_0(C_T) \qquad (3)$$

where η_H =Hull efficiency; and η_R= Relative rotating efficiency

As is obvious from the figure, measured propulsive efficiency values (η_D) are generally lower than estimated values (η_{D0}). The difference between measured values and estimated values is about 5 % on average and this value is difficult to be disregarded. The reduction of the propulsive efficiency under wind and wave disturbance seems not to be only due to the increase of propeller loading.

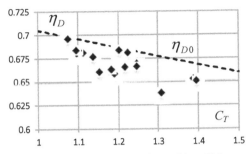

Figure 13. Comparison of measured propulsive efficiency and estimated propulsive efficiency

4.4 Wake coefficient at actual sea

The authors examined the relationship between wake coefficient $(1 - w_t)$ and propeller loading (C_T) in order to analyze the cause of propulsive efficiency reduction shown in Figure 13.

Figure 14 shows the relationship between propeller loading and wake coefficient in the region of 16 knots. Although some dispersion is observed in the data, it seems that wake coefficient increases in proportion to the propeller loading. As the propeller loading will be changed by wind, wave and displacement, the difference of the displacement is considered to be one of the causes of wake coefficient increasing.

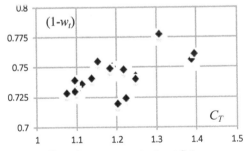

Figure 14. $(1 - w_t)$ and propeller loading (C_T)

The authors investigated the relationship between displacement and wake coefficient as shown in Figure 15. In the figure, the wake coefficients obtained by the model test are also displayed as the dotted line. The wake coefficients obtained by the full-scale experiments increase in proportion to the displacement. On the other hand, the wake coefficients obtained by the model tests in this region remain almost constant.

Therefore, there is a possibility to explain the cause of difference between measured (η_D) and estimated (η_{D0}) shown in Figure 13 by the effect of displacement on the wake coefficient. Further model tests seem to be necessary in order to clarify the effect of displacement on the wake coefficients.

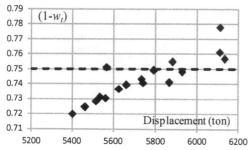

Figure 15. $(1 - w_t)$ and displacement

5 CONCLUSION

The authors performed an experimental study in order to clarify the characteristics of propulsion performance in the actual sea. Results obtained in this study are summarized as follows.

1 The propulsion performance of a ship in the actual sea with wind and wave disturbances decreases compared with the performances obtained by power prediction and a speed trial. The degree of propulsive power reduction depends on the wind direction, wind force and wave height.

2 As the propulsion performance decreases compared with the speed trial results even if a ship proceeds under the follow wind, the effect of wave is considered to be greater than that of wind.

3 Causes of propulsive efficiency reduction in the actual sea can be divided into the unavoidable reduction of propeller efficiency due to the resistance increase by wind and wave and others.

4 In order to determine the characteristics of Self Propulsion Factors in the actual sea, it is important to measure the thrust and calculate the propulsive efficiency directly.

5 The wake coefficient $(1 - w_t)$ in the actual sea can be estimated directly by measuring her thrust. However, as the accuracy of thrust measurement is generally inferior to that of the torque measurement, it is necessary to examine their mutual relation beforehand.

6 The effect of propeller loading on the wake coefficient in the actual sea is small and the ship's condition such as displacement and trim seems to have a larger impact on the wake coefficient. The authors consider that the above results are necessary to be examined by model tests.

7 In order to improve the performance of a weather routing, it is important to predict the Power Curve more precisely taking into account the characteristics of the effect of wind and wave obtained in the present study.

REFERENCES

Yokoi, T. et al. 2010. A mechanism on parallel processing to numerical weather prediction for weather routing – Accuracy evaluation and performance benchmark of the sea surface wind prediction-, *Proceeding of Asia Navigation Conference* 2010; 33-40

Sasaki, N. et al. 2009. Development of ship performance index (10 mode at sea), *Report of National Maritime Research Institute, Vol.9 No.4*; 1-46 (in Japanese)

Tsujimoto, M. et al. 2000. Evaluation of ship propulsive performance by analyzing full-scale data and time history of ship motion on actual seas, *Proceeding of 4th Osaka Colloquium on Seakeeping Performance of Ship* 2000; 1-9

Sasaki, N. et al. 2010. Development of basic design tools for high performance ships, *Report of National Maritime Research Institute, Vol.10 No.3*; 1-21 (in Japanese)

Propulsion and Mechanical Engineering
Maritime Transport & Shipping – Marine Navigation and Safety of Sea Transportation – Weintrit & Neumann (Eds)

Engine Room Simulator (ERS) Training Course: Practicability and Essentiality Onboard Ship

R.A. Alimen
John B. Lacson Foundation Maritime University-Molo, Iloilo City, Philippines

ABSTRACT: This study determined the essentiality and practicability of the Engine Room Simulator (ERS) Training Course onboard ship among special program cadets of JBLFMU-Molo, Iloilo City, Philippines. This employed the qualitative research method where data are gathered through an interview and the subjects were the special program cadets who had taken the ERSTC and had undergone apprenticeship onboard an international vessel. The participants were the ten (10) engine cadets of the special program specifically the Norwegian Ship-owners Association (NSA) Cadets of JBLFMU-Molo, Iloilo City, Philippines taking up marine engineering, which had taken the Engine Room Simulator Training Course (ERSTC) and had undergone apprenticeship onboard international vessel. As whole, the ERS Training Course is essential onboard ship/ in a manner, that most of the vessels are computer based or UMS. It gives basic idea and knowledge on the operations and functions of the machineries and equipment in a specific system onboard, gives experiences on how to trouble shoot and rectify and make the mastery of operating procedure easy like starting and stopping of the main engine, synchronizing of generators. Furthermore, ERS Training Course is very practicable on the UMS vessels and essential on the manned machinery space when taken as whole. As such, the machineries and equipment, operation and functions are the same onboard even though that the positions are complicated not on the simulator that it is fixed and organized. Lastly, it is applicable onboard regardless on the types of vessel, kind of cargo carried, and mode of operations.

1 BACKGROUND AND THEORETICAL FRAMEWORK OF THE STUDY

At the time that man started discovering things around him, many innocent people were amazed. Starting from the discovery of fire upon the ignition of two stones that creates flame, to the invention of gunpowder by the Chinese people.

After several years of evolution, technology had created a great change in life on land and in water. Men had created highly complicated gadgets, the development of machines, treatments in the field of medicine, and in science and technology.

Now, even on board the ship technology had really affected the life of many seafarers. Just imagine the kind of ships 30-40 years ago. During those times everything is being done manually. For example, when an alarm is heard everybody must go down the engine room to trace the exact place where a deficiency is spotted. But now, even inside one's cabin, one can immediately track the place where the alarm started. Because of the sophistication of technology, most international vessels are under a special operational system, which is widely known as the "UMS" or the Unmanned Machinery Space. This system helps most marine engineers do their work easier. Through this system, they are not obliged to monitor everything in the engine department from time to time. If the crew on duty can already stay inside his cabin while doing his duty at the same time and if the alarm is heard he can immediately determine where the alarm is coming from.

With the continuous development of technology, ships became complicated and highly powered with their machineries and gadgets. And so, in order to be competent in using these machineries one must undergo trainings, seminars and special courses that could comply with the standards of these vessels.

With the rapid development of technology, a new and better training course is introduced which gives the new generation of marine engineers the idea and knowledge with the usage of these technologies. Such course is known as the "Engine Room Simulator Training Course." The ERSTC is an upgrading course offered by a school to the future

Engine officers with the functions and usage of the machinery and equipment in the engine room and also enhances the abilities and competency of the engineers.

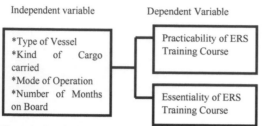

Independent variable   Dependent Variable

Figure 1. Practicability and Essentiality of ERS Training Course

2 STATEMENT OF THE PROBLEM

This study determined the essentiality and practicability of the Engine Room Simulator training Course onboard ship among special program cadets of JBLFMU-Molo Inc. This study also aimed to answer the following questions:

1 Is the Engine Room Simulator training Course essential to special program cadets of John B. Lacson Foundation Maritime University (JBLFMU)-Molo Inc. when taken as a whole?
2 Is the Engine Room Simulator training Course essential to special program cadets of JBLFMU-Molo Inc. when grouped according to a) Type of Vessel; b) Kind of Cargo Carried; c) Mode of Operation?
3 Is the Engine Room Simulator training Course practicable to special program cadets of JBLFMU-Molo Inc. when taken as a whole?
4 Is the Engine Room Simulator training Course practicable to special program cadets of JBLFMU-Molo Inc. when grouped according to a) Type of Vessel; b) Kind of Cargo Carried; c) Mode of Operation?
5 How does the Engine Room Simulator training Course help the special program cadets of JBLFMU-Molo?

3 SIGNIFICANCE OF THE STUDY

The researchers believed that this study is beneficial to the following:

JBLFMU community and Administrator. This shall be the basis in enhancing and promoting the quality of education and learning among marine engineering students, specially the special program cadets onboard ship.

JBLF Training Center. This will give an insight about the essentiality and practicability of the Engine Room Simulator training Course onboard a ship in improving the quality of learning and training.

Student. They will be able to appreciate and give more importance to the Engine Room Simulator training Course on its essentiality and practicability onboard ship. Also this will give an idea about the said course.

For Future Use. This will give information about the Engine Room Simulator training Course, on its essentiality and practicability onboard ship to the researchers who find interest to pursue the same study.

4 RESEARCH DESIGN

This employed the qualitative research method where data are gathered through an interview and the results were compared.

The purpose of this study was to determine the essentiality and practicability of the Engine Room Simulator Training Course onboard ship among the Special Program Cadets of JBLFMU-Molo Inc. The qualitative research method was employed in the study. This type of method employed an interview method in gathering of data.

The variables are: Essentiality and Practicability of Engine Room Simulator Training Course used as dependent variable and Type of Vessel, Kind of cargo, Numbers of month onboard, and Mode of Operation as independent variable.

The participants were the ten (10) engine cadets of the special program specifically the Norwegian Ship-owners Association (NSA) Cadets of JBLFMU-Molo Inc., Iloilo City taking up marine engineering, which have taken the Engine Room Simulator Training Course and have undergone apprenticeship onboard international vessel, during this second semester of the school year 2007-2008 and second semester of the school year 2008-2009.

5 THE PARTICIPANTS

The participants were the ten (10) engine cadets of the special program specifically the Norwegian Ship-owners Association (NSA) Cadets of JBLFMU-Molo Inc., Iloilo City taking up marine engineering, which have taken the Engine Room Simulator Training Course and have undergone apprenticeship onboard international vessel.

6 DATA-GATHERING INSTRUMENT AND STATISTICAL TOOLS

The research procedure involved the preparation of the study instrument, choosing the participants, data processing and analysis.

The participants were chosen randomly and the interview technique was employed because the researchers chose the qualitative type of research, using a qualitative-questionnaire made by the researchers and approved by the adviser.

The research procedure involved the preparation of the study instrument, choosing the participants, data processing and analysis.

The participants were chosen randomly and the interview technique was employed because the researchers chose the qualitative type of research, using a qualitative-questionnaire made by the researchers and approved by the adviser. The researchers asked the interviewees the following questions: (1) Is the Engine room Simulator training course essential onboard ship? Why? (2) Is the Engine Room Simulator Training Course practicable onboard ship? Why? (3) How does the Engine Room Simulator Training Course help you onboard ship? (4) Is there a difference on the machineries and equipment onboard and on the Engine Room Simulator Training Course? and (5) Why did you take the Engine Room Simulator training Course? The data gathered were then compared.

7 RESULTS OF THE STUDY

The interviewee number one (1) answers on the question asked by the researcher and was analyzed based on qualitative method. Interviewee number one (1) has already taken the ERSTC (Engine Room simulator Training Course) and boarded an Oil-Chemical tanker vessel for ten (10) months carrying finished products of oil like gasoline, LPG, LNG, etc. The mode of operation of machinery is unmanned machinery space (UMS). For him, the ERSTC is very essential and practicable onboard for the reason that most of the vessels today use UMS. The simulator gives him knowledge and basic ideas of the correct procedures in operating machineries and different systems which are carried onboard. It made him familiarize because the machineries onboard are the same in the simulator but some fittings are not found in his vessel.

The results gathered by the researcher on interviewee number two (2) were shown in the following sections. Interviewee number two (2) has already taken the ERSTC and boarded a General Cargo Vessel carrying bulk, ore, etc for almost eleven (11) months. The mode of operation of the machineries is UMS.

The interviewee number three (3) the data gathered were shown on the following sections. Interviewee number three (3) has already taken the ESRTC and boarded a Tanker vessel for almost ten (10) months carrying LPG, LNG, etc. The mode of operation of machineries is UMS.

Interviewee number four (4) has already taken the ERSTC and boarded a General Cargo Ship for almost ten (10) months carrying all forest products like lumber, wood, etc. The mode of operation of the machineries is manned machinery space.

The results gathered by the researcher from the interviewee number five (5) were shown in the following sections. Interviewee number five (5) has already undergone an ERSTC and boarded a Bulk vessel for almost 11 months carrying ore, bulk, etc. and the mode of operation of machineries is a manned machinery space.

The results gathered by the researcher on interviewee number six (6) were shown in the following sections. Interviewee number six (6) has already taken the ERSTC and boarded a General Cargo Vessel carrying bulk, ore, etc. for 12 months and 2 days. The mode of operation of the machineries is UMS.

The results gathered by the researcher from interviewee number seven (7) were shown in the following sections. Interviewee number seven (7) has already taken the ERSTC and boarded an Oil Chemical Tanker carrying palm oil, gas oil and molasses for almost ten (10) months. The mode of operation of the machineries is UMS.

The results gathered by the researcher on interviewee number nine (9) were shown in the following sections. Interviewee number nine (9) has already taken the ERSTC and boarded a General Cargo Vessel carrying pulp and different kinds of metals for 12 months and 8 days. The mode of operation of the machineries is manned machinery space.

The results gathered by the researcher on interviewee number ten (10) were shown in the following sections. Interviewee number ten (10) has already taken the ERSTC and boarded an Oil Chemical Tanker carrying various oils for almost eleven (11) months. The mode of operation of the machineries is UMS.

8 CONCLUSIONS

Based on the interviews, the qualitative results lead the researcher to conclude that:

As whole, the ERS Training Course is essential onboard ship/ in a manner, that most of the vessel is computer based or UMS. Also, it gives basic idea and knowledge on the operations and functions of the machineries and equipment in a specific system onboard. When the system fails, it gives experiences on how to trouble shoot and rectify and make the mastery of operating procedure easy like starting and stopping of the main engine, synchronizing of generators, etc.

The same qualitative findings shared by the respondents during the interview when they were grouped according to type of vessel, kind of cargo carried and mode of operations, the ERS is essential onboard ship.

Furthermore, ERS Training Course is very practicable on the UMS vessel and practicable on the manned machinery space when taken as whole. As such, the machineries and equipment, the operation and functions are the same onboard even though that the position are complicated not on the simulator that it is fixed and organized.

9 IMPLICATIONS FOR THEORY AND PRACTICE

The Engine Room Simulator (ERS) Training Course is applicable onboard regardless on what type of vessel, kind of cargo carried, and mode of operations.

10 RECOMMENDATIONS

Based on the findings of this study, the researchers arrived at the following recommendations:

The administrator and the head of the JBLF Training Center must give importance on the ERS Training Course. They should open the Engine Room Simulator to everybody, so that the student could practice on the operation of the machineries and equipment on board ship in the time they were available or must be added to the curriculum of the Marine Engineering Course.

For the school, they should maintain the computer and the equipment in good condition and additional computer to occupy more students.

To Instructors and Assessors of the ERSTC of JBLF Training Center, they should be strict to the student in assessing and must improve his teaching skills.

In addition, students must take the ERS seriously so that they could learn more about the operations and functions of the different machineries onboard.

REFERENCES

Baria, R.O. (2004). Electronics Technology As A Subject: Attitude and Performance Among Marine Engineering Students. Unpublished Thesis. John B. Lacson Foundation Maritime University-Molo, Iloilo City.
Careta, T. and Dunlap, R. (1998). Transfer of Training Effectiveness in Flight Simulation.
Doyle, E. (2009). Reconstructing A Marine Casualty: The Effectiveness of the Full-Mission Simulator As A Casualty Analysis Tool. Marine Navigation and Safety of Sea Transport. Taylor and Francis Group, London, U.K., pp.69-74.
Gattiker, J. (2005). "Using the Gaussian Process Model for Simulation Analysis Code", Los Alamos technical report LA-UR-05-5215.
Glimm, C., Higdon, G., Schultz, S. (2005). "Error Analysis in Simulation of Complex Phenomena", Los Alamos Science special issue on Science-Based Prediction for Complex Systems, no.29, pp.6-25.
Hu. Y. & Wan, B. (2000). A Simulation Study on Diesel Engine Performance Failure. Proc ISME Tokyo. Vol. 2., 797-803.
Jaleco, V. (2004). Teaching-Learning Situation in Maritime Schools in Western Visayas. Unpublished Dissertation. University of San Agustin, Molo, Iloilo City.
Lobaton, J. A. (2003). Students' Mathematics Performance As Moderated by Their Attitude Towards the Subject. Unpublished Thesis. John B. Lacson Foundation Maritime University-Molo, Iloilo City.
McKellop, H.A. & D'Lima, D. (2008). How have wear testing and joint simulator studies helped to discriminate among materials and designs? J Am Acad Orthop Surg, Vol 16, No suppl_1, July 2008, S111-S119.
Olsen, N., et al. (2006). The Swarm End-to-End mission simulator study: A demonstration of separating the various contributions to Earth's magnetic field using synthetic data Earth Planets Space, 58, 359–370.
Porras, E.P. (2004). Teaching Performance Towards Management, Work and Study Conditions, and Communication Climate Among the Faculty of a Maritime Institution. Unpublished Thesis. John B. Lacson Foundation Maritime University-Molo, Iloilo City.
Roof and Kapos Associates. (1996). Trade-Offs Between Live and Simulated Training.
RuppeL. F. and Wysor, W. (1997). Guidelines for Simulator-Based System Testing.
Towne, D. M. (1995). Learning and instruction in simulation environments. Englewood Cliffs, NJ:Educational Technology Publications.
Tumala, B., Trompeta, G., Evidente, L., and Montaño, R. (2008). Impact of Simulator Training on Cognition Among Marine Engineering Students. JBLFMU Research Review. Volume XVIII. Number 1. pp.65-87.
White, S. R., & Bodner, G. M. (1999). Evaluation of computer simulation experiments in a senior-level capstone course. Chemical Engineering Education, 33(1), 34-39.
Zalewski, P. (2009). Fuzzy Fast Time Simulation Model of Ship's Maneuvering. Marine Navigation and Safety of Sea Transport. Taylor and Francis Group, London, U.K. , pp. 75-78.

Contribution to Treatment System Deformed Highlighted a Network Connection Point of Medium and High Voltage

V. Ciucur

Constanta Maritime University, Constanta, Romania

ABSTRACT: Distorting regime is a permanent working AC networks, wherein at least one of the time variations of voltage or current is not sinusoidal. If voltage is sinusoidal, distorting regime appears only exists receiving non-linear (distorted), such as limiters, inductance, receivers flaming arc, discharge lamps gas or metal vapor, etc. If the voltage is harmonic, such as the saturated magnetic circuit transformers, distorting regime may be increased by the presence of capacitors in the circuit, which, with small reactance at high frequencies, can lead to significant current harmonics and can create resonant circuits capacitor-network.

1 INTRODUCTION

The study for distorting regime is done by decomposing the harmonics voltage and current in periodic non-sinusoidal Fourier series. In the distorting regime, besides active power, reactive power and apparent power interferes distorting power.

Distorting regime has the following negative effects on power system:

- Show of voltage and current distortion at all levels between generators and receivers
  Distortion coefficient, showing regular size deviation from the sinusoidal shape is defined as the ratio between U_{def} deforming residue and nominal voltage U_n.

$$\delta_U = \frac{U_{def}}{U_n} \qquad (1)$$

the residue distorted has an expression

$$U_{def} = \sqrt{\sum_{i=2}^{n} U_i^2}, \qquad (2)$$

where U_i is the real value of harmonic i.
Practical, for $\delta_U \leq 5\%$ the size can be considered sinusoidal.

- The many power losses and energy in networks and electric machines due to increased apparent power because power distorting.

- Resonance overvoltage caused by capacitor reactive power compensation and harmonic due to overheating of these capacitors.
- Couples parasitic at motors produced by harmonics 5, 7, 11, 13 etc..
- Additional measurement error of power measurement devices (up to 3%) and energy (up to 14%) due to different value under power factor deforming than sinusoidal, the same reactive power. The harmonics limitation do using filters with discharge or absorption. Pressing filter designed to avoid resonance between the network and existing capacitors to improve power factor using reactors in air (tuning coils) in series with the batteries.

In AC traction substations, capacitor banks are used in series with resistors, connected to bars that connect the substation to supply leaders contact wire, amending LC parameters of the contact line in order to avoid occurrence of resonance for any of superior traction current harmonics produced by the locomotive rectifiers (resonant frequency falls below 50 Hz).

Absorbent filter consisting of capacitors connected in series with the air reactors so calculated as to provide resonance for higher harmonics (5, 7, .) present in consumer bars distorted in this way, the filter will absorb all current harmonics for which it was granted. Calculation of these filters is based on the current or voltage harmonic analysis, performed using oscillograms.

2 CONTRIBUTION TO TREATMENT SYSTEM DEFORMED HIGHLIGHTED A NETWORK CONNECTION POINT OF MEDIUM AND HIGH VOLTAGE

2.1 *Indicators of the harmonics*

The AC wave form ideal voltage or current is sinusoidal, and dc is the ideal constant pressure to be perfect. Causes distortions can be found in some extent the producer or supplier of electricity, but more often they are consumers, especially those who are receivers and linear elements. Basically, distortion current and voltage waves would qualify as deviations from the ideal form, together with voltage pulses, but while distortions are periodic, non-periodic pulses usually are.

Deviations from ideal shape can be characterized on the basis of harmonic analysis, by highlighting harmonic oscillations with periods of full submultiples of the fundamental period, highlighted the real wave period.

Figure 1. The division for a period nonsinusoidal wave periodic in order to the harmonic analyze

Harmonic analysis of real wave voltage or current, obtained experimentally by with an oscillograph or registration involves identifying fundamental period T and divider it into equal parts 2p, so p parts on a half-period, as shown in Figure 1.

Of course, the number of divisions is limited below the maximum order of harmonics that must be identified: if it looking for p harmonics, the division must be at minimum 2p parties. Upper limiting the number of divisions is due to the possibilities of "reading" the function values analysis.

Where is noted by z (t) voltage or current wave subject to harmonic analysis, Fourier coefficients expressions known as:

$$A_N = \frac{1}{p}\sum_{k=1}^{2p} Y_k \cdot \sin Nk\frac{\pi}{p}, N \geq 1$$

$$B_N = \frac{1}{p}\sum_{k=1}^{2p} Y_k \cdot \cos Nk\frac{\pi}{p}, N \geq 1$$

(3)

and continuous component is calculated by the formula:

$$Y_0 = \frac{1}{2p}\sum_{k=1}^{2p} Y_k$$

(4)

where:

Y_k, k=1...2p, consecutive ordinates wave is analyzed as practiced divisions;

N–Order harmonics, already stressed that. N≤p.

Fourier development wave z(t) is presented first form:

$$y(t) = Y_0 + \sum_{N=1}^{p}(A_N \sin N\varpi t + B_N \cos N\varpi t)$$

(5)

or condensed expression of:

$$y(t) = Y_0 + \sum_{N=1}^{p} Y_N \sin(N\varpi t - \varphi_N)$$

(6)

the harmonic amplitude and phase are determined, respectively, by the relations:

$$Y_N = \sqrt{A_N^2 + B_N^2}$$

(7)

$$\varphi_N = arctg\left(\frac{-B_N}{A_N}\right)$$

(8)

and $\varpi = 2\pi/T$ is the angular frequency corresponding to the fundamental.

Number of divisions (2p) is important not only for maximum harmonic order that can determine but the accuracy each harmonic and for resultant wave y. If the number of divisions is too small, appears especially errors in the phases of harmonics. For an acceptable harmonic analysis is recommended sampling of voltage and current waves for a period of 80ms, so for four periods, with a minimum number of divisions for a period 2p = 64.

A harmonic analysis relevant and fast can be obtained as follows:

– the correction phase φN given by (8), the size

$$\Delta\varphi_N = \frac{\pi N}{p}$$

(9)

correction is of the form

$$\varphi'_N = \varphi_N - \Delta\varphi_N$$

(10)

where φ 'N are considered correct phase, which will be introduced in relation (6), instead φN phase affected by the error in calculating the relations (3), which approximates the integral relations;

– Calculated by summing harmonic wave and comparison with the results analyzed.

– By stopping calculating when the standard deviation of the values of the two waves, above, is minimal.

Establishing indicators for deviations wave characteristics from the ideal form, sine allowable values are based on either current or voltage harmonics of either the synthetic parameters of distortion.

There is tend to include the consequences of the harmonics characteristic sizes of telecommunications equipment, electromagnetic compatibility therefore, the deforming regime indicators.

If some indicators can be determined for both voltages and currents, there are indicators that make sense only for wave power.

2.2 The harmonics of order N

Is defined as the ratio, expressed as a percentage, between the Y_{eN} effective value harmonic and Ye effective value wave:

$$y_N = \frac{Y_{eN}}{Y_e} \cdot 100, [\%] \tag{11}$$

where Y_{eN} obtains by ratio with $\sqrt{2}$ on YN amplitude, calculated with relation (7) and effective value Ye for the analyzed wave can calculated from array by 2p values with relation:

$$Y_e = \sqrt{\frac{1}{2p} \sum_{k=1}^{2p} Y_k^2} \tag{12}$$

and using the result of a complete Fourier analysis

$$Y_e = \sqrt{\sum_{N=0}^{p} Y_N^2} \tag{13}$$

Sometimes, in the relation (11), instead of the effective use of Ye effective value of the fundamental Ye1.

2.3 Distorting residue

Is defined by the expression:

$$Y_{def} = \sqrt{\sum_{N=2}^{p} Y_{eN}^2} \tag{14}$$

can be easily calculated with the equation

$$Y_{def} = \sqrt{Y_e^2 - \left(Y_e^2 + Y_{el}^2\right)} \tag{15}$$

since the three values involved are already determined.

Without always considered directly, an indicator of non-sinusoidal load, deforming residue is used to determine the periodic wave distortion, harmonic. Is normal for certain types of equipment, it is natural to be considered however as a separate indicator.

2.4 Distortion factor (the total Harmonic Distortion THD) of a wave harmonic

Is defined as the ratio, expressed as a percentage of the residue deforming and effective value of the fundamental

$$\delta_Y = \frac{Y_{def}}{Y_{el}} \cdot 100, [\%] \tag{16}$$

Regime is now distorting aspect which is the largest breakdown limit values of indicators in regulations and standards. Factors are specified for the harmonic distortion for both input voltages and currents.

Internal rules limiting distortion factor of the voltage curves as follows:

$\delta_u \leq 8\%$ for low and medium voltage networks

$\delta_u \leq 3\%$ high voltage networks

Measured current (I) at the point of connection (PC) is the sum of harmonic current injected the actual consumer network (IHC) and from the mains (IHS).

To determine whether the consumer meets the harmonic emission regulations or standards necessary to separate the total current measured by the injected fuel system.

In general, the actual consumer injected current is predominantly but not always, in some cases from the current system can be important due to resonance phenomena that can appear in the modification of network parameters that occurs during such incidents short circuits.

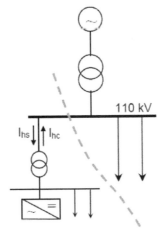

110 kV

I_{hs} I_{hc}

Harmonic voltage can be described by the relations:

$$\underline{V}_h(i) = \underline{Z}_s \cdot \left(\underline{J}_s - \underline{I}_h(i) \right) \qquad (17)$$

$$\underline{V}_h(i) = \underline{Z}_c \cdot \left(\underline{J}_c - \underline{I}_h(i) \right) \qquad (18)$$

where J_s and J_c is harmonic current from the source and the consumer, $\overline{Z}_s$ and $\overline{Z}_c$ is equivalent harmonic impedance on the source and the consumer.

One of the methods used to estimate the real part mark total harmonic impedance $Real \dfrac{\Delta V_h}{\Delta I_h}$ which will provide impedance value $\overline{Z}_s$ if the report is positive or value $\overline{Z}_c$ if the report is negative.

3 QUALITY CONDITIONS IN THE SUPPLY OF ELECTRICITY TO CONSUMERS

For proper operation of receivers, power supply must meet a number of conditions related to frequency, voltage, power and continuity. Detailed presentation of these conditions is systematizes below.

1 constant frequency voltage is a major challenge for both receivers functioning, maintenance precision measuring instruments and working machines driven by AC motors.

Frequency variations can be caused by variation in load or severe damage to the system, such cases can be originated and electricity consumers. Constant industrial frequency (50 Hz) is a system-level energy problem is related to the power reserve of power plant efficiency and dispatch system. In some cases, the possibility of generating electricity in power are limited, the supply tells consumers (killing distributors) in order to maintain the system frequency. Maximum permissible deviations by frequency are ± 0.5 Hz (1%).

2 constant voltage, the amount and form, is a precondition for any type of receivers.

It is recommended that the voltage across the receiver is constant and equal to the nominal or possible variations should be within the limits specified for each handset. In operation voltage electrical installations are variations caused by consuming due to changes in load or short circuits. Periodic variations can be slow due to load changes while receivers, and fast, also called fluctuations caused by rapid changes in pregnancy (e.g. arc furnaces, welding machines, rolling mills, compressors, machinery and others pulsating torque), including those due to connection - disconnection of receivers.

A permanent reduction in value voltage may result to go down in value for the conductors sections, a situation with negative consequences such as: electrical insulation damage, malfunction and over-heating equipment and piping receivers. Voltages higher than nominal causes overload operation of receivers reduce the life force and light receivers. Voltage decrease under nominal value attracts heat density, operation at lower parameters, or the receivers not working. As indicated voltage unilateral variation of voltage are drop-away voltage or over-potential of short duration. Is used as drop-away voltage for any decrease in voltage electrical network, with an amplitude in the range (0.2 ÷ 0.9) for nominal voltage and a duration of maximum 3 s.

Of receivers and sensitive installations drop-away voltage, the following:
− Motors and synchronous compensators;
− Asynchronous Motors (depending on torque resistant feature);
− Electronic equipment, including controlled rectifiers;
− Contactors of 0.4 kV and secondary circuits;
− Automatic or protecting circuits bottlenecks and technological adjustments.

The problem arises form the voltage shaped if DC powered receivers and if the supplied AC.

DC voltage across the DC receivers can have a series of harmonics, especially if the power supply is a rectifier ordered. Harmonic content is restricted according to their effects on the receiver, specifying permissible distortion coefficient.

Deviation from the sinusoidal voltage wave form determines receivers AC operation under distorted. While some receivers, such as induction ovens, wave voltage harmonics in mind not to others - including electric motors - voltage harmonics also be limited by specifying the allowed distortion coefficient.

3.1 Harmonics are a second product of modern equipment to control electric current

Propulsion systems are generally used for positioning and operation ship. In a great measure, these systems are electrically driven due to the need for precise control of speed. Propulsion systems consume a lot of energy and is often a significant part in the task of damp harmonics generator. Active filters have considerable sizes to meet required levels of maritime standards, and solutions are often economically attractive.

This ship, a cable-layer wind turbine was equipped with seven high-power inverter and harmonics attenuated by two active filters installed centrally. Flexible mounting with robust and compact assets filters can install into your black gang away from the frequency changer drive system.

By controlling the engine speed or pump water installations, you can save energy up to 50% and therefore variable frequency changer make their sense of efficiency. Most electrical energy saving have side effect to transform current in a non-sinusoidal current - known as harmonic distortion of current.

Harmonic distortion is therefore an ongoing concern. Harmonics are second product of modern equipment to control electric current. Using variable frequency inverter generates harmonics.

Harmonic currents gave as a result:
– increase energy consumption
– increase system losses
– request series equipment
– increase the distortions in the network

The problem for the distorted current is that it affects waveform voltage, which results in deformation of the supply voltage. If the network supply is affected by harmonic distortion, all equipment supplied from the network to operate abnormal and thus will deviate from their normal behavior.

This leads to:
– limitations on power and network use
– premature aging products
– higher losses
– vibration motor shaft
– production stops
– increase electromagnetic interference

Harmonics reduce reliability, increase downtime, affect product quality, increase operating costs and cause lower productivity. It will never be possible to completely eliminate harmonics, but reducing harmonics for individual tasks nonlinear voltage distortion can be reduced.

An alternative is to compensate individual harmonic active filter using a point common coupling to compensate for multiple tasks or all at once. Active filter can also be used for upgrading facilities supplied from sources with harmonics or later connected nonlinear loads to improve energy efficiency. An active filter works just as external noise canceling headphones. Using external current transformers, power active filter monitors, including any distortion.

From this signal, the control system identified that the compensation necessary and creates a characteristic switching. This creates a low impedance in filter and current flow harmonic in place to work towards electricity supply goes to the filter installed.

4 CONCLUSIONS

Disturbing phenomena occurring in power system may have different causes and different solutions can attain such resolution. Power quality monitoring is a complex process of information gathering, analysis and interpretation of measurement results as useful information.

An important aspect of monitoring is to establish points to make measurements. When power quality problems become obvious need metering equipment location as close as possible to the equipment affected by disturbances.

Station operating regimes are important for monitoring process because they represent a point of overlap for many variations of voltage. Some monitoring equipment may send data through a telecommunication line in a processing center for analysis and interpretation.

REFERENCES

Martinon I, Assessment of harmonic and flicker planning levels for high voltage, Conference on Harmonic and Quality of Power ICHQP, Athens 1998

Gonbeau O, a.o., Method to determine contribution of the customer and the power system to the harmonic disturbance, CIRED 2003

IEC 61000-1-4, Electromagnetic compatibility (EMC) Part 1-4: Historical rationale for limitation of power-frequency conducted harmonic current emission from the equipment, Ed.1, 2003

Ungureanu M., Rosca R., Power quality in high voltage network due to electrical utilities, WSCE Torino 2006

Power Quality in European Electricity Supply Networks, Eurelectric Rapport of NE Standardization Specialist Group EMC&Harmonics, 2004

*** Review of methods for measurement an evaluation of the harmonic emission level from an individual disturbing load, CIGRE 36.05 /CRED 2 Joint VG (voltage quality)

Chapter 8

Hydrodynamics and Ship Stability

Prognostic Estimation of Ship Stability in Extreme Navigation Conditions

S. Moiseenko, L. Meyler & O. Faustova
Baltic Fishing Fleet State Academy, Kaliningrad, Russia

ABSTRACT: The article presents an approach for preliminary estimating ship's stability when there is a forecast of extreme hydrometeorogical conditions at the area where navigation is supposed. Consequences of such conditions can be negative even if initial ship stability meets requirements of classification societies. Based on the analysis of the ship extreme navigation conditions the method of the ship's stability at high angles of heel estimating is suggested. The first stage is the calculation of the standard elements of the initial stability. At the second stage it is proposed to add the common algorithm of stability estimation with calculations of possible extreme situations of the ship navigation. It is recommended to tabulate the "Information on extreme navigation conditions" as a result of prognostic evaluation calculations. The paper presents a case study of ship stability analysis in extreme conditions. Data was taken from the real practice of navigation.

1 INTRODUCTION

There are cases in the practice of navigation when ships, having the sufficient stability according to requirements of the Registers of Shipping in various countries, have been capsized. The accidents happen due to different causes.

According to the Final Report... (Final Report and Recommendations 1999) stability against capsizing in heavy seas is one of the most fundamental requirements considered by naval architects when designing a ship. The purpose of studying capsizing is to establish an understanding of ship behavior in extreme seas and to relate this to the geometric and operational characteristics of the ship to achieve cost effective and safe operation. It was defined that the capsizing modes of an intact ship include loss of static and dynamic stability and broaching, where loss of dynamic stability may be associated with dynamic rolling, parametric excitation, resonant excitation, impact excitation and bifurcation. Some provisions of the Maritime Safety Committee, for instance (MSC.1/Circ.1228 2007, MSC.1/Circ.1281 2008) and many scientific papers and books, in particular (Dahle, E.A. & Myrhaug, D. 1996, Ananyev D.M. & Yarisov Y.Y. 1997, Kobylinski, L.K. & Kastner, S. 2003, Belenky, V.L. & Sevastianov, N.B. 2007) were devoted to the problem of providing ships stability and prevention of capsizing.

As a rule capsizing is the consequence of extreme navigation conditions such as a wind squall or deck flooding. For instance, when a ship has strong rolling and the roll resonance phenomenon is observed. In this case the amplitude of rolling can greatly exceed values calculated by the method of the Rules for Classification and Construction of ships (Rules... 2008) of the Russian Maritime Register of Shipping and taken to determine the moment of force tending to capsize.

According to these Rules it is necessary to calculate characteristics of the ship stability in atypical case of loading and to include them in the "Stability information".

General intact stability criteria for all ships are given in the Code on intact stability for all types of ships (Code on intact stability... 1993, International Code... 2008). According to this document each ship should be provided with a stability booklet which contains sufficient information to enable the master to operate the ship in compliance with the applicable requirements contained in the Code, including "any necessary guidance for the safe operation of the ship under normal and emergency conditions. For example, there is the guidance to the master for avoiding dangerous situations in following and quartering seas (MSC/Circ.707 1995).

But it contains recommendations to the master when the ship navigates already under extreme conditions.

It is necessary to have prognostic data when the ship has a threatening situation of capsizing, at which angles of heel the calculated heeling moment can become an upsetting one. That it is important to know which the rolling amplitude is dangerous. One of such guidance could be the "Information for extreme navigation conditions".

2 EXAMPLE OF NAVIGATION EXTREME CONDITION

As an example it is possible to consider the voyage of the transport refrigerator ship "Bering Strait" which followed to a fishing ground at the Bering Sea (Moiseenko & Meyler, 2011). The ship had the following main dimensions: maximum load displacement Δ_{max} = 19,600 t, length overall $L_{o.a.}$ = 172 m, length between perpendiculars $L_{b.p.}$ = 160 m, extreme breadth B_{ex} = 23 m, mean draught loaded d = 8.09 m. (Here and below designations are given in accordance with (Rules... 2008) but in parentheses they are presented according to IMO standards).

Stability of the ship was satisfied to the Russian Maritime Register of Shipping requirements. The value of the angle of roll $\theta_{1r} \approx 21°$ was calculated according to (Code...1993, Rules...2008):

$$\theta_{1r} = 109kX_1X_2\sqrt{rS} \qquad (1)$$

where k, X_1, X_2, r, S = coefficient and nondimensional multipliers depending on the ship main dimensions, the area of navigation and the rolling period.

As it is shown in Figure 1 the initial metacentric height was equal to h_0 (GM_0) = 0.8 m, the upsetting lever l_c = 0.36 m. Thus the upsetting moment had a value of $M_c \approx 70,600$ kNm.

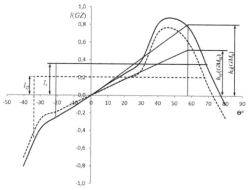

Figure 1. Statical stability curve (GZ curve)
(--- for flooded deck)

The ship runs into heavy weather being in the Sea of Okhotsk. The wind from the north-north-west reached the speed up to 30 m/s. The ship pointed the course close to the wind on port side with the speed of 8 knots. The rolling amplitude was no more than 10°. However, suddenly the ship received a list to port to 20°, and then up to 30°. At the same time the ship was under the influence of a wind squall and the list to starboard reached to 35°. The large mass of water flooded the deck. The ship's course was changed strictly on the wave, and then navigation continues without complications.

The heel resonance amplitude reached up to 30° and the ship could capsize if the angle of heel is equal to 35° because the value of the upsetting lever for the flooded deck l_{cf} = 0.21 m and the upsetting moment decreased to M_{cf} = 41,160 kNm. This example shows that the ship having the sufficient initial stability can capsize in the case of an unfavorable combination of various factors.

3 PROCEDURE FOR THE ANALYSIS OF THE SHIP'S STABILITY IN THE EXTREME CONDITIONS AT HIGH ANGLES

Based on the analysis of possible operational situations the following procedure for the analysis of the ship's stability in the extreme conditions at high angles of heel may be suggested in order to obtain prognostic estimations of the ship's capacity to counter to forces of nature. An algorithm for the first stage of analysis is well known (Makov Ju. 2005, Barrass C.B. & Derrett D.R. 2006) and includes:
– calculation of the standard elements of the initial stability;
– comparison an applicate of the ship's center of gravity $z_g(KG)$ with its critical value $z_{gcrit}(KG_{crit})$;
– plotting the statical stability curve and its checking on requirements (Rules...2008);
– calculation of the angle of roll;
– determination of the upsetting moment;
– calculation of the heeling moment in accordance with the adopted method (Rules...2008);
– calculation of the weather criterion K according to recommendations (Code... 1993, Rules...2008).

If the calculated elements of the stability meet these requirements and $z_g(KG) \leq z_{gcrit}(KG_{crit})$ then the ship can navigate safely with loads from external forces taken in calculating.

4 CONSIDERATION OF UNUSUAL SITUATIONS FOR SHIPS

At the second stage of analysis it is proposed to add a well-known algorithm with calculations relating to the consideration of unusual situations for the ship. These situations include: the phenomenon of resonance, when the amplitude of the roll can be much higher than the estimated value; the simultaneous impact on the ship of the dynamically applied heeling moment due to wind and the moment of cargo displacement or the heeling moment from the mass of water on the deck; the ship "stalling" off the course at the time of finding on the wave crest and the impact of the heeling moment due to external forces. Navigators should avoid the above situations but sometimes weather forecasts are unreliable, operators' errors in assessment of situations are also inevitable. Therefore there is a need to obtain prognostic estimates which would give a quantitative idea of the degree of a danger on the ship and on people of external forces with their worst combinations. This can be confirmed by many losses of cargo and fishing ships as a consequence of capsizing.

At the second stage it is recommended to add the common algorithm of stability estimation with calculations of possible extreme situations of the ship navigation, namely:

- to receive the forecast of the meteorological conditions of navigation and information on abnormal natural phenomena that have ever been observed in the area of navigation;
- to calculate the elements of stability at high angles of heel at the time of the approach to the area of unfavorable conditions (such work has to be done in advance);
- to check if the condition $z_g(KG) \leq z_{gcrit}(KG_{crit})$;
- to calculate the possible value of the heeling moment due to a squall wind impact;
- to determine on the static stability diagram at which angles of heel the dynamic heeling moment can capsize the ship;
- to calculate the value of the moment which may arise as a consequence of cargo shifting;
- to calculate the heeling moment from the joint impact of the wind squalls, water received on the deck and displacement of cargo;
- to determine by the statical stability curve angles of heel at which the total dynamic moment can capsize the ship;
- to consider the risk of icing if the ship is sailing in winter.

It is recommended to tabulate the "Information on extreme navigation conditions" as a result of prognostic evaluation calculations. Calculation and an analysis of prognostic estimates of the ship's stability under extreme conditions connected to certain time expenditures. It is clear that these calculations are performed or have to perform before the ship sailing. And, if there is even a smallest possibility to reduce the probability of the ship and people loss, any amount of time is reasonable. These calculations require only time for collection, processing and inputting information into the computer's memory.

The "Stability information" existing on ships at present does not contain all initial data that are required for the analysis of the stability, including the scheme described above. Therefore, it is possible to recommend to a navigator to complete himself the necessary information about the ship's stability with necessary initial data.

Let consider an example of the analysis of ship stability in extreme conditions (it is taken from the real practice). The ship m/v «Priboy» had the following main dimensions: length overall $L_{o.a.}$ = 157 m, length between perpendiculars $L_{b.p.}$ = 145 m, extreme breadth B_{ex} = 21.2 m. The ship leaves harbour with 30% of ship's stores and 7,500 tons of general cargo on board. According to the "Stability information" in this case of loading there are the following dimensions of the ship: displacement Δ = 15,400 t; mean draught d = 7.37 m; abscissa of the center of gravity $x_g(XG)$ = - 3.29 m; applicate the center of gravity $z_g(KG)$ = 8.47 m. The value of $z_{gcrit}(KG_{crit})$ = 8.49 m.

The statical stability curve (GZ curve) of the ship is shown in Figure 2a,b.

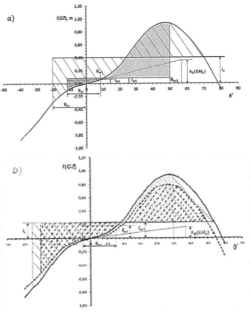

Figure 2. Statical stability curve of m/v "Priboy" (--- for flooded deck)

Some indicators of the GZ curve are given in Table 1. Parameters of the statical stability curve: the maximum lever $l_{max} \geq 0.2$ m when maximum angle of stability curve $\theta_m \geq 30°$, the angle of vanishing stability $\theta_v \geq 60°$, the initial metacentric height $h_0(GM_0) \geq 0.15$ m and the condition $z_g(KG) \leq z_{gcrit}(KG_{crit})$ meet requirements (Rules... 2008). Hence stability of the ship is provided.

According to the formula (1) the calculated angle amplitude of roll is $\theta_{1r} \approx 20°$. The weather criterion is defined according to the (Rules... 2008).

Table 1. Indicators of the ship "Priboy" GZ curve

Maximum righting lever	$l_{max}(GZ_m)$, m	0,94
Maximum angle of stability curve	θ_{gm}, deg	47
Angle of vanishing stability	θ_{vm}, deg	>70
Initial metacentric height	$h_0(GM_0)$, m	0,35

It is necessary to calculate a steady heeling lever:

$$l_{w1} = p_v A_v z_v / 1000 g \Delta, \text{ m} \qquad (2)$$

and a gust wind heeling lever

$$l_{w2} = 1.5 \, l_{w1}, \qquad (3)$$

where p_v = a steady wind pressure acting perpendicular to the ship's centerline which is defined depending on the ship navigation area (for this case $p_v = 504$ N/m^2 that corresponds to the mean wind speed $v = 25$ m/s), A_v = projected lateral area of the portion of the ship and deck cargo above the waterline ($A_v \approx 1,600$ m^2), z_v = vertical distance from the centre of A_v to the centre of the underwater lateral area or approximately to a point at one half the draught ($z_v \approx 9.5$ m), $g = 9.81$m/s^2.

For this ship values of the heeling levers are $l_{w1} \approx 0.051$ m, $l_{w2} \approx 0.075$ m. The heeling moments are $M_{w1} \approx 7,700$ kNm and $M_{w2} \approx 11,330$ kNm respectively. Comparing the fully shaded areas in Figure 2a it is seen that the weather criterion $K \approx 10$. Thus the ship's stability is sufficient according to this criterion. The value of the upsetting lever $l_c \approx 0.39$ m is found from equality of the dashed areas in Figure 2a. The value of the upsetting moment is $M_c \approx 58,920$ kNm for the calculated angle $\theta_{1r} \approx 20°$. The gust heeling moment can become upsetting one when the value of the angle of heel will be more than 50°. An analysis of the characteristics of the ship and the possible effects of external forces shows that stability of the ship is enough for safe navigation.

The next step in the analysis is the determination of hazardous heeling angles that may arise in the case of a resonance with the simultaneous squall action. It is known that the ship will consume 360 tons of fuel at the time on the way to the Bay of Biscay. The preliminary calculation shows that in this case the applicate of gravity center: $z_g(KG) = 8.54$ m, which is more than the critical value $z_{gcrit}(KG_{crit}) = 8.49$ m. Therefore, requirements to the ship's stability are not satisfied. Taking into account that at the moment of the voyage beginning the value of the calculated applicate of the center of gravity was too near to its critical value the ship should be ballasted before navigation, because $z_g(KG)$ becomes equal to $z_{gcrit}(KG_{crit})$ already on the third day of the voyage.

Let's assume further that the ship took the ballast and at the time on the way to the Bay of Biscay its stability was characterized by the same values of the elements as at the beginning of the voyage.

Let's assume that a forecast about a hard storm (even a hurricane) with the wind speed up to 35 - 45 m/s at the Gulf of Biscay has been received. According to the Beaufort scale the mean wind pressure at the height of 6 m above the sea surface is about $p_v = 878$ N/m^2 but under a squall action it is up to 1,265 N/m^2 when the wind speed value is more than 29 m/s. The values of the wind pressure can be calculated using the formula (Blagoveschenskiy S.N. & Kholodilin A.N. 1975):

$$p_v = C_y \rho v^2 / 2, \text{ H/m}^2 \qquad (4)$$

where C_y = nondimensional aerodynamic coefficient of the area A_v ($C_y \approx 1.3$), ρ = mass density of air, v = wind speed.

Thus for the ship stability assessment it is necessary to take the mean value $p_v \approx 1.700$ N/m^2 when the wind speed is equal to 45 m/s and $p_v \approx 2.240$ N/m^2 under a squall action. The value of a steady heeling lever $l_{w1} \approx 0.17$ m and a gust wind heeling lever $l_{w2} \approx 0.23$ m. According to fully shaded areas on the GZ curve shown in Figure 2b the weather criterion decreased to the value $K \approx 3.5$. It is seen comparing areas shaded with solid lines that the value of the heeling moment can become the gust upsetting moment $M_c \approx 34,740$ kNm when angles of heel can reach values $\theta_v \geq 35°$. In this case the ship will be at a critical situation because such the heeling moment can capsize the ship.

In such stormy conditions the ship takes on the deck a significant quantity of water at big angles of heel. It was observed cases when ships of this type took up to 200 tons of water. Consequently, the ship's stability is deteriorated in such conditions and an additional heeling moment is created. Deterioration of stability when flooding the deck can be estimated approximately, making a mistake in the direction of the ship safety. Thus, when quantity of water on the deck is about 200 tons, the calculated values are: $z_g(KG) = 8.64$ m that is more than $z_{gcrit}(KG_{crit})$. The initial metacentric height $h_0(GM_0) = 0.18$ m as it is shown in Figure 2b for the dotted curve. The calculated gust wind heeling lever is $l_{w2} \approx 0.22$ m and the heeling moment with the increased ship's displacement is $M_v \approx 33,660$ kNm. Comparing areas shaded with dotted lines on the corrected statical stability diagram shown in Figure 2b, it is possible to find that the ship can capsize under the impact of this upsetting moment when angles of heel can reach values $\theta_v \geq 29°$.

Let's further assume that the ship navigates in winter at the definite area and icing is possible from the forecast of hydrometeorological conditions. Carrying out the ship's stability assessment it is necessary to take into consideration changes of displacement, applicate of the ship's center of gravity and projected lateral area of the portion of the ship and deck cargo above the waterline. The added mass of ice is equal to 30 kg on 1 m^2 of horizontal area of the upper deck and 15 kg on 1 m^2 of the projected lateral area (Rules…2008). The mass of ice on the decks, superstructures, cabins and masts of the ship can reach up to 150 t. The calculated applicate of the gravity center of the ship in icing conditions is equal to $z_g(KG) = 8.54$ m. Thus, in this case of icing $z_g(KG) > z_{gcrit}(KG_{crit})$ and the ship does not meet the requirements of stability. The upsetting moment will have practically the same value as in case of the flooding deck. If there is a threat of icing ballast should be taken in order to make the condition $z_g(KG) \leq z_{gcrit}(KG_{crit})$.

The example of the mentioned above "Information on extreme navigation conditions" for this ship as a result of preliminary calculations and possible actions of the ship's crew are given in Table 2.

Table 2. Information on extreme navigation conditions

Conditions of navigation	M_c, kNm	Critical angle of heel θ_v	Possible actions
Wind speed up to 45 m/s, squalls, observed resonance conditions	34,740	> 35	Turn the ship's nose to/on the wave and to ride out a storm or leave to harbour
Wind speed up to 45 m/s, the ship takes water on the deck, strong roll, possible resonance phenomenon	33,660	≥ 29	Take the extra ballast. Turn the ship's nose to/on the wave and to ride out a storm or leave to harbour
The ship is in icing conditions. Wind speed is up to 45 m/s, squalls	33,660	≥ 29	Take the extra ballast, to ride out a storm on the wave

5 CONCLUSION

Based on the analysis of the ship extreme navigation conditions the method of the ship's stability at high angles of heel estimating is suggested. Such an analysis It is recommended to analyze do such conditions in advance basing on the of meteorological forecast conditions at the area of the proposed prospective voyage.

The presented method allows the master to calculate predictive assessments of the ship stability under extreme navigation conditions at the stage of the voyage planning. Based on this preliminary assessment it is possible to develop measures improving the safety of the ship, to choose a rational cargo plan and cargo securing in addition to the schemes recommended by IMO.

The proposed "Information on extreme navigation conditions" may be an addition to the existing on each ship "Stability information" and provide operational management of the master in the event of extreme conditions.

A simulation model can be developed on the base of the suggested method and used in the educational process for naval officers' training and upgrading.

The preliminary assessment of the ship stability was performed repeatedly by one of the authors (the master) when carriage of the deck cargo (mahogany) from ports in Africa to ports in Europe and Malaysia. It allowed the successful navigation in extreme meteorological conditions in areas of the Bay of Biscay, the Cape of Good Hope and the Philippine Islands.

REFERENCES

Ananyev D.M. & Yarisov Y.Y. 1997. Heeling and capsizing of small ships flooded by following waves. In Trans. Russian Maritime Register of Shipping, Issue 20, Part 1, 43-56 (in Russian).

Barrass C.B. & Derrett D.R. 2006. *Ship Stability for Masters and Mates*. Sixth edition. Burlington: Elsevier.

Belenky, V.L. & Sevastianov, N.B. 2007. Stability and safety of ships. Risk of capsizing. The society of Naval Architects and Marine Engineers, 2nd ed. Jersey City: USA.

Blagoveschenskiy S.N. & Kholodilin A.N. 1975. Handbook on ship's statics and dynamics. Vol. 1. Statics of a ship. Leningrad. Sudostroenie. (in Russian).

Code on intact stability for all types of ships covered by IMO instruments: Resolution A.749(18) as amended by resolution MSC.75(69) 1993. London: IMO.

Dahle, E.A. & Myrhaug, D. 1996. Capsize risk of fishing vessels, Schiffstechnik, 43.

Final Report and Recommendations 1999. The Specialist Committee on Stability. Proc. 22nd International Towing Tank Conference (ITTC), Seoul

International Code on intact stability 2008. London: IMO

Kobylinski, L.K. & Kastner, S. 2003. Stability and Safety of Ships. Oxford UK: Elsevier.

Makov Ju. 2005. *Stability… What it is? (Dialogs with a Master)*. Saint-Petersburg: Sudostroenie. (in Russian)

Moyseenko S. & Meyler L. 2011. *Safety of Shipping*. Kaliningrad: BFFSA (in Russian).

MSC/Circ.707 1995. Guidance to the master for avoiding dangerous situations in following and quartering seas.

MSC.1/Circ.1228 2007. Revised guidance to the master for avoiding dangerous situations in adverse weather and sea conditions.

MSC.1/Circ.1281 2008. Explanatory notes to the international code on intact stability.

Rules for classification and construction of ships 2008. Saint-Petersburg: Russian Maritime Register of Shipping (in Russian)

The Values and Locations of the Hydrostatic and Hydrodynamic Forces at Hull of the Ship in Transitional Mode

O.O. Kanifolskyi
Odessa National Maritime University, Odessa, Ukraine

ABSTRACT: The transitional mode isn't studied fully. The data about value and coordinates of locations of hydrostatic and hydrodynamic forces are absent. This knowledge is necessary for calculation stability and strength of high-speed ships, at first steps of project. In article, it is trial to define some parameters for transitional mode ships, with apply modern program's pack. In duration of study, the values of forces and their location at ship's hull were found.

1 INTRODUCTION

In theory and designing of high-speed ships, there are much unexplored hitherto parts. Particularly, this fact is actually for ship of the transitional mode. At hull of such craft, the hydrostatic and hydrodynamic forces are appearing, when ship is moving. The questions about value of these powers and their location are not studied. For making efficient and safe ship, shipbuilder must have data about value and coordinates of locations of these forces. This is a goal of this research. The objective and subject of research are ship of the transitional mode and hydrostatic, hydrodynamic forces, accordingly.

2 MAIN PART

Usually, shipbuilder considers three main regime of motion for high-speed ships. For definition parameters of these modes, the engineers use Froude numbers based on volume $Fr_V = v / \sqrt{g \sqrt[3]{V}}$.

The "displacement mode" means the regime, whether at rest or in motion, where the weight of the craft is fully or predominantly supported by hydrostatic forces $\Delta = \gamma V$, $Fr_V < 1$. The "non-displacement mode" is the normal operational regime of a craft when non-hydrostatic forces substantially or predominantly support the weight of the craft $\Delta = Y$, $Fr_V > 3$. The "transitional mode" is defined as the regime between displacement and non-displacement modes $\Delta = \gamma V_1 + Y$, $1 < Fr_V < 3$.

The transitional mode is not research fully now. In this regime, the calculation of forces is difficult; because many factors are unknown. But this knowledge is very important, for calculation of the strength and stability of ship. We can define two types of forces which are doing to hull at motion of craft in transitional mode. There are hydrostatic and hydrodynamic forces.

For calculation of these forces, Bunkov's experimental data were used [1]. The data are including experiments about ships in transitional mode. The parameters of models are: relative length 5,25; average deadrise angle 7,25°; average location of relative center of gravity, from transom 0,4; static beam-loading coefficient $Cd = \Delta / \gamma B^3$ from 0,427 to 0,854. The results of search showed in view of relative wetted surface area $\Omega / V^{2/3}$; relative wetted length $L_{aw} = (L_{ch} + L_k)/2$ (L_{ch} - length at chine; L_k - length at keel) and trim φ. The experimental data is available on www. http://www.twirpx.com.

Some calculation methods, for definition of hydrodynamic and hydrostatic forces in planning mode, exist, but we haven't data about values of these forces and their location at ship's hull in transitional mode.

The two computer programs were used for definition of parameters forces: AutoCAD and Dialog Statika. The program AutoCAD is US's product and is using for creating drawings. The program Dialog Statika was developed in Russia and was made for calculation in sphere of ship's theory. This program gives, for engineer shipbuilder, wide possibilities for calculations. With help of this program, the design gyrostatics, cross cavers, trim and many other parameters can be calculated.

In AutoCAD the shape of Bunkov's model was drawn and the positions of waterlines for every relative speed were defined, Figure 1. These positions are characterized with trim and relative wetted length. After that, the bow and aft draft were measured and the displacement and coordinates center of buoyancy x_{V1}, z_{V1} were calculated, with help of program Dialog Statika. The variant of such calculations is showed on Figure 2.

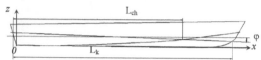

Figure 1. Definition of the relative wetted length and two draft in AutoCAD.

Bow draft = 0.07 m

Aft draft, m	Δ, t	X_{V1} m	Z_{V1} m
0.16	0.08	-0.543	0.085
0.17	0.08	-0.547	0.089
0.18	0.09	-0.550	0.093
0.19	0.10	-0.553	0.096

Figure 2. The variant of calculation displacement and coordinates center of buoyancy with Dialog Statika (x_{V1} measurement from middle).

The relative buoyancy (hydrostatic force), on Figure 3, was calculated with two methods. The first of them is value from Dialog Statika and second is defined with Peter du Cane's formula [2]:

$$W_1 = \gamma \frac{L_{aw}^2 B}{4} sin 2\varphi \qquad (1)$$

where φ - trim; B - width of the ship; γ - density of water; L_{aw} – relative wetted length.

The relative locations of center of buoyancy are showed on Figure 4 and Figure 5.

The hydrodynamic force is difference between weight of ship and hydrostatic force. The significance of this force and coordinate of point its location x_Y / L_{ch}, are presented on Figure 6 and 7. On this figure, the results of research ETT towing tank is showed also. The coordinate z_Y is at shell of ship's bottom.

The longitudinal location of force Y, if point of calculation is transom, was defined with formula of moments. The parameters x_g and z_g are equal $0.4L_{ch}$ and $0,7D$, accordingly.

$$W_1(x_{V1} cos\varphi - z_{V1} sin\varphi) - \Delta(x_g cos\varphi - z_g sin\varphi) + \\ + Y(x_Y cos\varphi - z_Y sin\varphi) = 0 \qquad (2)$$

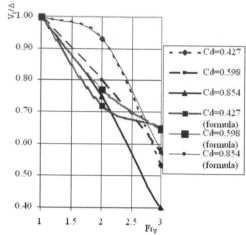

Figure 3. The relative buoyancy (hydrostatic force) calculated with two metods.

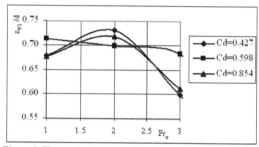

Figure 4. The vertical location of center of buoyancy.

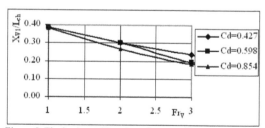

Figure 5. The horizontal location of center of buoyancy.

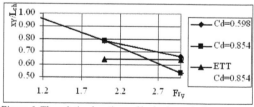

Figure 6. The relative locations of hydrodynamic force.

278

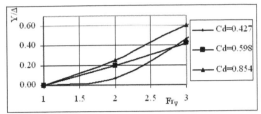

Figure 7. The relative value of hydrodynamic force.

3 CONCLUSIONS

For definition of location forces at ship's hull, in transitional mode, engineer can use modern program's pack AutoCAD and Dialog Statika.

The presented figures were obtained by the author, are available, in this article, for all shipbuilders and could take, for calculations of ship's stability and strength, in early steps of project.

REFERENCES

Zinkin, V. 1980. Propulsion high-speed ships. LSI
Peter du Cane. 1960. High-speed small craft. Temple Press Limited

Contrary Hydrodynamical Interactions Between the Model and Prototype of Boats

A. Şalci
Gedik University, Faculty of Engineering, Kartal, Istanbul, Turkey

ABSTRACT: In this study presents the comparison of analysis results obtained from the calculation and resistance tests results obtained from the "Geosim Tests" on the Lake Köyceğiz and Dalyan Delta and the resistance tests results obtained from a model infantry type of fishing boats in I.T.U. Ata Nutku Ship Model Towing Tank and the analysis results arrived by calculation. The geosim tests were carried out under single loading with the geometrically similar infantry type of fishing boats with their size varying 7-12 meters.

This paper consists of six sections. The first section gives information about the resistance tests of prototype boats and the built model which produced in laboratory. The second section outlines the principal particulars of the model and prototypes and geometric details. The third section explains the similarity and model theory for ship resistance. In the fourth section, results of experimental researches for model and prototypes are presented. In the fifth section, contrary hydrodynamical interactions and the form factors developed with the author's method (modified or transformed Hughes method) and Prohaska's method are presented. Finally, in the last section the comparison of the result in the model and prototype scales are mentioned.

1 INTRODUCTION

Empirical formula that have been developed by statistical evaluations, results of sea trials of a similar ship, graphs that are based on experiment results of systematic model experiments and results of ship model towing tank experiments are used in the "Ship Resistance and Power" calculations that are one of the most important calculations that are related to ship design. The last method, which is the most scientific of them all, produces results that are very close to accurate. In this method, the ratio between the size of the model experiment laboratory-towing tank and the model is important and the importance of an "optimum model" which does not create "scale effect" or "blockage effect" increases. These two effects affect each other in the opposite direction.

In this study, the experiment results of a model and six prototypes of this model under real conditions have been put through a resistance analysis mutually and the interaction between the model and its prototypes are studied.

In the routine studies that have been done in the past, there has always been one prototype and many models. However, in this study, there was only one model and many prototypes. The type of boat that was used in the study was a fishing boat. The type of boat is used a lot especially in the Aegean and Mediterranean Seas.

These wooden boats are produced completely by hand using molds. The scientific properties of such boats have not been discovered, yet, due to the fact that their production was never a part of a project. In this research, the geometry and the form plans of these boats have been determined and by the help of these experiments that were performed using this information, virtually the "hydrodynamic anatomy" of these boats have been discovered.

Prior to this study, development of these boats were guided by years of experience at sea and the skills of the masters. The name **"infantry"** was given to these boats, which have gone through approximately a century long process, by the local people who lived in the region where these boats were used. These boats are used as fishing boats all year and as touristic tour boats only during the summer.

Table 1. Main Dimensions of Boats

Principal Particulars	Symbol	Unit	Model	Ship (prototype)					
				7	8	9	10	11	12
Length of overall	L_{OA}	m.	3	7	8	9	10	11	12
Length of betw. Perp.	L_{BP}	m.	2,640	6,160	7,040	7,920	8,800	9,680	10,560
Length of load water line	L_{WL}	m.	2,694	6,286	7,184	8,082	8,980	9,878	10,776
Breadth	B	m.	0,690	2,140	2,370	2,620	2,830	3,180	3,610
Height	H	m.	0,345	1,000	0,800	0,800	0,940	0,900	1,370
Draught (midship)	T	m.	0,168	0,320	0,420	0,380	0,520	0,550	0,600

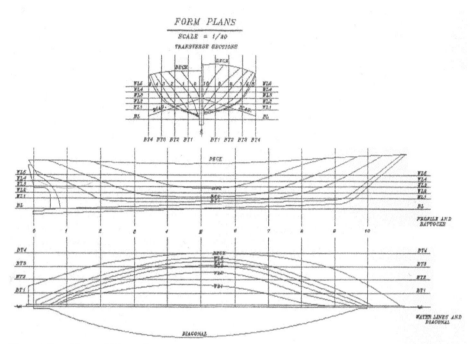

Figure 1. Form Plans of Infantry Type of Fishing Boats

2 GEOMETRIC CHARACTERISTICS

The length of overall of infantry type fishing boats range from 7 to 12 meters. Their cross sections are in the (V) form shape, keels are llama keels, stem forms are of Clipper type and their stern forms consist of a heel and a small transom. There is a rudder and a 3-bladed propeller at the stern of these boats. The main dimensions of boats that are the subject of research can be seen in Table 1 and their form plans are shown in Figure 1.

3 SIMILARITY AND MODEL THEORY FOR SHIP RESISTANCE

According to Hughes Hypothesis, ship resistance can be written

$$R_T = R_V + R_W = (1+K) \cdot R_F + R_W, \quad (1)$$

where $R_V = f(Re)$ and $R_W = f(Fr)$

$$Re = \frac{\upsilon \cdot L}{\nu}, \quad \nu = \frac{\mu}{\rho} \qquad \text{Reynolds number} \quad (2)$$

$$Fr = \frac{\upsilon}{\sqrt{g \cdot L}} \qquad \text{Froude number} \quad (3)$$

On the other hand, frictional resistance of ship

$$R_F = C_F \cdot \frac{1}{2}\rho S \upsilon^2 \quad (4)$$

and wave resistance of ship

$$R_W = C_W \cdot \frac{1}{2}\rho S \upsilon^2 \quad (5)$$

and also total resistance of ship

$$R_T = C_T \cdot \frac{1}{2}\rho S \upsilon^2 \quad (6)$$

In ship model experiments, given the speed υ_m and the total resistance of the model R_{Tm}, the equivalent speed and the total resistance of a ship are calculated in the following way. According to the similarity laws, in other words Froude similarity,

$$Fr_m = Fr_s \rightarrow \frac{\upsilon_m}{\sqrt{g \cdot L_m}} = \frac{\upsilon_s}{\sqrt{g \cdot L_s}} \rightarrow \upsilon_s = \upsilon_m \sqrt{\frac{L_s}{L_m}}$$

Here, $\alpha = \dfrac{L_s}{L_m}$ is the geometric similarity ratio and came be written as

$$\upsilon_s = \upsilon_m \sqrt{\alpha} \quad \text{and} \quad V_s \cong \frac{\upsilon_s}{0.5144} \tag{7}$$

$$R_{T_s} = C_{T_s} \cdot \frac{1}{2} \rho_s \cdot S_s \cdot \upsilon_s^2 \tag{8}$$

where, $\rho_s = \dfrac{\gamma_s}{g}$ and $S_s = \alpha^2 \cdot S_m$

Total resistance coefficient for the ship can be written as

$$C_{T_s} = C_{V_s} + C_{W_s} = C_{F_s}(1+K) + C_{W_s} \tag{9}$$

According to the (ITTC – 1957) formula, friction resistance coefficients for the ship and the model are calculated in the following way,

$$C_{Fs,m} = f(Re_{s,m}) = \frac{0.075}{(Log\,Re_{s,m} - 2)^2} \tag{10}$$

The most important stage of the calculation is the determination of the form factor K and according to the results of the model experiment, the form factor becomes,

$$K = \frac{R_V}{R_F} - 1 = \frac{C_V}{C_F} - 1 = \frac{\overline{AC}}{\overline{AB}} - 1 = \frac{\overline{BC}}{\overline{AB}} \tag{11}$$

At the point where the form factor is determined, C_W and Re possess the following values,

$$C_W \rightarrow 0, \quad Re = Re^*$$

These values can be seen in Figure 2. Viscous resistance coefficient curve is similar to the friction resistance coefficient curve and is an envelope curve of the total resistance coefficient curves.

$$C_V = (1+K) \cdot C_F = \lambda C_F \quad (\lambda > 1) \tag{12}$$

For this reason, it is necessary to perform many experiments for the models and prototypes in order to draw this envelope curve. On the other hand, this necessity can be done away with by the use of an inverse transformation technique.

$$C_F = \frac{0.075}{(Log\,Re - 2)^2} = 0.075X \tag{13}$$

and

$$X = \frac{1}{(Log\,Re - 2)^2} \tag{14}$$

This way each of the C_F and $C_V = \lambda\, C_F$ curves can turn into a straight line as a result of this inverse transformation. This situation leads to the determination of a straight line C_V that goes through the origin and runs tangent to a curve C_T leads to the calculation of K form factor. In Figure 3, the extrapolation diagram which is obtained as a result of inverse transformation can be seen.

Here,

$$\alpha = \tan^{-1}(0.075) \cong 4.289° \quad \text{ve}$$

$$\beta = \tan^{-1}(0.075 \cdot \lambda), \quad \beta > \alpha \ (\lambda > 1).$$

Another method for finding the form factor is Prohaska's method. According to that method,

$$C_{T_m} = C_{F_m}(1+K) + C_{W_m} \quad \text{and} \quad C_{W_m} = a \cdot Fr^n \tag{15}$$

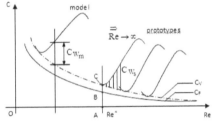

Figure 2. Normal Transformation

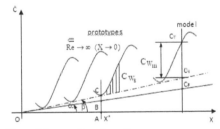

Figure 3. Inverse Transformation

Here, (a) is given as a constant and n=4. According to this,

$$C_{T_m} = C_{F_m}(1+K) + a \cdot Fr^4 \tag{16}$$

and

$$\frac{C_{T_m}}{C_{F_m}} = a \cdot \frac{Fr^4}{C_{F_m}} + (1+K) \ \Rightarrow \ y = ax + b \tag{17}$$

linear relation is obtained.

Where, $x = \dfrac{Fr^4}{C_{Fm}}$ and b=1+K. This relation shows a straight line that has a slope (a) and which crosses the ordinate axis (b).

According to Hughes hypothesis, for $Fr_m = Fr_s \rightarrow C_{W_m} = C_{W_s}$ can be written. Given this information, we can arrive at the formula that are listed below,

$$C_{W_s} = C_{W_m} = C_{T_m} - (1+K) \cdot C_{F_m} \qquad (18)$$

$$C_{T_m} = \frac{R_{T_m}}{\frac{1}{2}\rho_m \cdot S_m \cdot \upsilon_m^2} \qquad (19)$$

$$C_{T_s} = C_{F_s} \cdot (1+K) + C_{T_m} - (1+K) \cdot C_{F_m} \quad \text{or}$$

$$C_{T_s} = (1+K) \cdot (C_{F_s} - C_{F_m}) + C_{T_m} \qquad (20)$$

4 EXPERIMENTAL RESEARCHS

4.1 Model Tests

Model experiments have been performed in an I.T.U. Towing Tank using a model that has a length of overall 3 meters. The model is made out of wood and it was tested at speeds of 0,0 to 2,3 m/s. at the laboratory using different draughts. The characteristics of the main model experiment are shown in Table 2.

Table 2. Specifications of Model Experiment (main experiment, test number:4) M 204

Loading Condition		WL 3	
		Value	Unit
Geometric Similarity Ratio	α	10/3	-
Draught	T	0,168	m.
Length of water line	L_{WL}	2,694	m.
Wetted surface area	S	1,260	m².
Displacement	Δ	90,865	kg.
Water temperature	t	13,0	°C
Mass density	ρ	101,91	kg.s²/m⁴.
Kinematic viscosity	ν	1,2036.10⁻⁶	m²/s.
Form factor	K	0,8073	
Roughness		None	
Trim		None	
Appendages		None	
Turbulence generator		Trip wire (φ 3)	

In experimental researches, resistances of model number M 204 at the water lines that correspond to six different loading situations were measured and experiment data was functionally defined using the curve-fitting method. This way, $R_{Tm} = f(v_m)$ resistance characteristics of the model were determined. Data of experiment number 13 and resistance characteristic which has been determined are shown Figure 4 as an example. The (R_{Tm} - v_m - T_m) variation is shown in two-dimensional and three-dimensional forms in Figures 5 and 6, respectively. This was accomplished by determining total resistances R_{Tm} that correspond to the draught T_m values for each speed value of the model within the range of 0,8 to 2,3 m/s.

Model Number: M 204   (L_{BP})m = 2,640 m.
φ3 Trip wire S_m = 1,121 m².
LWL : WL 2,5 Δ_m = 68,00 kg.
t_m= 20°C B_m = 0,630 m.
(L_{WL})m = 2,670 m. T_m = 0,146 m.

Figure 4. Results of Test Number 13 and its Curve Fitting.

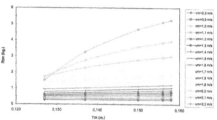

Figure 5. Variation of Resistance with Draught and Model Forward Speed (2-dimensional representation)

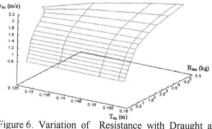

Figure 6. Variation of Resistance with Draught and Model Forward Speed (3-dimensional representation)

The resistance characteristics of the model were determined by using speed and resistance values that were obtained in the draughts of a model scale. These draughts are the equivalent of draughts of prototypes which have lengths of overall that range from 7 to 12 meters. When resistance analysis was being performed for the model that was geometrically similar to each and every prototype, corrected data was used in the calculations.

(ITTC – 1957) formula was used in the friction resistance calculations. The method that was developed by the author (modified or transformed Hughes method) and the method of Prohaska were used comparatively in the determination of the form factor. According to model experiment results, form factors range from K = 0,339 to K = 0,496 based on the loading condition.

4.2 Prototype Tests

In this section, resistance experiment results of prototype boats that were obtained within the scope of the research Project which was supported by the Republic of Turkey Office of the Prime Minister Environmental Protection Agency for Special Areas of which the author is the executive, German Technological Cooperation agency and Turkish Scientific and Technical Research Council.

Boats that have been selected by means of sampling for being geometrically similar to each other have length of overall that range from 7 to 12 meters, inclusively. Lengths of overall of these boats are 7, 8, 9, 10, 11 and 12 meters. Experiments were performed at a lake in an area where there is still water and no wind, 100 meters off the coast under smooth-water conditions.

Boats that are involved in the experiment were pulled by another boat that functioned as a tugboat at a constant forward speed while winding through signal flags which were separated from each other by a pre-determined distance. The forces that act upon the towing line are measured by using a dynamometer. Sufficient space was left between the boat that was pulling and the boats that were being pulled. Each experiment was performed after the waves that were formed during the previous experiment were completely damped. The **C** and **P** values of each boat were calculated and checked against the literature to confirm the reliability of the experiments. It was observed that the results matched those that were found in the literature.

Here,

$$C = 427.1 \frac{EHP}{\Delta^{2/3} \cdot V^3} \quad \text{and}$$

$$P = 0.746 \frac{V}{\sqrt{C_p} \, L} = \frac{0.746}{C_p} \cdot Fr$$

is being provided along with it.

The resistance characteristic of the 10 meters long boat is shown in Figure 7 and the "Resistance Analysis" that was performed based on it is shown in Table 3. Resistance coefficients that were determined as a result of the analysis are shown graphically in Figures 8 and 9. The determination of a form factor using the Prohaska method is shown in Figure 10.

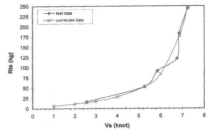

Figure 7. Prototype Resistance Characteristics $(L_{OA})_s = 10$ m., $T_s = 0.520$ m.

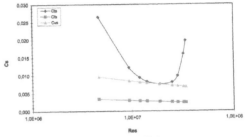

Figure 8. Prototype Resistance Coefficients $(L_{OA})_s = 10$ m., $T_s = 0.520$ m.

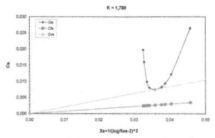

Figure 9. Calculation of Form Factor According to Modified Hughes Method

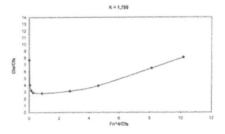

Figure 10. Calculation of Form Factor According to Prohaska's Method

Table 3. Resistance analysis for prototype and model: $\alpha = 10/3$, $(L_{oa})_s = 10$ m., $T_s = 0.520$ m., $(L_{oa})_m = 3$ m., $T_m = 0.156$ m.

V_s(knot)	V_m (m/s.)	Fn	Re_s	C_{fs}	C_{ts}	C_{vs}	C_w	Re_m	C_{fm}	C_{vm}	C_{tm}
1,00	0,514	0,055	$4,70.10^6$	$3,44.10^{-3}$	$2,65.10^{-2}$	$9,58.10^{-3}$	$1,69.10^{-2}$	$7,28.10^5$	$5,03.10^{-3}$	$1,40.10^{-2}$	$3,09.10^{-2}$
2,00	1,029	0,110	$9,40.10^6$	$3,0,3.10^{-3}$	$1,22.10^{-2}$	$8,46.10^{-3}$	$3,77.10^{-3}$	$1,46.10^6$	$4,33.10^{-3}$	$1,21.10^{-2}$	$1,58.10^{-2}$
2,56	1,317	0,140	$1,20.10^7$	$2,91.10^{-3}$	$9,41.10^{-3}$	$8,10.10^{-3}$	$1,30.10^{-3}$	$1,86.10^6$	$4,11.10^{-3}$	$1,15.10^{-2}$	$1,28.10^{-2}$
3,00	1,543	0,164	$1,41.10^7$	$2,83.10^{-3}$	$8,27.10^{-3}$	$7,89.10^{-3}$	$3,77.10^{-4}$	$2,18.10^6$	$3,98.10^{-3}$	$1,11.10^{-2}$	$1,15.10^{-2}$
4,00	2,058	0,219	$1,88.10^7$	$2,70.10^{-3}$	$7,55.10^{-3}$	$7,52.10^{-3}$	$2,64.10^{-5}$	$2,91.10^6$	$3,76.10^{-3}$	$9,95.10^{-3}$	$1,05.10^{-2}$
5,28	2,716	0,289	$2,48.10^7$	$2,58.10^{-3}$	$8,10.10^{-3}$	$7,19.10^{-3}$	$9,10.10^{-4}$	$3,85.10^6$	$3,57.10^{-3}$	$9,71.10^{-3}$	$1,09.10^{-2}$
6,00	3,086	0,329	$2,82.10^7$	$2,52.10^{-3}$	$9,91.10^{-3}$	$7,04.10^{-3}$	$2,87.10^{-3}$	$4,37.10^6$	$3,48.10^{-3}$	$9,71.10^{-3}$	$1,26.10^{-2}$

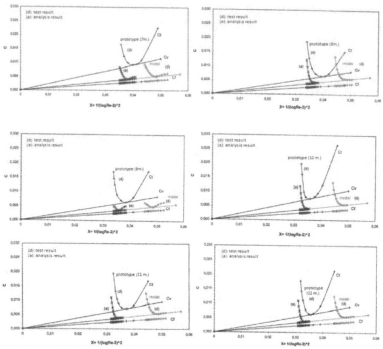

Figure 11. Test and Analysis Results for Model (Prototype → Model)

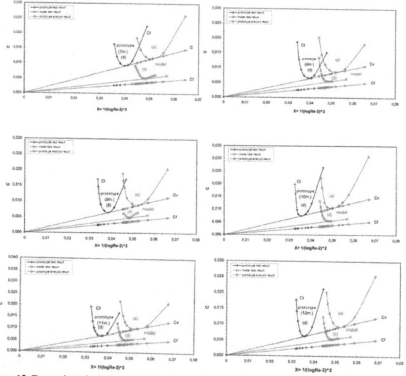

Figure 12. Test and Analysis Results for Prototype (Model → Prototype)

Figure 13. Normal and Inverse Transformed Extrapolator Curves Finding from Prototype Tests.

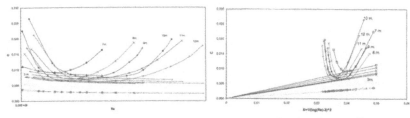

Figure 14. Prototype (12 m.) and its model (3 m.)

5 CONTRARY HYDRODYNAMICAL INTERACTIONS

In this section, the relationship between the data which is obtained from the resistance experiment of model and prototype boats and data that is obtained from resistance analysis was studied. The data of the resistance experiment of prototype boats was compared with analysis data was obtained from the model draught values. These model draught values were those that were equivalent to the data of resistance experiment of the prototype boats. In Figure 11, analysis results that were obtained from the experiment of each prototype are compared with the results of model experiments one by one. In Figure 12, the analysis results that were obtained from model experiments are compared with the results of the prototype experiments one by one. When the geometrical, kinematic and dynamic similarities are fully realized, the analysis and experiment results of both the prototype and model boats are expected to be the same.

The flow must be checked to see if it is laminar flow or turbulence flow to confirm the similarity of the flow. For this reason, turbulence generators are used in model experiments. On the other hand, turbulence flow is usually present in prototype experiments.

Extrapolator curves which display the experiment results of boat the model boat and the six prototype boats together and the image of these curves after the inverse transformation are shown in Figure 13.

6 CONCLUSIONS

Model resistance experiment is the most realistic and reliable method in the determination of the resistance of a ship. Results that have been obtained by resistance analysis that was done in accordance with similarity laws during the transformation from model to prototype, in other words, during normal transformation using corrected data with curve-fitting that is obtained from the data of model experiments are the following :

1 Minimum resistance values were obtained in the model draught which is equivalent to the draught of a 9 meters long boat.
2 In this study, it has been observed that form factors which have been calculated using the method that was mentioned by the author (Hughes method in which transformation is used) and Prohaska methods are highly approximately the same.
3 The average of form factors was determined to be K = 0,441. These form factors are those that were obtained in the model draught which is equivalent to that of prototype vessels.

Results that have been obtained by resistance analysis that was done in accordance with similarity laws during the transformation from the prototype to the model, in other words, during inverse transformation using corrected data which was obtained from resistance experiments of prototype boats are listed below :

1 In this study, it has been observed that the form factors of prototype boats which have been

calculated using the method of the study and the Prohaska method were highly approximately the same.

2 It has also been determined that the form factors of prototype boats did not change proportionally with draughts.

3 The harmony that is expected under ideal conditions in extrapolation curves cannot be realized in cases where complete geometric similarity cannot be observed amongst prototype boats that were chosen by means of sampling.

Conclusions that can be made after comparing the results that were obtained from model and prototype scales are listed below :

1 When the extrapolation diagram is studied, it can be said that prototype which has a total length of 8 meters is the most equivalent prototype for a model that has a total length of 3 meters.

2 Virtually the "resistance anatomy" of boats have been revealed as a result of this study. This situation makes the optimization of these boats in terms of systematic propeller design and the planning of developments in form design possible.

3 The method that is generally used in **Geosim Analysis** is the one in which at the least two models are tested in a model towing tank with each model being of different size and the results of this experiment and that of the experiment at sea involving a prototype are compared.

4 In this study, an unusual method that is not routine and for which there are no examples in the literature is used. Only a single model and corresponding six prototypes are used. An increase in the number of models can be considered in order to deepen the research a lot further.

REFERENCES

Lammeren, V., Troost, L., Koning, J.G., Resistance, Propulsion and Steering of Ships, 1950.

Kafalı, K., Static and Dynamic Fundamentals of Ship Forms (in Turkish), vol.2 (Ship Resistance and propulsion), 1972.

Harvald, Sv. Aa., Resistance and Propulsion of Ships, 1983.,

Şalcı, A., Power Calculation of Fishing Boats, Seminar Notes (in Turkish), 9 September University, Izmir, 1985.

Lewis Edward, V. (Editor), Principles of Naval Architecture, vol.2 (Resistance, Propulsion and Vibration) SNAME, 1988.

Kafalı, K., Form, Stability, Resistance and Propulsion of Fishing Boats (in Turkish), 9 September University, Izmir, 1989.

Şalcı, A., Investigation of Hydromechanical Characteristics of Köyceğiz-Dalyan Boats, Final Report III (in Turkish/English), T.R. Prime Ministry Ö.Ç.K.K. – GTZ (Germany), 1991.

Garcia, G.A., On the Form Factor Scale Effect, locate/ocean eng. Madrid-Spain, www.elsevier.com., 1992.

Şalcı, A., Boats Form Design Which Suitable GAP (Southeastern Anatolia Project) Waters, TUBITAK Marmara Research Center, Technical Report Nr. T3-1, Gebze-Izmit, 1993

Kalıpçı, S., Systematic Resistance Analysis of Infantry Type Fishing Boats, B.Sc. Thesis (supervisor: Şalcı, A.), Istanbul Technical University, Istanbul, 1995.

Şalcı, A., Hydrodynamic Design Evaluation of Multi-Purpose Container Ship, Istanbul Technical University-Ship Model Towing Tank, Project Nr. 98-03, Istanbul, 1998.

Kalıpçı, S., Hydromechanic Analysis of Infantry Type Fishing Boats, M.Sc. Thesis (supervisor : Şalcı, A.), Istanbul Technical University, Istanbul, 1999.

Şalcı, A., Kalıpçı, S., Systematic Drag Analysis of Infantry Type Fishing Boats (in Turkish), Technical Congress of Naval Architecture Proceedings Book, Istanbul, 1999.

Aydın, M., Computer-Aided Design of Fishing Boats Suitable For Turkish Waters, Ph.D. Thesis (supervisor : Şalcı, A.), Istanbul Technical University, Istanbul, 2001.

Yıldırım, S., Contrary Hydrodynamical Interactions Between The Model And Prototype Of Fishing Boats. "Geosim Analysis" (in Turkish), M.Sc. Thesis (supervisor : Şalcı, A.) Kocaeli University, Kocaeli, 2004.

New Methods of Measuring the Motion (6DOF) and Deformation of Container Vessels in the Sea

D. Kowalewski & F. Heinen
Geo.IT Systeme GmbH, Berlin, Germany

R. Galas
Technische Universität Berlin, Department for Geodesy and Geoinformation Sciences, Precise Navigation and Positioning, Berlin, Germany

ABSTRACT: The state-aided project MoDeSh (Motion and Deformation of Ships) lasted for three years and was completed in March 2012. In this project, six highly precise GNSS receivers were placed on 300 metres long container vessel "Kobe Express" from Hapag Lloyd. In these three years the vessel travelled around the world, allowing us to collect raw data about the motion (all six degrees of freedom) and the deformation of the ship during all of this time. In the past months we have evaluated the collected data and can now present the results. We have found a new method of calculating moving baselines and we achieved to calculate the deformation with a very high accuracy.

1 INTRODUCTION OF THE PROJECT

The project MoDeSh (Motion and Deformation of Ships) was promoted by the German government. The two main project partners are the "Germanischer Lloyd (GL)" and "The Hamburg Ship Model Basin (HSVA)". One part of the project was the development of a 6-DOF (degree of freedom) GNSS system for the measurement of movements and deformation. For this part the company Geo.IT System acted as subcontractor of HSVA, together with the Technical University of Berlin (TU Berlin), Department of Geodesy, Chair of Precise Navigation and Positioning. Using three GNSS systems at the bug of the ship and three ones on the bridge, we made a long-term measurement of the movements as well as of the deformation of the container vessel.

1.1 Container vessel "Kobe Express"

Subject of the project was the container vessel "Kobe Express", which belongs to the German Hapag-Lloyd and whose port of registry is Hamburg. It is the first time that this vessel was used for research projects and the installation of wave radar and acceleration sensors helped us to validate the measurements with the GNSS system. Another advantage of the Kobe Express is its use as a liner. The ship starts from Hamburg to Rotterdam over the Atlantic to Halifax, New York, through the Panama Channel, to San Francisco, over the Pacific to Tokyo, Beijing and Shanghai and the same way back. Every three month the Kobe Express was in Hamburg or Bremerhaven and we had the chance to check our system.

Facts about Kobe Express:

Dimensions:	
Total lengths:	293.94 m
Total width:	32.31 m
Total height:	21.40 m
Max. water depth:	13,63 m
Max. height:	54.25 m
Dead weight:	67.631 t
Gross load weight:	89.129 t
Max. payload:	21304 t
Container:	4612 TEU
Machine:	
Manufacturer:	MAN
Model:	B&W 9 K90 MC MK 6
Output power:	41.130 kW (55.937 PS)
Engine speed:	94 RPM
Max. speed:	24,5 kts (45,4 km/h)
Range:	21.500 nm (39.818 km)
Rudder area:	61 m²
Propeller diameter:	8.30 m
Bridge over water level:	43.30 m

2 PROJECT START

2.1 *The selection of GNSS receivers and antennas*

The six GNSS receivers and antennas needed, meant high investments. One of the main criteria for the selection was therefore the price. The company Topcon gave us very good conditions for their 6 GB-1000 and 6 PG-A1 antennas. Positive was also the fact that we could work with GPS and GLONASS systems and that we had 20 Hz raw data on output. Unfortunately, the quality of the PG-A1 antennas was sub-optimal, as after one year they were out of order. The seawater had penetrated the cover and destroyed the electronics of the antenna. As a result, we swapped the six antennas for the high-end GNSS antennas 3G+C from the company navXperience. These antennas proved to be absolutely waterproof as their cover are sealed with a new laser welding technology. The 3G+C antennas also have the highest military standard (MIL-STD) 810g. After two years on the vessel the antennas were functioning as if they were new.

2.2 *The selection of the operating system and the development environment*

With a data-sampling rate of 20 Hz, using GPS and GLONASS, we received 60 MB of raw data per receiver and per hour. Having six receivers in total and an operating time of three months we ended up having 777 GB of data. In order to keep the risk of data loss low, we used three 500 GB hard disks and the operating system Ubuntu Linux 10.04. As a result, no data was lost during the project. As the project partners GL and HSVA preferred to work with Windows OS, we used the software Sharpdevelop 2.0 and 3.0 and Monodevelop 2.0 as a development environment. The advantage of this was that we could work with one source code for Windows, Linux and Mac operating systems.

2.3 *Installation of the instrumentation*

For the measurement of the 6 DOF we used the starboard side of the bridge, because there we had the most space and the lowest shadowing effects. Regarding the geometry, it would have been better to install the GNSS antennas in the middle of the ship. However, this was not an option, because this space is usually reserved for the compass, where it is not allowed to install any other electronic devices. At the bow however, we were able to use the mast in the middle of the ship. The following graphics show the installation plan of GNSS antennas:

Installation Map GNSS Antennas at the Kobe Express

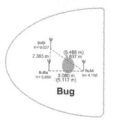

All units m und the slope distance

Bug

Bridge

2.4 *Installation of the network*

The expected raw data transmission from the GNSS receivers necessitated the installation of a Local Area Network (LAN). One problem was the ship length of 300 metres. We had to install a glass fibre cable reaching from under the mast to the machine room. From there, we used the normal network installation of the vessel with our own number group of the IP address. The reason for doing this, was to ensure the network security of the vessel. The next graphic shows the network plan:

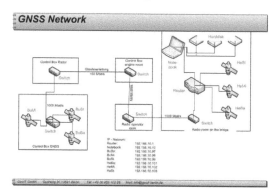

3 SOFTWARE DEVELOPMENT

3.1 *Graphical user interface (GUI)*

We developed the GUI according to the project work-flow. The software is usually used for ship testing and for manoeuvre ways. The first line presents general data, for example about the antenna installation, and data about the ship. The second line shows the raw data recording and the post processing of the data. The last line provides some tools. The software is very user-friendly and one does not need to be a computer expert to work with this software.

3.2 *Raw data recording*

The requested antenna arrangements in the equipment configuration can be found again in the raw data recordings. One arrangement is, that in the GUI the user will see three antennas at the same time. The user can see the status of each GNSS receiver and he or she can start operating all receivers of one arrangement at the same time. The user has got the option to start the receivers with different parameters or, alternatively, all receivers of one arrangement with the same script file. He or she can also observe how many satellites are available and with what data sampling rate the observations are recorded.

3.3 *Post processing*

A number of papers describing algorithms for GPS-based precise attitude control in marine environment have been published in the last years. Among others: Kleusberg (1995), Lachapelle et al (1996), Andree et al (2000). Our processing algorithm, used in the experiment, is based on the so called relative moving baseline approach.

The post processing takes place in only a few steps. At first we calculated the position as well as the position variations of the master antenna. Then, we calculated the position variation of the two slave receivers. With these data we went to the next step and calculated the nautical data: roll, pitch and yaw and the translational movements: heave, sway, surge as well as the speed and the acceleration in the three main axles. In the third and final step the graphic analysis of the nautical data was displayed.

4 RESULTS

For the validation of the results we used a GNSS assisted inertial system. This system uses the GNSS receiver only for the initial coordinates and for the first time synchronization. The IMU had an accuracy of 0.1° per hour. The application software was developed by HSVA and has been in use for years with good results. In all comparative measurements we obtained exactly the same results in the first ten minutes. After that time span the IMU started to drift, as expected. These experimental results showed that our 6 DOF GNSS system fulfils the requirements for long-term measurements better than an IMU with a very poor GNSS support. This was of course expected.

4.1 *Accuracy of the relative positioning*

The following screen dump shows the accuracy in the relative positioning. This data was collected when the sea was calm. The slightly wave-like course is typical for Kobe Express. So we can be sure that our accuracy is higher than ± 2 cm.

4.1 Accuracy of the movements

The next graphic shows the roll movements from the ship in driving a curve with a rudder position of 20°. The nearly 30 seconds that the vessel needs from **portside to starboard** and back are absolutely significant. The bumpy trend of the graph is not only due to inaccuracy. It also presents the vacillations of the ship. The roll angle is at only 1.5°, which is negligible. Our client, the HSVA, was really surprised about these good results.

4.2 *Measurement of the deformation*

The deformation is the difference between the angles from stern to bow of the vessel. For example, in the following screen dumps you can see the two roll angles. Mostly, the roll angles show a similar graph line, but there are also deviations, showing the differences in the twist. The result can simply be calculated by taking the difference.

5 SUMMARY

This project gave us an opportunity of a special scientific experiment – which is very rare in our normal GNSS engineering applications. Working together with the ship engineers was a good experience and we learned a lot about roll, pitch and heading under wave effect and about the problems with weather, wind and the loading of a container vessel. The ship engineers on the contrary learned a lot about coordinate systems, and working with precise GNSS technology and on the geoid undulation.

REFERENCES

[1] Andree, P., Läger, R., Schmitz, M., Wübbena, G. (2000). Bestimmung von Schiffsbewegungen und anderer hochfrequenter Bewegungen mittels GPS. In Lechner, W., editor, DGON Symposium Ortung + Navigation. Freising-Weihenstephan, Germany, Oct.17-19, 2000.
[2] Kleusberg, A. (1995). Mathematics of attitude determination with GPS. GPS Word, Vol. 6, No. 9, pp. 72-78, September, 1995.
[3] Lachapelle, P., Cannon, M.E., Lu, G., Loncarevic, B. (1996). Shipborne GPS Attitude Determination During MMST-93. IEEE Journal of oceanic Engineering, Vol. 21, No. 1, pp. 100-1005. January 1996.

Hybrid Bayesian Wave Estimation for Actual Merchant Vessels

T. Iseki
Tokyo University of Marine Science and Technology, Tokyo, Japan

M. Baba & K. Hirayama
Japan Radio Co, Ltd., Tokyo, Japan

ABSTRACT: A new algorithm of Bayesian wave estimation is proposed in which the information obtained from the radar wave observation system are introduced to estimate the directional wave spectrum around the ship. Radar wave observation systems can measure the wave directions and periods precisely but cannot provide any information about the wave heights theoretically. The Bayesian method can estimate wave heights according to the wave buoy analogy. Therefore, the combination of the two methods can be considered as a mutually complementary system. In order to investigate the effectiveness of the proposed method, on-board experiments using an actual container vessel which was operated on the north pacific route was carried out. The response functions of the ship motion were estimated by using the dummy hull-form which can be generated by only the principal particulars. The estimated directional spectra are compared to NOAA data and the usefulness is shown.

1 INTRODUCTION

To reduce the GHG (Greenhouse Gas) emissions from shipping, the Energy Efficiency Design Index (EEDI) and the Ship Energy Efficiency Management Plan (SEEMP) were made mandatory at MEPC 62 (July 2011). Among the shipping companies and the operators, the Energy Efficiency Operational Indicator (EEOI) was defined as a tool of SEEMP and "slow steaming", "weather routing" and "just in time arrival" are considered as the most effective SEEMP related measures. It is very important but difficult for those measures to obtain the wave information around the ship in a real seaway. Of course, it is possible to get the wave data from forecast or hindcast but their resolution is not sufficient for such purposes. The on-site wave estimation based on the buoy analogy can be considered as the most possible measure for obtaining the wave information. Based on the assumption of linearity between waves and ship motions, the directional wave spectrum can be estimated by regarding the ship's hull as a wave rider buoy (e.g. Webster & Dillingham 1981). In order to improve the accuracy, Bayesian modeling procedure was introduced to the analogy and the usefulness as an on-board guidance system was investigated (Iseki & Ohtsu 2000, Iseki & Terada 2002, Nielsen 2006).

The buoy analogy requires detailed hull-form data of the ship. Nowadays, however, it is very difficult to get the body plan or the off-set data that represents the hull shape in detail. That is the most crucial problem for the actual application to merchant vessels. Therefore, one of the authors investigated the possibility of introducing a dummy hull-form to the Bayesian wave estimation, using T.S. Shioji-maru of Tokyo University of Marine Science and Technology (Iseki 2010). The dummy hull-form can be generated by the 2-parameter Lewis forms (Lewis 1929) and the ship's principal particulars. Introducing the Lewis forms, there are two advantageous points. First, the Lewis forms require only the distributions of cross sectional area, sectional breadth and draught to approximate the hull-forms. Second, the hydrodynamic forces can be calculated by Ursell-Tasai's method (Tasai 1969). Using the generated data, the response functions are evaluated and wave parameters are estimated.

The investigation was followed by the on-board experiments using an actual container vessel which was operated on the north pacific route (Iseki et al. 2012). The estimated wave parameters were compared with a radar wave observation system which was also installed in the vessel. It is widely known that radar wave observation systems can measure the wave directions and periods precisely but cannot provide any information about the wave heights theoretically. The Bayesian method can estimate wave heights based on the wave buoy analogy. Therefore, the combination of the two methods can be considered as a mutually complementary system. In the report, a simple algorithm was investigated in which the wave direction and the period measured by the radar wave observation system were transferred to the Bayesian wave estimation. The simple algorithm was confirmed very effective but it was also pointed out that more effective and consistent algorithm should be developed.

In this report, a new algorithm of Bayesian wave estimation is proposed in which the directional spectrum obtained by the radar wave observation system are effectively introduced to the modeling. In order to investigate the effectiveness of the proposed method, the estimated directional spectra are compared to NOAA data and the usefulness is discussed.

2 NEW ALGORITHM FOR BAYESIAN WAVE ESTIMATION

It is widely known that radar wave observation systems can precisely measure wave directions and wave periods but cannot provide any information about the wave heights. The reflection intensity of the radar wave is not related to the wave heights but just roughness of the sea surface. Therefore, the method that estimates the power of the directional wave spectrum estimated by radar wave observation systems can be considered as a mutually complementary algorithm. In this section, the new algorithm of the Bayesian wave estimation.

If ship motions are considered to be linear responses to incident waves, the cross spectrum of ship motions and the directional wave spectra are related by frequency response functions as follows:

$$\varphi_{ij}(f_e) = \int_{-\pi}^{\pi} H_i(f_e, \chi) H_j^*(f_e, \chi) E(f_e, \chi) d\chi \quad (1)$$

where f_e denotes encounter frequency; $E(f_e, \chi)$ is the directional wave spectra based on encounter frequency; $\varphi_{ij}(f_e)$ is the cross spectrum between the i-th and j-th components; and $H_i(f_e, \chi)$ is the response function of the i-th component of the time series.

As the directional wave spectra should be expressed based on true wave frequencies for convenience, Equation (1) must be transformed into true wave frequencies from encounter frequencies. Considering the triple valued function problem in the following seas, the discrete form of Equation (1) can be expressed by the following matrix expression:

$$\Phi(f_e) = H(f_{01})E(f_{01})H(f_{01})^{*T} + H(f_{02})E(f_{02})H(f_{02})^{*T} \\ + H(f_{03})E(f_{03})H(f_{03})^{*T} \quad (2)$$

where f_{01}, f_{02} and f_{03} are the true wave frequencies that correspond to the encounter frequency f_e. is the measured cross spectrum matrix, and denotes encounter frequency; $E(f_e, \chi)$ is the directional wave spectra based on encounter frequency; $\Phi(f_e)$ is the measured cross spectrum matrix, and $H(f_{0i})$ and $E(f_{0i})$ for ($i=1,2,3$) denote the matrices of response functions of the ship motions and directional wave spectrum at f_{01}, f_{02} and f_{03}, respectively.

As $\Phi(f_e)$ is a Hermitian matrix, Equation (2) can be reduced to a multivariate regressive model expression using only the upper triangular matrix:

$$B = AF(x) + W \quad (3)$$

where B denotes the cross spectrum vector, which is composed of real and imaginary parts of each element of $\Phi(f_e)$, A denotes the coefficient matrix composed of products of the ship motion response functions, W is a Gaussian white noise sequence vector introduced for stochastic treatment and $F(x)$ denotes the unknown coefficient vector that is composed of the discretized directional wave spectrum.

In this model, since the number of unknown coefficients is much larger than that of equations, the fitting problem of Equation (3) cannot be solved alone. Based on the Bayesian modeling procedure, which was formulated by Akaike (1980), the unknown coefficients of Equation (3) can be evaluated by maximization of the product of the likelihood function and appropriate prior distributions that are introduced as the stochastic constraints. Here, the prior distribution can be recognized as a general character of the model. In case of wave estimations, two Gaussian smoothness prior distributions are usually used for smoothness of the directional wave spectrum in the directional-wise and frequency-wise.

$$\sum_{m=1}^{M}\sum_{n=1}^{N}\varepsilon_{1mn}^2 = \sum_{m=1}^{M}\sum_{n=1}^{N}(x_{m,n-1} - 2x_{mn} + x_{m,n+1})^2 \quad (4)$$

$$\sum_{m=1}^{M}\sum_{n=1}^{N}\varepsilon_{2mn}^2 = \sum_{n=1}^{N}\sum_{m=1}^{M}(x_{m-1,n} - 2x_{mn} + x_{m+1,n})^2 \quad (5)$$

where, M and N are the number of discrete wave frequencies and angles of encounter.

In order to take into account the information obtained from the radar wave observation system, another type of prior distributions can be introduced.

$$\sum_{m=1}^{M}\sum_{n=1}^{N}\varepsilon_{(1 \, or \, 2)mn}^2 = \sum_{m=1}^{M}\sum_{n=1}^{N}(x_{m,n} - x_0)^2, \quad (6)$$

where x_0 denotes small value to restrain the spectrum from having power. Replacing Equation (4) or (5) with Equation (6) at the direction and the frequency where any power was estimated by the radar wave observation system, the power of the remaining points can be estimated by Bayesian modeling procedure.

Assuming ε_{imn} $(i=1, 2)$ to be normal distributions with mean zero and variance σ^2/u_i^2, the probability density function of the prior distributions are written as:

$$P_i(\mathbf{x}) = \left(\frac{u_i^2}{2\pi\sigma^2}\right)^{MN/2} \exp\left[-\frac{u_i^2}{2\sigma^2}\sum_{m=1}^{M}\sum_{n=1}^{N}\varepsilon_{imn}^2\right], \quad (i=1,2) \quad (7)$$

The cost function to be minimized can be expressed as follows (Nielsen 2008):

$$J(\mathbf{x}) = \|\mathbf{AF}(\mathbf{x}) - \mathbf{B}\|^2 + \mathbf{x}^T\left(u_1^2\mathbf{D}_1^T\mathbf{D}_1 + u_2^2\mathbf{D}_2^T\mathbf{D}_2\right)\mathbf{x} \quad (8)$$

where $\|\bullet\|$ represents the L_2 norm and u_1 and u_2 are the hyperparameters that control the trade-off between the good fit to the data and the prior distributions. $\mathbf{D}_1$ and $\mathbf{D}_2$ are the Gaussian smoothness prior distributions for directional-wise and frequency-wise.

In the actual calculations, the unknown vector should be expressed in the following form to avoid a negative wave spectrum:

$$\mathbf{F}(\mathbf{x})^T = (\exp(x_1)\cdots\exp(x_J)), \quad \exp(x_j) = E_j(f_0) \quad (9)$$

As a consequence of the substitution of Equation (9), the cost function becomes a non-linear equation. Equation (8) is, therefore, linearized for numerical calculation as follows:

$$J(\mathbf{x}) \cong \|\hat{\mathbf{A}}\mathbf{x} - \hat{\mathbf{B}}\|^2 + \mathbf{x}^T\left(u^2\mathbf{D}_1^T\mathbf{D}_1 + v^2\mathbf{D}_2^T\mathbf{D}_2\right)\mathbf{x}$$

$$= \left\|\begin{pmatrix}\hat{\mathbf{A}}\\u\mathbf{D}_1\\v\mathbf{D}_2\end{pmatrix}\mathbf{x} - \begin{pmatrix}\hat{\mathbf{B}}\\0\\0\end{pmatrix}\right\|^2 \quad (10)$$

where

$$\hat{\mathbf{A}} = \mathbf{A}\mathbf{E}(\mathbf{x}_0), \ \hat{\mathbf{B}} = \mathbf{B} - \mathbf{AF}(\mathbf{x}_0) + \hat{\mathbf{A}}\mathbf{x}_0, \ \mathbf{E}(\mathbf{x}) \equiv \frac{\partial\mathbf{F}(\mathbf{x})}{\partial\mathbf{x}} \quad (11)$$

and $\mathbf{x}_0$ is the initial value of unknown vector $\mathbf{x}$.

Finally, the directional wave spectrum can be obtained by minimizing Equation (10) and the optimum hyperparameters can be obtained by minimizing Akaike's Bayesian information criterion (ABIC), which can be expressed as:

$$ABIC = -2\log\int L(\mathbf{x}|\sigma^2)P_1(\mathbf{x})P_2(\mathbf{x})d\mathbf{x} \quad (12)$$

where $L(\mathbf{x}|\sigma^2)$ denotes the likelihood function of the model.

3 GENERATION OF DUMMY HULL-FORM

Response functions of ship motions $\mathbf{H}(f_{0i})$ in Equation (2) are calculated using the actual hull-form of the ship. In the previous study (Iseki & Ohtsu 2000), the response functions are estimated by New Strip Method (NSM) in which added masses and damping coefficients are calculated by the direct boundary element method known as the "Close Fit method". In case of merchant ships, however, it is very difficult to obtain the body plan or the off-set data that represents the detailed hull shape. Therefore, in this section, a simple procedure based on the Lewis transformation (Lewis 1929) is proposed by only referencing the ship's principal particulars for making up the dummy hull-form data. Using the Lewis transformation, the cross section of the ship's hull can be mapped conformally to an unit circle. The 2-parameter Lewis conformal mapping is expressed by:

$$\frac{\mathbf{z}}{M} = \mathbf{w} + \frac{a_1}{\mathbf{w}} + \frac{a_3}{\mathbf{w}^3} \quad (13)$$

where $\mathbf{z}$ and $\mathbf{w}$ are complex coordinates of the plane of the ship's cross section and the unit circle, respectively. The conformal coefficients a_1 and a_3 are called Lewis coefficients, while M is the scale factor. The coefficients and the factor will be defined when the half breadth to draught ratio H_0 and the sectional area coefficient σ are provided.

$$H_0 = \frac{B}{2T}, \quad \sigma = \frac{S}{BT} \quad (14)$$

where B, T and S are the sectional breadth on the waterline, the sectional draught and the area of the cross section, respectively.

The added masses and damping coefficients are calculated by Ursell-Tasai's method (Tasai 1969). Therefore, if we represent the distributions of B, T and S along the ship length as the functions $B(x)$, $T(x)$ and $S(x)$, the generation of dummy hull-form can be changed into a problem of approximation of the functions.

3.1 Distribution of Cross Sectional Area S(x)

In this study, the quartic polynomial is used to approximate $S(x)$.

$$S(x)/S_0 = ax^4 + bx^3 + cx^2 + dx + e, \quad (-0.5 \leq x \leq 0.5) \quad (15)$$

where S_0 is the midship cross sectional area under the waterline. If we consider the following conditions:

$$S(-0.5) = 0, \quad S(0) = 1, \quad S(0.5) = 0, \quad (16)$$

each coefficient can be related to each other.

$$c = -a/4 - 4, \quad d = -b/4, \quad e = 1 \quad (17)$$

Furthermore, we can introduce two conditions to determine the coefficients as follows:

$$\int_{-0.5}^{0.5} \frac{S(x)}{S_0} dx = C_P, \quad \int_{-0.5}^{0.5} \frac{S(x)}{S_0} x dx = C_P \frac{l_{cb}}{L_{PP}} \quad (18)$$

where C_P, l_{cb} and L_{PP} are the prismatic coefficient, longitudinal distance of the centre of buoyancy from the midship and the length between perpendiculars. Applying these conditions, we obtain the remaining coefficients.

$$a = 120\left(\frac{2}{3} - C_P\right), \quad b = -120 C_P \frac{l_{cb}}{L_{PP}} \quad (19)$$

3.2 Distribution of Sectional Breadth B(x)

The quartic polynomial approximation is not suitable for $B(x)$ because the actual ships have some parallel body amidships. The cross section in which its breadth is wider than that of midship section may appear based on the location of centre of buoyancy. In this study, $B(x)$ is approximated by a trapezoidal distribution.

$$B(x)/B_0 = \begin{cases} (x+0.5)/l_s & \cdots (-0.5 \le x \le l_s - 0.5) \\ 1 & \cdots (l_s - 0.5 \le x \le 0.5 - l_b) \\ (0.5 - x)/l_b & \cdots (0.5 - l_b \le x \le 0.5) \end{cases} \quad (20)$$

where B_0 denotes the breadth at the midship section on the waterline l_s and l_b denote the length of triangular part near the stern and the bow. Let l_{pb} be the non-dimensional length of the parallel body and let x_{pb} be the coordinate of the center, the l_s and l_b can be represented by the following relationship.

$$l_s = \frac{1}{2}(1 - l_{pb}) + x_{pb}, \quad l_b = \frac{1}{2}(1 - l_{pb}) - x_{pb} \quad (21)$$

4 ON-BOARD MEASUREMENT

Ship motion time histories were measured using the 6,500 TEU class container carrier. The photo and principal particulars of the ship are shown in Figure 1 and Table 1. In order to obtain the actual wave condition around the ship, a radar wave observation system (JMA-242, Japan Radio Co, Ltd.) was also installed in the wheel house. The system consists of a desk-top computer and communication boards and the specific software. The successive images are obtained from the x-band marine radar and the directional wave spectra are estimated by Fourier analysis. The wave parameters, such as wave direction, speed and length, are also provided as the outputs in NMEA format.

The measurement was started from the winter of 2010 and is still going on. The ship motions were measured by the attitude measurement units (AMU-1802BR Sumitomo Precision Products Co. Ltd.) installed in the wheel house and the bosun store. The data acquisition PC was set in the wheel house and the data measured at the bow were obtained through the wireless LAN. Both of the measured data were stored in the PC as 10 minutes time histories every one hour.

Table 1. Principal particulars of the container carrier

Length (P.P.)	292.00 (m)
Breadth (MLD)	40.00 (m)
Depth (MLD)	24.80 (m)
Draught (MLD)	13.00 (m)

Figure 1. The 6,500 TEU class Container carrier.

5 DUMMY HULL-FORM AND RESPONSE FUNCTIONS

Figure 2 shows the body plan generated by the proposed dummy hull-form. The cross sections were calculated by the 2-parameter Lewis conformal mapping and the dummy B(x) and S(x) described in the section 3. The left and right side of the figure are representing the after and the fore parts of the ship's hull. As shown in the figure, rather roundish hull shapes are obtained while the modern merchant vessels have square midship section.

Figure 3 shows the response functions of heaving, pitching and rolling motions in 15 knots. The horizontal axes are true wave frequencies in Hz and the vertical axes are response amplitude operators (RAOs). The heaving RAO is normalized by the incident wave amplitude and the pitching and rolling RAOs are normalized by the wave slope. In this study, transfer functions were calculated using the New Strip Method (NSM) in which added masses and damping coefficients are calculated by the Ursell-Tasai's method with the 2-parameter Lewis form, and wave exciting forces are evaluated by a sum of the Froude-Krylov forces, and diffraction forces that are approximately calculated based on the concept of relative motion. In this study, the response functions were transformed into vertical acceleration, pitch rate and roll rate in order to be consistent with the measured time histories.

6 COMPARISONS OF THE RESULTS

The time histories of the ship motions analyzed here were measured during the 12 days voyage on the north pacific route in the wintertime of 2010. The trajectory is indicated in Figure 4. The wave parameters estimated by the radar wave observation system and the proposed Bayesian method were compared with NOAA data. The sea conditions at the container carrier were evaluated by referencing the nearest three points of NOAA data. Those data could be obtained every 3 hours, therefore, 82 time histories were analyzed for the comparisons of wave parameters.

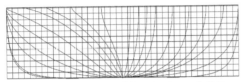

Figure 2. Body plan generated from the dummy hull-form.

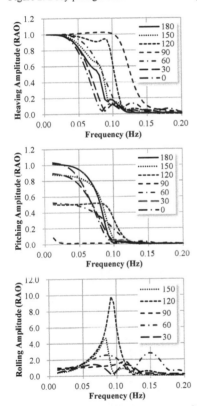

Figure 3. Response amplitude operators of heaving, pitching and rolling that were calculated from the dummy hull-form.

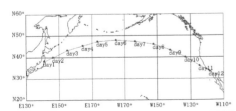

Figure 4. Trajectory of the voyage in the winter of 2010.

Figure 5 shows comparisons of the wave directions, periods of the waves and the significant wave heights. The horizontal axes denote the day of the voyage and three kinds of markers are used to represent the data of NOAA, the radar wave observation system and the proposed Bayesian new algorithm.

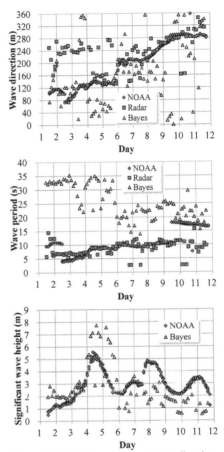

Figure 5. Scatter diagrams of the wave direction, the period and the significant wave height.

The NOAA data were chosen from the nearest point to the sailing ship and it should be noted that the wave conditions don't coincide with the actual sea conditions experienced by the ship. The radar wave observation system equipped to the ship estimates directional wave spectra every 2 minutes and the plotted points in the figures are the mean value of a certain length of the time. The system, however, cannot estimate the wave height theoretically. Therefore, the comparison of the significant wave height indicates only results of NOAA and the Bayesian estimation. Concerning the Bayesian estimates, the directions, periods and wave heights were evaluated by harmonic analysis, volumes and moments of the spectra. The estimated results of the radar wave observation system agree well with data of NOAA in case of wave directions and wave period. This is indicating that the radar wave observation system has high accuracy and validity for estimating the wave parameter excepting the wave heights.

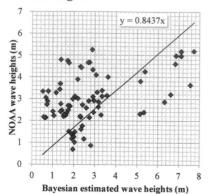

Figure 6. Scatter diagram of the wave height estimated NOAA and the new Bayesian algorithm.

On the other hand, looking at the results of wave directions and periods estimated by the Bayesian estimation, some of the results agree well but other results are dispersed. The reasons can be considered that the natural period of the ship motion is longer than 10 seconds and the wave periods were estimated longer. It is confirmed by Figure 3 that the response functions don't have large amplitude in the frequency range that is higher than 0.1 Hz. However, the comparison of the significant wave height rather agrees well with the data of NOAA. It should be emphasized here that the Bayesian wave estimation is based on the data of vertical acceleration, pitch rate and roll rate, moreover, the dummy hull-form was used for estimating calculations. Therefore, the idea combining the radar wave observation system and the Bayesian wave estimation seems to be quite effective.

Figure 6 shows the scatter diagram of the wave height estimated NOAA and the Bayesian wave estimation. The graph shows good correlation in the high wave height region. It should be noted again here that the NOAA data was chosen from the nearest point to the sailing ship. Furthermore, the sailing ship may be experiencing the higher waves than the NOAA data. This is pointing out the on-site wave estimation is very important for the actual ship navigation.

7 CONCLUSION

In this report, the Bayesian wave estimation applied to the actual merchant vessel introducing a new algorithm. In the algorithm, the dummy hull-form was generated to estimate the response functions of the ship and estimates of a radar wave observation system were taken into account. The estimated wave parameters were compared with NOAA data. The results are summarized below:

1 The radar wave observation system can provide wave directions and the periods with reasonable accuracy in comparisons with NOAA data.
2 The wave height estimated NOAA and the Bayesian new algorithm showed good correlation in the high wave height region.
3 The new algorithm for improving the Bayesian wave estimation is confirmed very effective and pointed out the importance of the on-site wave estimation for the actual ship navigation.

ACKNOWLEDGEMENTS

The measurement of the ship motion data was fully supported by NYK Line and Monohakobi Technology Institute. This work is partly supported by Grant-in-Aid for Scientific Research of the Japan Society for Promotion of Science (No. 23560966). The authors express sincere gratitude to the above organizations and thank the captain and crew of the container carrier.

REFERENCES

Akaike, H. 1980. Likelihood and Bayes procedure, Bayesian Statistics: 143-166. Valencia: University Press.
Iseki, T. & Ohtsu, K. 2000. Bayesian estimation of directional wave spectra based on ship motions, Control Engineering Practice 8, 215-219.
Iseki, T. & Terada, D. 2002. Bayesian Estimation of Directional Wave Spectra for Ship Guidance Systems, International Journal of Offshore and Polar Engineering 12, 25-30.
Iseki, T. 2010. Sensitivity Study of Bayesian Wave Estimation to Ship Motion Response Functions, Proc. Asia Navigation Conference 2010, 602-610.
Iseki, T., Baba, M. & Hirayama, K. 2012, Study on Bayesian Wave Estimation for Actual Merchant Vessels, Proc. The 14th IAIN Congress 2012, 1-8.
Lewis, F.M. 1929. The Inertia of Water Surrounding a Vibrating Ship, Transactions SNAME 37, 352-254.
Nielsen, U.D. 2006. Estimations of on-site directional wave spectra from measured ship responses, Marine Structures 19, 33-69.
Nielsen, U.D. 2008. Introducing two hyperparameters in Bayesian estimation of wave spectra, Probabilistic Engineering Mechanics 23, 84-94.
Tasai, F. 1969. Improvements in the Theory of Ship Motions in Longitudinal Waves, Proc. 12th I.T.T.C., 677-687.
Webster, W.C. & Dillingham,J.T. 1981, Determination of directional seas from ship motions, Proc. Directional Wave Spectra Application '81, 1-20.

Modelling Studies of the Roll and the Pitch Training Ship

W. Mironiuk
Polish Naval Academy, Gdynia, Poland

A. Pawlędzio
Det Norske Veritas, Gdynia, Poland

ABSTRACT: Results of tests of school-ship model's free rolling have been presented in the elaboration. The research was conducted in model basin of the Naval Academy. Based on the measurement results, periods of the model's rolling and pitching were determined for the vessel model. Moreover, general mass of accompanying water and respective terms of restoring forces matrix were calculated – with empirical dependencies applied. Next, the determined values were used to calculate periods of the model rolling and pitching and compared with results of the tests.

1 INTRODUCTION

Problem of prognosing seakeeping ship's qualities is an important area of concern in a process of the ship designing, as a complex technical system. This task is mainly realized by means of model tests run by specialized research teams. In parallel to the experimental tests, calculations of the parameters in question are made by means of dedicated software using modern numerical methods in numerous scientific centres. It seems that both methods double each other because they are used for almost the same aims. In such a case, theoretically, it would be the best solution to elaborate one universal, yet accurate and inexpensive, method possible for application in tests of ships of any type. Model tests mostly meet the above said conditions provided that one has appropriate (expensive) back-up laboratory facilities at disposal - facilities prepared for tests of various ship types. In case of calculation methods, implemented in specialist computer software, one may conclude that they are many times cheaper and yet – universal. However, they have one important defect due to which they have not squeezed the experimental tests out. Namely, from various tests of physic's problems, it is difficult to obtain results comparable to experimental tests. In spite of this defect, almost every research centre decides to apply computer calculation methods. Usually, companies having appropriate experience in programming numerical methods are manufacturers of the software. Many research centres undertake difficulty to elaborate their own independent computer

programmes which are focused on one specific problem. Possession of back-up laboratory facilities is a trump of such centres - by means of the facilities they may verify and correct code of the computer programme and assumptions for the physical as well as for the mathematical models. Following such institutions, the Naval Academy undertook steps towards research on seakeeping qualities of ships based on both the research facility and the numerical methods. Several years ago, a research facility was built in the Academy enabling conduction of tests with ship models in a scope of both statics and dynamics of vessels. Then, aside from the model tests, there are computer programmes being elaborated in the Academy, executing calculations regarding ship's hydromechanics. This elaboration is a result of initial experiments run in the research facility. It contains results of laboratory tests on rolling, pitching and heaving of a school-ship model on calm water and calculations of the rolling and pitching periods, executed with a use of analytical and empirical dependencies.

2 AIM AND EXECUTION OF TESTS

Determination of heaving, pitching and rolling periods of ship model is a target of the tests' stage under description - based on measurements done in the laboratory facility, and then – conduction of analogical calculations with a use of analytical and empirical formulas. The research was executed in the facility for tests on statics and unsinkability located in the Naval Academy in Gdynia, Poland. Model of school-vessel was an object of the tests, made in a scale of 1:50. Main dimensions of the real ship are: length L-72 m, breadth B-12 m, draught T-4,2 m and displacement 1750 t. This warship is used for training of Polish seafarers taking part in numerous international cruises. The basic data of the model are as the following:

- length between perpendiculars of the model

  L = 1,284 m
- breadth of the model   B = 0,232 m
- displacement of the model   D = 13,15 kos
- avg. draught of the model   T = 0,079 m

Hull of the model was made out of polyester-glass laminate, based on a drawing of body lines, and plywood was used to build the model's superstructure. Compartmentation of the model hull's internal space roughly corresponds with a real vessel. One compartment that is lengthwise divided into three smaller ones constitutes an exemption. Similar to a real vessel, the model under description has a bilge keel, two screw propellers and rudder. The ship model is equipped, among the others, with an angle of heel and trim sensor and a sensor of average draught [4]. Determination of the model's draught is done indirectly by measurement of hydrostatic pressure in respect to water surface. The sensor registers measurements with accuracy of 0.0001 m, while inclinometer makes the measurements with accuracy of 0.01 degree. Accuracy of the measuring equipment was evaluated during many experiments run with the model. Electric signals are emitted from the ship model to the computer via cable of insignificant weight, and the results – in a form of given parameters – are registered and displayed on monitor in a real time. The ship model, together with connected cables, is given in Figure 1.

Figure 1. School-ship model in model basin
- length between perpendiculars   L = 1,284 m
- breadth of the model   B = 0,232 m
- displacement of the model   D = 13,15 kos
- avg. draught of the model   T = 0,079 m

The tests are conducted in the model basin of the dimensions of l = 3 m, b= 2 m, h =0.5 m, filled with water up to the level of 0.4 m. Such dimensions of the tank result in some problems with waves that rebound from its edge and upset the measurements.

Therefore, several first seconds of the model movement, when the wave rebound was not visible, were taken into account in the result analysis.

The model used to be placed in parallel to the longer side of the basin during measurements of the parameters under discussion. All compartments of the model were empty. The model was capable of moving free in all degrees of freedom (in the required ranges). The free movement of the model was caused by application of starting conditions different from zero. Also, at the same time, efforts were made in order to minimize a phenomenon of individual movements' coupling (mainly between heaving and pitching). Measurements of the parameters under discussion were taken every 0.05 second.

In the initial period of the facility operation, tests on rolling were made many times. Based on them, location of centre of the ship model mass was determined precisely and also her metacentric height was determined which now is 0.0083 m.

3 RESULTS OF MEASUREMENTS OF SHIP MODEL REPLACEMENTS

3.1 Measurement of ship model rolling

Rolling of the ship model was registered by means of a sensor that simultaneously measures two angles: angle of heel and angle of trim. Results of the measurements are presented in Figure 2.

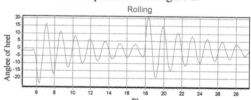

Figure 2. Results of measurements of ship model rolling

At the beginning, the ship model was tilted to an angle of some 23 degrees port side and then it was rolling free for some 17 seconds. Next, it was again tilted to an angle of some 21 degrees but towards opposite side. From the analysis of the presented course of rolling it results that its period is 1.64 second.

3.2 Measurement of ship model pitching

Measurements of the angles of trim were made after the wavy motion caused by rolling of the ship model stopped [10]. The obtained results are given in Figure 3.

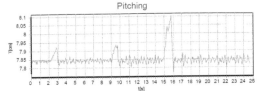

Figure 3. Results of measurements of ship model pitching

The initial model's angle of trim was about 4.5 degree. After some 6 seconds, effect of the waves rebound from the basin's walls occurred. Its impact is visible in the next test on measurement of pitching. Analysis of several courses of trim angles (not included in the elaboration) did show that the period of pitching is 0.6 second.

3.3 Results of measurements of ship model heaving

Because of the applied mode of measurement of the ship model heaving, conduction of correct tests on heaving in the facility under description in its current state is made difficult. The used sensor (converter) of average draught operates in such a way that it measures hydrostatic pressure in respect to water surface. There is an opening in the sensor containing air during execution of measurements. The air is being compressed by water getting inside during increase of draught. Compressibility of air is of insignificant importance at the time of execution of static measurement, but there are pulsations occurring there during execution of dynamic measurements upsetting the test. Moreover, in the course of the research conducted in the facility, it was observed that water "wanders" upwards together with the hull, i.e. the water surface undergoes deformation. Hence, measurement of hydrostatic pressure is burdened with additional error. Execution, for instance, of measurement of the model's distance from the basin's bottom could solve the problem. In current state, the laboratory facility does not allow to perform such tests. Results of the heaving measurements registered by means of the described sensor of heaving are presented in Figure 4.

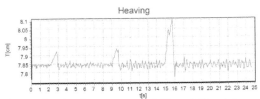

Figure 4. Results of measurements of ship model heaving

4 CALCULATIONS OF SHIP MODEL OSCILLATIONS' PERIODS

4.1 Results of calculations of model's rolling

Calculations of the ship model's rolling period have been done with a use of the following dependency [1, 3, 7,9,10]:

$$T_\phi = 2\pi \sqrt{\frac{I_x(1 + k_{xx})}{W\,\overline{GM}}} \tag{1}$$

where:
T_φ – period of rolling [s];
I_x – moment of inertia in respect to roll axis running through the vessel's centre of gravity [kos m^2];
k_{xx} – coefficient of mass of accompanying water during rolling;
W – buoyancy of the vessel [N];
GM – transverse metacentric height [m].

Pawlenko's empirical equation [3, 7] has been applied to determine the model's moment of inertia in respect to her roll axis:

$$I_x = D\frac{B^2 + H^2}{16} \tag{2}$$

where:
D – displacement of the vessel [kos];
B – breadth of the vessel [m];
H – side height of the vessel [m].

As a result, the period of rolling has been calculated with the following dependency applied:

$$T_\phi = 2\pi \sqrt{\frac{B^2 + H^2}{16g}\frac{(1 + k_{xx})}{GM}} \tag{3}$$

Coefficient of mass of accompanying water has been read from a graph [3] based on a block coefficient of the hull, taking presence of bilge keels in the vessel's model into consideration at the same time. After providing the following data:
$B = 0.2326$ m;
$H = 0.1112$ m;
$GM = 0.0083$ m;
$k_{xx} = 0.35$.
the rolling period has been calculated, resulting in a value of 1.64 s.

4.2 Results of calculations of pitching

The period of pitching of the vessel's model has been calculated with the following dependency applied [3, 7]:

$$T_\psi = 2\pi \sqrt{\frac{I_y\left(1 + k_{yy}\right)}{W\,GM_L}} \tag{4}$$

where:

I_y – moment of inertia in respect to transverse axis running through the vessel's centre of gravity [kos m^2];

k_{yy} – coefficient of mass of accompanying water during pitching;

$\underline{W}$ – buoyancy of the vessel [N];

GM_L – longitudinal metacentric height [m].

The following empirical dependency [5] has been used to calculate the transverse moment of inertia:

$$I_y = D(0,24L)^2 \tag{5}$$

The coefficient of accompanying mass k_{yy} has been accepted as equal 1.1. Its value has been evaluated based on a comparative analysis made for vessels of similar shapes and block coefficients. After entering with the following data:

$I_y = 1,248$ kos m^2;

$k_{yy} = 1.1$;

$\overline{W} = 129$ N;

$GM_L = 1.72$ m,

into the dependency (4), the pitching period equal 0.68 s has been obtained. Statement of the measurements and calculations made for periods of vessel model tested movements are given in the table 1.

Table 1. Comparison of values of oscillation periods measurements' and calculations' results for 888 project vessel's model

Tested movement of model	Periods of vessel model's oscillations [s]		Relative error [%]
	Results of measurements	Results of calculations	
Rolling	1.64	1.64	-
Pitching	0.6	0.68	13

5 SUMMARY

Based on the comparison of measurements' and calculations' results of oscillation periods for movements of the tested vessel's model, one may conclude that satisfactory results can be obtained already after application of simplified analytical formulas and empirical dependencies. From experience we do know that this conclusion may not be generalized and ascribed to all water crafts. Moreover, variety of water crafts' constructions causes that broad research has not been conducted for all of them; hence adequate empirical dependencies have not been formulated for them. Therefore, aside from model tests, it is recommended – among the others – to elaborate and to use modern calculation techniques in the shipbuilding industry.

So far, model tests on vessel's seakeeping qualities have been run in the Naval Academy in a very limited range and for specific constructions such as submarine model [2] and pinnace model [6]. Tests described in this paper were conducted in a totally new facility, for a model very accurately reflecting geometry of the Polish Navy vessel's hull being still operated. Thanks to the new laboratory facility, equipped w specialist sensors and vessel models executed in accordance with body lines, technical level and quality of the tests have been seriously improved.

The described tests are a preliminary stage for a broader analysis of phenomena regarding hydromechanics of vessels. In the future, results of measurements included in this paper shall be used for verification of computer programmes engaging numerical methods and being elaborated by the authors of this paper.

REFERENCES

[1] Dudziak J., *Ship theory*, Foundation for Promotion of Shipbuilding and Maritime Industries, Gdańsk, Poland 2008.;

[2] Gniewszew J.: *Impact of free water surface on accompanying masses and inertia moments of submarine operating in underwater location*, Naval Academy, Gdynia, Poland 1982;

[3] Kabaciński J., *Ship's stability and unsinkability*, Marine Academy, Szczecin, Poland 1993;

[4] Mironiuk W., Pawlędzio A., Wróbel R., Zacharewicz M., Report on scientific-research work entitled *Model tests on stability and unsinkability of vessels*, Gdynia, Poland 2004;

[5] Pawłowski M., *Linear model of ship movements on irregular waves*, Technical Report No. 41, Polish Register of Shipping, Gdańsk, Poland 2001;

[6] Pawlędzio A.: *Method of determination of water masses accompanying vessel during free rolling*, Scientific Papers of Naval Academy no. 2 (173)/2008, Gdynia, Poland 2008;

[7] Staliński J.: *Ship theory*, Maritime Publication House, Gdańsk, Poland 1969;

[8] Stahlberg K., Goerlandt F., Montewka J., Kujala P.: *Uncertainty in Analytical Collision Dynamics Model Due to Assumptions in Dynamic Parameters*, TransNav No1-March 2012, Gdynia, Poland;

[9] Soda T., Shiotani S., Makino H., Shimada Y.: *Research on Ship Navigation in Numerical Simulation of Weather and Ocean in a Bay*, 9-th International Symposium on Marine Navigation And Safety of Sea Transportation, Gdynia, Poland 2011;

[10] Song K., Kim Y., *Quantitative analysis of parametric roll and operational guidance*, 11-th International Conference Stability of Ships And Ocean Vehicles, Athens, Greece, 2012

Hydrodynamics and Ship Stability
Maritime Transport & Shipping – Marine Navigation and Safety of Sea Transportation – Weintrit & Neumann (Eds)

The Dynamic Heeling Moment Due to Liquid Sloshing in Partly Filled Wing Tanks for Varying Rolling Period of Seagoing Vessels

P. Krata, J. Jachowski, W. Wawrzyński & W. Więckiewicz
Gdynia Maritime University, Poland

ABSTRACT: Liquid sloshing phenomenon taking place in partly filled ships' tanks directly affects the stability of the vessel. However, only static calculations are carried out onboard ships nowadys and static transfer of liquid weight is taken into account in the course of routine stability calculation and assessment. Since previous researches reveals the necessity of dynamic approach towards liquid movement onboard ships, the presented investigation is focused on problems related to time dependent wave-type phenomena. This aspect is omitted in the course of standard ship stability calculations. The set of numerical simulations of liquid sloshing taking place in moving tanks is carried out. Among many obtained characteristics the heeling moment due to sloshing is emphasized and investigated. The realistic range of possible rolling periods is examined. The influence of ship's rolling period on the heeling moment due to liquid sloshing is analyzed for a representative geometry of seagoing vessels.

1 INTRODUCTION

At the time of globalization the importance of overseas transportation of goods is vital. The commonly discussed main features of maritime transport are usually its safety and effectiveness. However, the ships safety issues are crucial from the operational point of view and they ought to be considered as a prospective technical affair. One of the most critical features of seagoing ships related to her safety is stability influencing ship's overall seakeeping performance.

Stability against capsizing and excessive heeling is one of the most fundamental requirements considered by naval architects when designing commercial vessels and by their operators in the course of sailing and cargo handling (ITTC 1999). The stability of a vessel belongs to operational characteristics enabling cost effective and safe operation (ITTC 1999). The accuracy of ship's transverse stability assessment is an important problem in vessels' operation process. The ship loading condition of insufficient stability may induce a list, a strong heel and even her capsizing. Contrary to such state, the excessive stability causes high values of mass forces acting on cargoes and machineries due to strong accelerations.

One of the problem related to ship stability and her overall performance is an effect of moving liquids in partly filled tanks. Generally moving masses need to be avoided however it is impossible to evade them at all. The cargo securing procedures ensure an extinction of loose cargo onboard but some free surfaces of liquids in ships' tanks are inevitable. The crucial group of tanks onboard ships which may be partly filled are ballast tanks. They can be classified according to their purpose as follows (Krata, Wawrzyński, Więckiewicz, Jachowski 2012):

- trimming tanks (fore and after peaks) which are utilized very often as partly filled due to the need for precise trimming of a ship, so the ballast water level is adjusted according to the variable requirements, thus providing free surface of liquid;
- stability tanks improving ship's stability performance due to a decrease in the vertical center of gravity (usually double bottom tanks located between an engine room and a fore peak creates this group); quite often the breadth of these tanks equals half breadth of a ship or even sometimes it equals full ship's breadth (in ship's fore region) therefore the free surface of liquid can be massive in these tanks so generally they should be full or empty during voyage;
- list control tanks (side tanks) which are usually located amidships and due to their function quite often partly filled with free surface;

- strength control tanks utilized to adjust longitudinal weight distribution (fore and after peaks, double bottom tanks, side tanks and sometimes even cargo holds prepared for ballasting) which are very often partly filled to reduce excessive sheering force and bending moment and routinely they provide free surface of liquid;
- special purpose tanks like for instance anti-rolling tanks (flume) or anti-heeling tanks, which are usually filled up to the 50% level, providing free surface.

Although most of the mentioned tanks are usually filled up to their top or alternatively, they are empty during ship voyage, there are some relatively long periods when they are only partly filled. The most obvious is ballast water exchange applied to meet the requirements of ballast water management instructions providing natural environment protection. At the time of such operation the sufficient stability of a ship ought to be maintained.

Regardless the exact purpose of ballast tanks onboard, their total volume and resulting from it total weight of ballast water is significant which is shown in figure 1. This justifies focusing on this group of tanks in the course of the conducted research.

energy to induce and sustain a fluid motion. Under external large amplitude excitations or an excitation near the natural frequency of sloshing, the liquid inside tank is in violent oscillations which is of great practical importance to the safety of the liquid transport (Zeineb S., Chokri M., Zouhaier H., Khlifa M. 2010). Both the liquid motion and its effects are called sloshing. The interaction between ship's tank structure and water sloshing inside the tank consists in the constant transmission of energy (Akyildiz & Unal 2005).

The characteristics of heeling moment due to liquid sloshing depend on a variety of parameters, for instance tank's geometry, its filling level, location of a tank within a hull of a ship, rolling period and others. The influence of ship rolling period is the main subject of the paper, thus, the research is focused on this matter. However, a number of features needs to be taken into account.

The pre-study carried out in the course of the research enabled the classification of typical shapes and dimensions ship tanks (Krata, Wawrzyński, Więckiewicz, Jachowski 2012). The most often shapes are shown in figures 2, 3 and 4. The double bottom tanks belongs to standard arrangement of all seagoing ships, while the side tanks and wing tanks are typical for same types only.

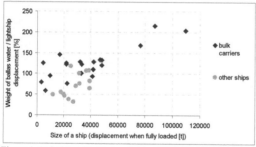

Figure 1. Relation of ballast water weight to the lightship disp. (Krata, Wawrzyński, Więckiewicz, Jachowski 2012)

The graph (Fig. 1) reveals that the total weight of ballast water carried onboard may reach and even exceed the lightship weight among ships other than bulk carriers. Moreover, ballast water weight can be twice a lightship in case of bulk carriers (Krata, Wawrzyński, Więckiewicz, Jachowski 2012). Obviously not all the tanks are partly filled at the same time but some of them can be which creates the need for stability calculations comprising the phenomenon of moving liquids in tanks.

2 SHIP AND TANK GEOMETRY APPLIED IN THE COURSE OF THE RESEARCH

The liquid sloshing phenomenon takes place in partly filled ships tanks. As a tank moves, it supplies

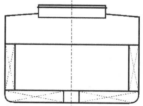

Figure 2. Typical side tanks and wide double bottom tanks

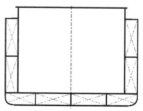

Figure 3. Division of side tanks and double bottom tanks typical for large ships

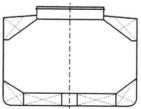

Figure 4. Wing tanks, bilge tanks and double bottom tanks of bulk carriers

The previous authors' researches reveals that liquid sloshing dynamics in ships side tanks can be neglected actually. The natural period of liquid sloshing is short enough to justify the quasi static calculation of the free surface effect. This approach is well known and routinely applied in the course of the stability assessment (ISC 2009). A liquid contained in partly filled side tanks remains actually horizontal and flat within the all ship rolling cycle, which is shown in figure 5. The remaining tanks, e.g. double bottom tanks and wing tanks need to be the subject to examine in terms of possible sloshing characteristics.

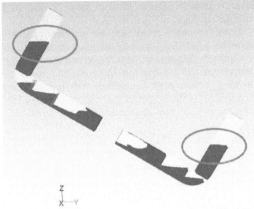

Figure 5. The surface of liquid sloshing in partly filled high side tanks (Krata, Jachowski 2010)

In the course of the study two typical ship dimensions were taken into consideration. In case of quasi static approach to the free surface effect the location of a partly filled tank does not play any role. Reversely, the dynamic approach is related to ship rolling and the location of considered tank is crucial. Therefore not only the dimensions of a model ship need to be specified but her rolling axis as well. The particulars applied in the research are given in the table 1.

Table 1. Main dimensions of considered ships

Ship particulars [m]	ship 1	ship 2
breadth	32,00	20,00
height	20,00	12,50
elevation of rolling axis	9,00	5,62

For the sake of simplicity the considered model tank was the same cuboid for both ships. Its breadth is 10 m and the height 2,5 m. Equal tanks dimensions enable the direct comparison of the obtained results of computation of the heeling moment due to liquid sloshing. The filling level of the tank equals 50% in all considered cases. The location of the tank was typical for wing tanks just beneath deck and close to the ship side.

3 COMPUTATION OF HEELING MOMENT DUE TO LIQUID SLOSHING IN SHIP'S TANK

The heeling moment due to liquid sloshing in a partly filled tank was computed with the use of CFD technique. The software FlowVision was applied. The simulations of liquid sloshing were carried out in 3D mode for the most typical rectangular ship ballast tank. The rolling period was variable according to the research assumptions and the range of angular motion reflects the very heavy sea conditions in extremely stormy weather. The rolling period depends on the stability performance of a considered ship therefore the wide range of such rolling periods is taken into account in the presented study.

The computational mesh applied in the course of the simulations was hexahedral type and related to two coupled reference frames, the stationary and a moving ones which is shown in figure 6. The Sub-Grid Geometry Resolution (SGGR) was applied where the triangulated surfaces naturally cut Cartesian cells and reconstructing the free surface (FlowVision 2010).

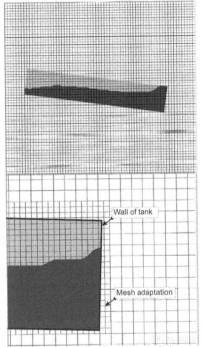

Figure 6. Mesh applied for CDF calculations

The SGGR method is intended for an approximation of curvilinear boundaries on a hexahedral mesh. The method consists in natural splitting of the boundary cells by the triangulated boundaries which is shown in figure 7.

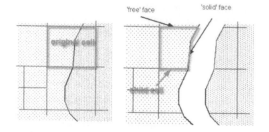

Figure 7. Sub-grid resolution of curvilinear wall (FlowVision 2010)

The number of the obtained child cells depends on the geometry peculiarities. The child cells are arbitrary polyhedrons. The equations of a given mathematical model are approximated on the polyhedrons without simplifications. The approach enables accurate calculations in a complex domain on a reasonably coarse mesh (FlowVision 2010). The FlowVision code is based on the finite volume method (FVM) and uses the VOF method for free surface problems.

The *FlowVision* code is based on the finite volume method (FVM) and uses the VOF method for free surface problems which is presented in figure 8 (FlowVision 2010).

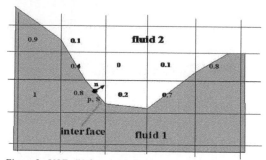

Figure 8. VOF (Volume Of Fluid) variable is the volume fraction of fluid 1 in a cell; VOF=1 - the cell contains only fluid 1; VOF=0 - the cell contains only fluid 2; 0<VOF<1 the cell contains fluid 1 and fluid 2 (FlowVision 2010)

High accuracy of computation is achieved by solving the governing equations in the 'free surface' cells (the cells partly filled with liquid) (FlowVision 2010). The RANS (Reynolds-averaged Navier–Stokes) equation is implemented and the simulation of turbulent flows is based on the eddy viscosity concept. The semi-empirical k-ε model turbulence model was applied.

The result of the simulation comprises the general flow pattern, the velocity and pressure fields and the user-predefined heeling moment due to liquid sloshing being the most important for the conducted study.

4 INFLUENCE OF SHIP'S ROLLING PERIOD ON HEELING MOMENT DUE TO LIQUID SLOSHING

The obtained heeling moment due to liquid sloshing in tanks was decomposed into two components. The first one comprises the moment due to dynamic action of solid-like liquid (i.e. 'frozen') at an angle of heel equal 0 degrees. The second component of the dynamic heeling moment due to liquid sloshing covers only the moment resulting from letting free the liquid to slosh inside the tank. All moments (components) are computed about the ship rolling axis which is fixed at the symmetry plane of a vessel at an elevation given in the table 1 for both considered ships.

The component containing the moment resulting from the solid-like liquid is included in the weight distribution calculation. And the remaining dynamic component of the heeling moment due to liquid sloshing which may be called 'the free floating component' is the matter of this paper. The core idea of this approach may be expressed by the formula:

$$M_{Total\_dyn} = M_{FL\_dyn} + M_{Ff} \qquad (1)$$

where:

M_{Total_dyn} – total dynamic moment due to liquid sloshing in a tank;

M_{FL_dyn} – dynamic heeling moment due to the weight of solid-like liquid in a tank;

M_{Ff} – free floating component of the dynamic moment due to liquid sloshing.

The exemplary result of CFD computations is presented in figure 9. The total dynamic moment due to liquid sloshing in a tank M_{Total_dyn} is obtained in time domain.

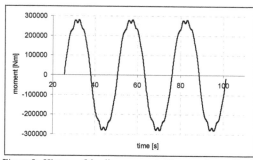

Figure 9. History of heeling moment M_{Total_dyn} due to liquid sloshing in the considered tank – exemplary case

As the momentary angle of ship's heel is know for every time step of CFD computations, the heeling moment may be plotted versus an angle of heel as well. This is a convenient approach which is shown in figure 10.

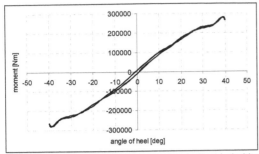

Figure 10. Heeling moment M_{Total_dyn} due to liquid sloshing versus an angle of ship's heel – exemplary case

The next heeling moment existing in the formula (1) is the dynamic heeling moment due to the weight of solid-like liquid in a tank. This moment is shown for an exemplary case in figure 11.

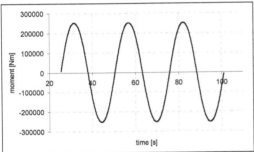

Figure 11. History of the dynamic heeling moment M_{FL_dyn} due to the weight of solid-like liquid in a tank – exemplary case

Similarly to the total heeling moment, this moment due to the solid-like weight in a tank can be plotted versus and angle of heel – like in figure 12.

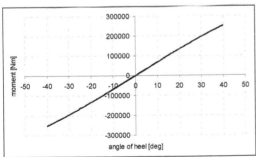

Figure 12. Dynamic heeling moment M_{FL_dyn} due to the weight of solid-like liquid in a tank versus an angle of ship's heel – exemplary case

According to the formula (1) the core component of the heeling moment due to sloshing is a difference between the total dynamic moment due to liquid sloshing and the dynamic moment due to solid-like weight in a tanks. The result is shown in figure 13.

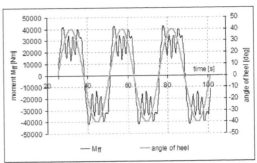

Figure 13. History of the free floating component M_{Ff} of the dynamic moment due to liquid sloshing – exemplary case

However, the most convenient way of presentation of the free floating moment is a graph plotted versus an angle of ship's heel which is presented in figure 14.

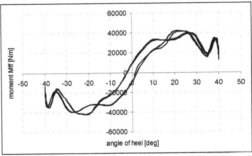

Figure 14. Free floating component M_{Ff} of the dynamic moment due to liquid sloshing plotted versus an angle of ship's heel – exemplary case

Although the heeling moment was computed in the course of time-domain calculations, the considered free floating component of the moment was plotted versus an angle of ship's heel. Then thanks to the application of the decomposition of the heeling moment (formula 1), the resultant hysteresis loop of the free floating component may be simplified by the use of a previously worked out linearization procedure (Krata 2009). The criterion of an equivalent work of a moment was adopted because of the main long-term purpose of the research which is the modification of the weather criterion of ship stability assessment. This criterion is just based on the work of both the heeling moment and righting moment which justifies the proposed linearization procedure (Krata 2009). The linearization formula can be concisely shown in following notation (Krata 2009):

$$\int\limits_{0}^{\varphi_A} M_{Ff(\varphi)} \cdot d\varphi - \int\limits_{\varphi_A}^{0} M_{Ff(\varphi)} \cdot d\varphi + \int\limits_{0}^{-\varphi_A} M_{Ff(\varphi)} \cdot d\varphi - ...$$

$$... - \int\limits_{-\varphi_A}^{0} M_{Ff(\varphi)} \cdot d\varphi = 4 \int\limits_{0}^{\varphi_A} M_{Ff\_LA} \cdot d\varphi = \varphi_A \cdot M_{FfA} \quad (2)$$

where:

M_{Ff} – free floating component of the dynamic moment due to liquid sloshing;

φ – angle of ship's heel;

φ_A – ship's rolling amplitude;

M_{Ff_LA} – linear approximation of the free floating component of the heeling moment for a given ship's rolling amplitude;

M_{FfA} – the value of the linear moment M_{Ff_LA} for an angle of heel equal rolling amplitude φ_A.

The linear function of heeling moment can be determined by the fixing of two in-line points having the coordinates (φ, M). One of them is the point (0,0) and the second one the point (φ_A, M_{FfA}). Therefore the complete description of the linear approximation of the moment M_{Ff} may be done by one scalar only. Such a scalar is the value M_{FfA} of the linear free floating component due to sloshing for the angle of heel equal φ_A and obtained according to the given formula (2) (Krata 2009).

The results of computation of the linear approximation of the free floating component of the heeling moment due to liquid sloshing in partly filled ship's tank are presented in figures 15 to 22. The linear value of the free floating component of the heeling moment are the subject of further analysis therefore all consecutive steps of calculation lead to these values obtained for all considered cases.

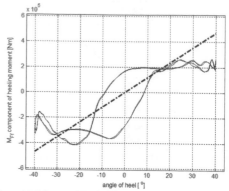

Figure 16. Moment M_{fFf} for the ship 1 and her rolling period 25,3 seconds

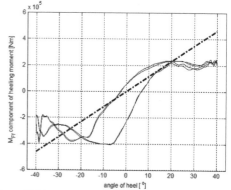

Figure 17. Moment M_{fFf} for the ship 1 and her rolling period 33 seconds

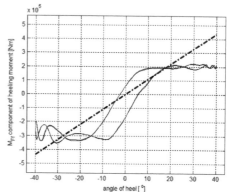

Figure 18. Moment M_{fFf} for the ship 1 and her rolling period 40 seconds

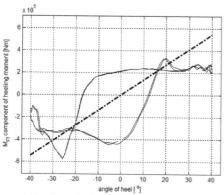

Figure 15. Moment M_{fFf} for the ship 1 and her rolling period 18 seconds

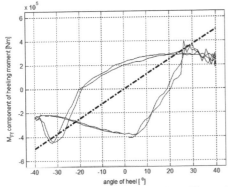

Figure 19. Moment M_{fFf} for the ship 2 and her rolling period 14 seconds

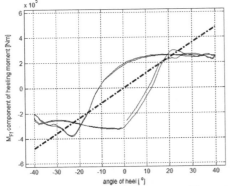

Figure 20. Moment M_{fFf} for the ship 2 and her rolling period 20 seconds

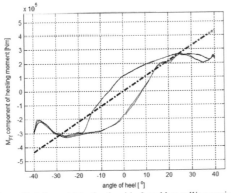

Figure 21. Moment M_{fFf} for the ship 2 and her rolling period 26 seconds

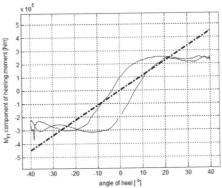

Figure 22. Moment M_{fFf} for the ship 2 and her rolling period 32 seconds

The resultant characteristic of an influence of the ship's rolling period on the free floating component of the heeling moment due to liquid sloshing in double bottom tanks is shown in figure 23. The intensity of effect of the free floating component of the heeling moment due to liquid sloshing is represented by the value M_{FfA} which is plotted versus a rolling period of a ship.

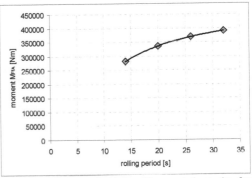

Figure 23. Influence of ship's rolling period on the free floating component of the heeling moment due to liquid sloshing in ship's double bottom tank – the value M_{FfA} for an angle of ship's heel equal φ_A (Jachowski, Krata ,Wawrzyński, Więckiewicz 2012)

The graph 23 is a result of earlier research and presents an important tendency (Jachowski, Krata, Wawrzyński, Więckiewicz 2012). The wave kind phenomena taking place in partly filled double bottom tanks result in asymptotic increasing in the M_{Ff} component of the heeling moment when ship's rolling period increases. Thus, the dynamic behavior of sloshing liquid less effects ships stability the static one. Such a remark opens a crucial question about the sloshing impact obtained for highly located wing tanks.

The characteristics of an influence of the ship's rolling period on the free floating component of the heeling moment due to liquid sloshing in wing tanks of both ships considered in this study are based on the linear values of free floating moments presented in figures 15-22. The values of M_{FfA} moments for all considered cases are gathered and plotted versus ship's rolling period which is the main objective of the study. The obtained tendencies are different then for double bottom tanks which is shown in figure 24.

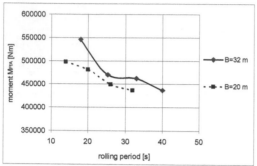

Figure 24. Influence of ship's rolling period on the free floating component of the heeling moment due to liquid sloshing in ship's wing tanks (for both considered ships – see table 1) – the value M_{FfA} for an angle of ship's heel equal φ_A versus ship's rolling period

The characteristics presented in the figure 24 reveals that the linear value of the heeling moment's free floating component gradually drops with the increase in rolling period. Theoretically the infinite rolling period reflects the quasi-static approach towards free surface calculation, so greater values of heeling moment for shorter then infinite ship's rolling period reveal more severe impact of liquid sloshing in wing tanks then it is assumed in contemporary standard stability calculations.

5 CONCLUSION

The study presented in the paper is focused on the dynamic effects of liquid sloshing in partly filled ship tanks. As the free surface effect is taken into account in the course of a routine stability calculations, the authors find a question about dynamic approach as justified.

The novel approach for the decomposition of the dynamic heeling moment was worked out and applied and together with the linearization procedure it enabled the convenient way of heeling moment presentation and analysis.

The results of the study enables an assessment of the influence of the ship's rolling period onto the heeling moment due to liquid sloshing in tanks. It shows that the actual heeling moment due to liquid sloshing may be larger then it is supposed on the basis of contemporary quasi-static approach. Such a case is found when ship's wing tanks are partly filled.

The conducted research reveals that the linear approximation of the free floating component of the considered heeling moment due to liquid sloshing in wing tanks gradually drops with the increase in ship's rolling period. Thus the dynamic and quasi-static approach meet their results for an infinite ship's rolling period which is only theoretically experienced. Practically for the realistic range of rolling periods the heeling moment is grater than expected one which exposes a ship to the potential danger. Moreover, the question about meeting ship stability criteria may be asked. The contemporary stability standards are based on the quasi-static approach towards free surface calculation (ISC 2009) which is underestimated in the light of the conducted research, while the general presumption is to keep the "safe side of ship stability in any doubt cases".

The study described in the paper may be a contribution to the ship safety estimation methods. The results suggest the need for further research work on the liquid sloshing phenomenon onboard seagoing ships.

REFERENCES

Akyildiz H., Unal E., „Experimental investigation of pressure distribution on a rectangular tank due to the liquid sloshing", Ocean Engineering 32 (2005), www.sciencedirect.com

Final Report and Recommendations to the 22nd ITTC, The Specialist Committee on Stability, Trondheim, Osaka, Heraklion, St. John's, Launceston 1996 – 1999.

FlowVision HPC Manual, Capvidia, Belgium 2010.

International Code on Intact Stability 2008, edition 2009, IMO 2009.

Jachowski J., Krata P., Wawrzyński W., Więckiewicz W., Analysis of dynamic heeling moment due to liquid sloshing in partly filled ship's tanks for realistic range of rolling periods – a case study, Journal of KONES Powertrain and Transport, Vol. 19/3, pp. 177-184, 2012.

Krata P., Linear characteristics of sloshing phenomenon for the purpose of on-board ship's stability assessment, Journal of Theoretical and Applied Mechanics (JTAM), Vol. 47, No 2, pp. 307-320, Warsaw 2009.

Krata P., Jachowski J.: 3D CFD modeling of ship's heeling moment due to liquid sloshing in tanks – a case study. Journal of KONES Powertrain and Transport, Vol 17, No. 4, pp. 245-251, 2010.

Krata P., Wawrzyński W., Więckiewicz W., Jachowski J., Ship's ballast tanks size and dimensions review for the purpose of model research into the liquid sloshing phenomenon, Scientific Journals of Maritime University of Szczecin, 2012.

Zeineb S., Chokri M., Zouhaier H., Khlifa M., Standing wave induced by free liquid sloshing in rectangular tank, Proceedings of the International Renewable Energy Congress, Sousse, Tunisia, November 5-7, 2010.

Safety Studies for Laker Bulker Trans-pacific Delivery Voyage

G. Mazerski
Deltamarin sp. z o.o., Gdańsk, Poland

ABSTRACT: Self Unloading Bulk Carrier for operations in the US/Canada Great Lakes was ordered in a Chinese shipyard. A risk assessment for delivery voyage across Pacific and Panama Canal was made. It showed that several studies are needed to ensure safe passage. First step consists of loading condition optimization with regards to longitudinal strength. Still water bending moment is taken into account and multi-objective genetic algorithm is utilized. Second step includes calculations of wave-induced bending moment for several stages of the delivery voyage, as function of initial loading condition and wave conditions in each month. The main purpose of the third stage is to provide motions and accelerations for design of equipment foundations onboard the vessel, as well as for assessment of habitability for crew onboard. Two methodologies are used, one considering whole route and all headings, another included different stages of delivery voyage and most probable wave directions in the area. Last step covers greenwater risk assessment and preparation of polar plots presenting risk of green water. This is to indicate greenwater probability and severity and to provide crew onboard with easy to use tool for avoiding water entering the deck.

1 INTRODUCTION

1.1 Problem definition

Whole series of Laker bulk carriers were ordered in a Chinese yard. An issue was raised of ship's sea-worthiness for the delivery voyage from China to North America East coast. Several risks were identified. These can be summarized in three points:
- Wave bending moment will exceed capacity of midship section
- Motions in waves will damage the equipment and cause discomfort for crew
- Greenwater on deck will present danger to equipment

1.2 Solution plan

Following action plan was prepared to ensure that all the problems are properly addressed. The actions include:
- Optimizing loading conditions to obtain minimum possible still-water bending moments
- Executing route-specific analysis for wave bending moment for the loading conditions optimized in the previous action
- Decide if delivery voyage strengthenings (DVS) is needed to enhance bending capacity of the ships' structure
- Update wave bending moment analysis (if DVS installed)
- Execute route-specific motion analysis in order to obtain motions and accelerations expected along the route in key areas of the vessel
- Execute green-water risk assessment study to check probability and severity of possible waves breaking over the deck

The actions described above were executed for both sub-types in the bulker series – namely self-unloading (SUL) and gearless (GLB) options. In the following chapters only results for SUL will be presented, since the analyses were identical and results turned up to be very similar.

2 CALCULATION MODEL

2.1 Hydrostatics and loading conditions

Shape and compartment definitions of the ship were modeled in NAPA stability software suite. This enabled calculation of hydrostatic properties of the hull, defining main compartments of the ship as well as allowed for preparation of numerous loading conditions that the ship is expected to experience. Also optimization of loading conditions, as described in chapter 3, was performed using macros within NAPA suite.

2.2 Hydrodynamics

Hydrodynamic model was prepared in ANSYS AQWA software package. This suite allows for potential, linear as well as non-linear, time- or frequency-domain calculation of hydrodynamic forces, based on response amplitude operators (RAOs) calculated using radiation-diffraction theory. The response values calculated this way include motion amplitudes in 6 degrees of freedom, accelerations in 3 degrees of freedom and longitudinal wave bending moments. Panel representation of ships' hull (there are about 7000 panels in the model, about half of them used in diffraction calculations) was used, together with weight data from loading conditions (see chapter 2.1). Utilising simulations of roll decay experiment in AQWA environment following values of damping between 3 and 5 % of critical roll damping were assumed for selected loading conditions.

2.3 Metocean analysis

The vessels' route from China to St. Lawrence River via Panama Canal was analyzed, utilizing Global Waves Statistics (BMT 1986) data and Fisheries and Oceans Canada. This way a "corridor" across Pacific and along North America east coast is created, see Figure 3-1 below

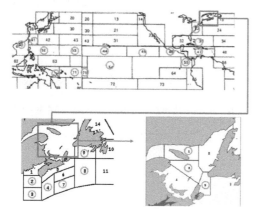

Figure 1. Analyzed delivery voyage route

The corridor consists of 15 different section. For each section of the ocean voyage 12 different monthly significant wave heights (for 3h storm duration of 400-day return period) were produced based on the seasonal GWS data. In addition the "all route" values were produced for the voyage route.

Several sea areas have seasonal warnings for hurricanes or typhoons. The analysis assumes that hurricanes/typhoons are avoided during the transport. The present ocean voyage scatter data is considered a conservative approximation. With careful route planning the route scatter data severity can be reduced.

Two approached were utilized for motion and greenwater analysis. Simpler one assumed the most severe of "all route" wave conditions for analysis, while the more elaborate one included regional variations and assumed that the most severe conditions for each of the legs of the route shall be taken into account for better accuracy.

2.4 Postprocessing and presentation

Various ways of post-processing the results were utilised. These include direct plots from AQWA Graphical Supervisor as well as macros used to export data from result files to MS Excel or Matlab environment. In most cases calculation of values for the given points onboard were made outside AQWA, the same applies to most figures and plots.

3 LOADING CONDITION OPTIMISATION

3.1 Requirements for loading conditions for delivery voyage

The purpose of this activity was to find loading conditions with minimum still water bending moment. This will allow for more wave bending moment and in turn will allow for transit in less favorable conditions. Two different conditions have been examined. Delivery condition only with ballast water (ballast condition) and delivery condition with ballast water and cargo (cargo condition) were checked. Cargo is utilized to ensure as much constant weight distribution along the vessel – this will help minimize still water bending moment. Arrival and departure conditions are included to account for varying amounts of consumables (fuel, water etc.)

Next draught limiting values are used to generate feasible loading conditions:
- Maximum mean draught should be less than 9.5 m
- Minimum forward draught should be 4.5 m or above (due to bottom slamming risk)
- Minimum aft draught should be 5.7 m or above (due to propeller immersion)

Please note that propeller is not completely immersed at aft draught of 5.7 m. Forward draught of 4.5 m doesn't give much freedom to generate feasible loading conditions when using only ballast water.

3.2 Optimization factors

The calculations have been carried out with NAPA program (Release 2009.1-1) and with General Optimisation Manager. The NAPA Manager application uses Multi-Objective Genetic Algorithm (MOGA). The details of the process are described below. (based on NAPA 2009)

3.2.1 Variables

For ballast condition the variables are the amount of ballast water in 5 pairs of ballast water tanks along cargo holds.

For cargo conditions the variables are the amount of cargo in 5 cargo holds. Ballast capacity is kept unchanged – it is fixed at the volumes resulting from ballast condition optimization.

3.2.2 Objectives

There are two objectives defined, namely hogging and sagging moment in the current loading condition.

3.2.3 Constraints

There are three constraints, as described in 3.1:
- Maximum mean draught should be less than 9.5 m
- Minimum forward draught should be 4.5 m or above (due to bottom slamming risk)
- Minimum aft draught should be 5.7 m or above (due to propeller immersion)

3.3 Results

The optimization process produced significant number of solutions, out of which only feasible ones (i.e. those that fulfill constraints requirements) were considered for post-processing (Aalto 2010, Dawn 1995, Goldberg 1989). The scatter diagram of various loading conditions showing how they perform against two objectives is shown below:

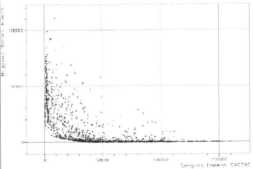

Figure 2. Scatter diagram of feasible loading conditions (cargo conditions) and bending moments they produce

A Pareto front can be clearly seen, showing that one cannot bring both bending moments to zero for a single loading condition. Due to equal weight of both objectives the optimum solution is the one with approx. 7000 tm for both sagging and hogging.

3.4 Final loading conditions

Following loading conditions were selected as optimum and taken for further analysis. These are presented in Table below:

Table 1. Final loading conditions

Loading condition	DBD	DBA	DCD	DCA
T aft [m]	5,816	5,834	8,181	7,522
T fore [m]	4,514	4,513	4,857	4,504
GM [m]	5,37	5,44	4,40	5,11
Displ. [t]	24703	24749	31720	29049
BM max [tm]	42136	31690	5292	6439

Loading conditions are described by following abbreviations:

DBD   Delivery Ballast Departure (SUL LC01)
DBA   Delivery Ballast Arrival (SUL LC02)
DCD   Delivery Cargo Departure (SUL LC03)
DCA   Delivery Cargo Arrival (SUL LC04)

All four loading conditions described above were checked in wave bending moment analysis.

For the motion and greenwater analysis it was noticed that within cargo and ballast loading conditions groups there were small differences, therefore only one of the pair is selected for further studies. Since motion response is assumed to be very similar (within pairs) the loading condition with larger draft is selected as this will produce more conservative results with regards to green water risks (to be covered in other parts of this project). Therefore following descriptions are used in further analysis:

DDD   dropped
DBA   −SUL1
DCD   =SUL2
DCA   dropped

These names will be used throughout this paper.

4 WAVE-INDUCED BENDING MOMENT

4.1 Procedure

In order to check longitudinal strength of the vessel during trans-Pacific voyage wave bending moments for wave scatter for each month of the year was performed. This allows for estimation of the delivery voyage time window, i.e. during which months the delivery voyage is feasible from the longitudinal strength point of view.

The calculations can be divided into following steps:

- Exporting longitudinal weight distribution from loading conditions in NAPA hydrostatic tool to AQWA hydrodynamic suite
- Calculation of wave bending moments RAOs for wide range of wave periods
- Short term response for irregular waves is performed to arrive at the probability of bending moment response for each sea-state/loading condition combination. Bending moment RAOs and Pierson-Moskowitz wave energy spectrum as well as environment directional spreading function is utilized
- Combined wave scatter for each month and whole year is prepared by means of combining monthly/yearly scatter diagrams from each leg of the delivery voyage (wave bending moment is calculated for whole route, no regional variations included)
- Long term response based on sea-state probabilities (from combined scatter diagrams) and short-term responses (from short-term analysis) is calculated for each loading conditions and each month of the year. Equal probability of all headings is assumed for simplicity
- The design wave bending moment values are evaluated such that they have a return period of 10 voyages. The theoretical probability of exceeding the resulting design wave bending moment value during the ocean voyage becomes about 10% (duration of each voyage is taken as 40days)
- Simplified springing moment calculations are performed. Based on full-scale measurement from similar vessel wave bending moments are increased by 35% to account for this hydroelastic response. This is done despite the fact that natural frequency in bending of the ship (incl. added mass) is much smaller that wave period – some significant response for wave periods which are multiplications of natural periods was identified during full-scale measurements.
- Total bending moments is calculated as a sum of still-water, wave-induced and springing bending moments

4.2 Results

As a first step a comparison of wave bending moments coming from utilizing world-wide navigation requirements (which corresponds to winter North Atlantic conditions, Lloyd's Register 2011):

Table 2. Summary of results for world-wide navigation

Studied still water moments (see Appendix 1)		
LC 01	+448	MNm
LC 02	+291	MNm
LC 03	+53	MNm
LC 04	+60	MNm
Wave moments for unrestricted navigation (based on winter North Atlantic)		
rule wave bending moment +1969		MNm
Total (still water + wave) moments for unrestricted navigation		
LC 01	+2417	MNm
LC 02	-2260	MNm
LC 03	+2022	MNm
LC 04	-2029	MNm
Level of excedance of total moments vs. design moments for Great Lakes (1824 MNm)		
LC 01		32,5 %
LC 02		23,9 %
LC 03		10,9 %
LC 04		11,2 %

As can be seen in case it is necessary to utilize world-wide navigation wave bending moments (still assuming that smaller-than-rule still water bending moments are allowed by authorities) the bending capacity of the hull is significantly exceeded, both for cargo and ballast loading conditions. This conclusion justifies the route specific analysis, which should result in loads smaller than for world-wide navigation.

The ocean voyage WBM was first calculated for the four loading conditions utilizing "All Year" ocean voyage scatter data. The cargo condition (LC 03 and 04) gave more favorable total bending moment compared to the ballast condition, see Figure 3 below:

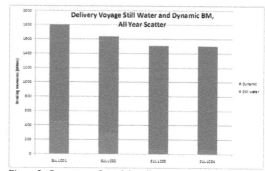

Figure 3. Summary of total bending moments for "all year" wave conditions.

As can be seen even for ballast-only conditions the bending moments do not exceed design bending moments for Great Lakes ships (Lloyd's Register 2003), which was calculated to be 1824 MNm.

Once more precise analysis is performed seasonal variations in bending moment can be utilized. The summary is presented in the Figure 4 below:

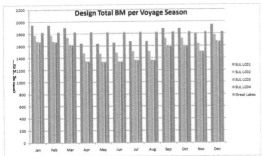

Figure 4. Summary of total bending moments for monthly wave conditions (right-most bar represent Great Lakes conditions)

4.3 Conclusions

As can be seen Great Lakes design bending moments are exceed during winter months for one ballast conditions (LC 01), remaining loading conditions can be utilized whole year. This means delivery voyage in ballast conditions is restricted to April-August summer window, while the same voyage with cargo is permitted irrespective of the season.

All these conclusions are valid assuming that hurricane/typhoon conditions will be avoided during the voyage.

5 MOTIONS AND ACCELERATIONS

5.1 Procedure

In order to assess risk level from excessive ship motions, that can be experienced during delivery voyage following procedure was followed to arrive at motions and accelerations results (based on Bertram 2000):

5.1.1 Sea state definition

Two different approaches were utilized. One of them assumed one, most severe sea-state for each loading condition, based on whole route wave data and time window defined by longitudinal strength (see chapter 3). Therefore following sea states are taken for further analysis:
- SUL1 Hs=7,7 m (as delivery window is April-August)
- SUL2 Hs=8,6 m (as delivery is possible whole year)

A range of wave periods is used following the formula below:

$$\sqrt{13 \cdot Hs} \le Tp \le \sqrt{30 \cdot Hs}$$

This can be re-calculated for the two sea states:

SUL1 – 10 < Tp < 15

SUL2 – 10,5 < Tp < 16

This is considered to be conservative approach as it assumes equal probability of any heading and that the most severe storm can happen during any stage of the delivery voyage.

Alternative approach is presented, which takes into account five sections of the delivery voyage. The values of significant wave height represent a storm of 3h duration, which is the most likely most severe storm to occur within 80-day period (20% or 1/5th of a total time of 10 voyages in each section of the route). Following sections were identified:

Table 3. Route sections for delivery voyage

Area	Env. direction	Wave height Hs	Relative ship course
East China Sea	NE	7,7 m	135
Equatorial Pacific	E	4,7 m	0
Caribbean	E	5,5 m	60
US East Coast	S	6,0 m	150
Canadian East Coast	SW	6,6 m	180

Range of wave periods is assigned to each of the areas, in line with $\sqrt{13 \cdot Hs} \le Tp \le \sqrt{30 \cdot Hs}$ formula.

5.1.2 Ship relative velocity to waves

Due to the fact the one is interested in the actual motions and accelerations for the sailing ship (and not stationary one) the speed of the ship need to be included when examining the range of wave zero-crossing period. Therefore following relation is identified and utilized for that purpose:

$$\omega_e = \left| \omega - \frac{\omega^2 V}{g} \cos \mu \right|$$

$\omega_e = \left| \omega - \frac{\omega V}{} \cos \mu \right|$ where:
ω_e – encounter frequency (as experienced by the ship)
ω – wave frequency
μ – relative wave heading
V – ship's speed
g – acceleration of gravity

Therefore e.g. for SUL1 all-route wave of 10 s< Tp < 15 s and ship speed of 14 knots wave (encounter) periods are in the range of 5,94 s and 22,52 s. Additionally, a fact that only some of the periods apply to the given direction need to be taken into account. For this reason the intermediate tables will include empty cells for the direction/period combination that is not relevant. This will result in tables where cells away from diagonal will be left empty.

315

5.1.3 *Motions and accelerations RAOs*

Once range of periods is established RAOs for motions and accelerations in generated. These operators cover whole range of encounter periods to allow for complete and accurate calculations. The RAOs provide linear response to a single wave of any height – the operator is denominated in degrees (for rotations) or meters (for linear motions) per 1 m of wave height.

Example of roll RAO for SUL1 condition is shown in Figure 5 below:

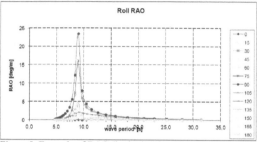

Figure 5. Example of RAO (in deg/m) for roll motion.

5.1.4 *Short term responses*

The RAOs are post-processed for short term response that describe extreme values for a 3h storm of given significant wave height and range of periods (to represent real period variations and account for the relative heading and encounter period. The motion amplitude and values of linear accelerations are calculated for a range of points onboard, to cover location of major equipment and places onboard where crew is working for extended periods of time. Figure 6 below and Table 4 below identify the points used in the motion analysis:

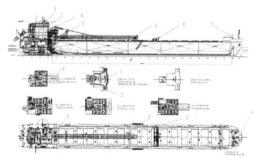

Figure 6. Calculation points onboard

Table 4. Coordinates of calculation points onboard

No	Description	X (m)	Y (m)	Z (m)
1	Bridge-conning	21,41	0,00	30,33
2	Bridge-wing	23,81	8,99	30,33
3	SU foundation	33,16	0,00	17,94
4	SU top	33,16	0,00	28,95
5	SU support	64,16	0,00	16,38
6	gantry midship	135,02	0,00	15,32
7	forecastle-winch	216,94	0,00	15,00

5.1.5 *Verification against habitability criteria*

Selected values are compared against limits recommended in industrial publications. Example of such comparison in presented in Table 5:

Table 5. Comparison of vessel motions with habitability limits for whole route approach (see 5.1.1)

Loading condition	Location	vert acc	hor acc x	hor acc y	roll
SUL1	Bridge conn.	0,96	0,08	1,44	10,08
SUL1	Bridge wing	1,01	0,11	1,44	10,08
SUL2	Bridge conn.	0,81	0,07	1,05	7,26
SUL2	Bridge wing	0,88	0,09	1,05	7,26
LIMIT		1,47 [m/s2]	1,18 [m/s2]	1,18 [m/s2]	6,00 [deg]

5.2 *Results and conclusions*

Extreme motions and accelerations were calculated for two delivery voyage loading conditions, using two different approaches – simplified whole route approach and more elaborate approach that divided the route into 5 sections. Calculations points were located ship's center of gravity and 7 different points onboard. Example of results is presented in Tables 6 and 7 below:

Table 6. Example of motion results (here results for SUL2, Canada East Coast region)

No	Description	MOTIONS		
		X (m)	Y (m)	Z (m)
1	Bridge-conning	0,39	2,14	2,42
2	Bridge-wing	0,39	2,12	2,35
3	SU foundation	0,38	1,29	2,13
4	SU top	0,38	1,93	2,13
5	SU support	0,40	0,90	1,42
6	gantry midship	0,41	0,29	0,99
7	forecastle-winch	0,40	1,00	2,82

Table 7. Example of acceleration results (here results for SUL2, Canada East Coast region)

No	Description	ACCELERATIONS		
(m/s^2)		X (m/s^2)	Y (m/s^2)	Z
1	Bridge-conning	0,24	1,06	1,19
2	Bridge-wing	0,25	1,05	1,17
3	SU foundation	0,17	0,58	1,05
4	SU top	0,23	0,97	1,05
5	SU support	0,17	0,41	0,71
6	gantry midship	0,17	0,15	0,51
7	forecastle-winch	0,17	0,41	1,44

All these values were transferred to equipment manufacturers and foundation designers so that all elements are properly attached onboard during delivery voyage. The results for the bridge locations were compared with habitability criteria and were found to be exceeding those in East China Sea for ballast conditions and in the Caribbean for both loading conditions. This provided navigational guidance so as to avoid stormy conditions for the sake of proper living and working conditions onboard.

6 GREENWATER RISK ASSESSMENT

6.1 *Calculating risk*

In similar manner to motion analysis specific check of relative distance from water surface to various points on the edge of the main deck was performed. The analysis utilized RAOs of this relative distance and probability of encountering an event where the on-deck measurement point is below water surface was estimated.

Following points, divided into three groups, were analysed:

Table 8. Coordinates of green-water measurement points along the deck edge.

No	Description	X (m)	Y (m)	Z (m)	
1	bow CL forecastle	222,06	0,00	17,83	fore
2	bow openings	222,06	0,00	14,70	
3	fcstl edge	211,65	11,26	14,70	
4	bow quarter	177,65	11,89	14,70	
5	midship	109,65	11,89	14,70	mid
6	stern quarter	65,45	11,89	14,70	aft
7	liferaft davit	13,08	11,87	14,70	
8	transom	-3,49	0,00	14,84	

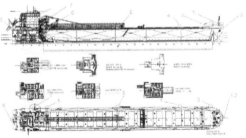

Figure 7. Location of green-water measurement points along the deck edge.

The probability of submersion was calculated using short-term response method. Combining that with storm duration and wave period one can calculate most probable number of greenwater occurrence during e.g. 3h storm. Example of results is presented in the Table 9 below.

Table 9. Number of occurrences of greenwater in aft part of the ship for SUL2 loading condition (Hs = 7,7 m for East China Sea section of the delivery voyage)

Tz	105 deg	120 deg	135 deg	150 deg	165 deg
5,50					
6,50					
7,50					
8,50	0	0	0	0	0
9,50	0	0	0	0	0
10,50	0	0	0	0	0
11,50	0	0	0	0	0
12,50	0	0	0	0	0
13,50	0	0	0	0	0
14,50	0	0	0	0	0
15,50	0	0	0	0	0
16,50	0	1	0	0	0
17,50	0	2	0	0	0
18,50		4	1	0	0
19,50		7	2	1	1
20,50			3	2	1
21,50					2
22,50					
23,50					

6.2 *Conclusion for route sections approach*

The probability of any green-water occurrence for SUL1 case (no cargo) is very low (below 0,06%). The consequences of such event are also very small, as it is likely to occur outside the areas where hatch covers are located. Therefore the risk of such event can be described as neglectable.

The probability of any green-water occurrence for SUL2 case (with cargo) is low (below 1,2%). The consequences of such event are also small - while some water will likely occur on deck next to hatch openings the relative height of the maximum water column is well below the design loads for hatch covers (0,5 m as compared to 2,95 m of design load water column). Therefore the risk of such event can be described as very low.

6.3 *Conclusion for route sections approach.*

The probability of any green-water occurrence for SUL1 case (no cargo) is very low (below 0,04%). The consequences of such event are also very small, as it is likely to occur outside the areas where hatch covers are located. Therefore the risk of such event can be described as neglectable.

The probability of any green-water occurrence for SUL2 case (with cargo) is low (below 2,0%). The consequences of such event are also small - while some water will likely occur on deck next to hatch openings the relative height of the maximum water column is well below the design loads for hatch covers (0,3 m as compared to 2,95 m of design load water column). Therefore the risk of such event can be described as very low.

6.4 *Preparing polar plots*

In order to assist crew onboard the ship during delivery voyage a set of polar plots was prepared. These polar plots visualize dangerous combinations of relative wave direction and encounter period of the waves.

Polar plots were prepared for three seastates:
- Beaufort scale sea state 5 (BF5)
  Hs=2,5m   5,7 < Tp < 8,7
- Beaufort scale sea state 7 (BF7)
  Hs=4,7m   7,8 < Tp < 11,9
- Beaufort scale sea state 9 (BF9)
  Hs=8,5m   10,5 < Tp < 16,0

A range of wave periods was used following the formula below:

$$\sqrt{13 \cdot Hs} \leq Tp \leq \sqrt{30 \cdot Hs}$$

Peak periods were corrected for forward speed (up to 14 kn) as presented described in 5.1.2. This is shown as "most probable period/direction range" in the plots.

Some simplified explanations how to read the plots is presented below:
- Circular direction show relative heading between ship and waves
- Radial direction show observed wave period (e.g. time separation between crests)
- Red area marked as "most probable period/direction combination range" represent the observed wave periods for given wave height, heading and range of speeds between 0 and 14 knots
- Areas marked in magenta, green, blue and dark blue represent wave/heading combination for which greenwater will likely occur

Therefore dangerous areas are those where red markings touch or overlap with color markings. These combinations of heading and relative have period should be avoided.

Most probable relative headings for each leg of the voyage are shown as arcs outside the circular plot. This is aimed at providing better understanding of the risks of greenwater in each voyage leg.

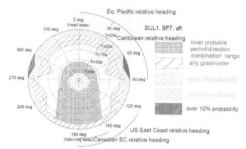

Figure 8. Example of the polar plot (here for SUL1, aft area of the deck and Equatorial Pacific area)

7 SUMMARY AND CONCLUSION

Elaborate set of safety studies was performed for the Laker bulk carrier due to planned delivery voyage from China to Canadian East Coast. Many aspects of the safety of transit were analyzed utilizing state-of-the-art hydrodynamic analysis software. Methodology of the analysis was presented together with some sample results and general conclusions from the studies.

The methods presented in this paper could be utilized for:
- studies concerning extreme weather events or special ship types
- safety assessment of marine transportation of heavy equipment (for offshore or land-based facilities like platforms or refineries)

REFERENCES

Aalto University, School of Engineering, 2010. Course material for Kul-24.4511 Optimization of Structures, Espoo

Bertram V., 2000. Practical Ship Hydromechanics, Oxford, Butterworth-Heinemann

BMT Ltd, 1986. Global Wave Statistics. Old Woking : Unwin

Dawn, T., 1995. Nature Shows the Way to Discover Better Answers Scientific Computing World 6 (1995), 23-27.

Goldberg, D.E., 1989. Genetic Algorithms in Search, Optimization & Machine Learning. Reading, Addison-Wesley

Lloyd's Register, 2011, Rules and Regulations for the Classification of Ships, London

Lloyd's Register, 2003, Classification of Ships for Service on the Great Lakes and River St. Lawrence, London

NAPA Oy, 2009. Manual, Optimization chapter, Helsinki

Milton Keynes UK
Ingram Content Group UK Ltd.
UKHW051853071024
449327UK00025B/1932

9 780367 576417